Chemically Modified Nanopores and Nanochannels

Chemically Modified Nanopores and Nanochannels

Edited by

Mario Tagliazucchi

Universidad de Buenos Aires—CONICET, Buenos Aires, Argentina

Igal Szleifer

Northwestern University, Evanston, IL, United States

ELSEVIER

AMSTERDAM • BOSTON • HEIDELBERG • LONDON
NEW YORK • OXFORD • PARIS • SAN DIEGO
SAN FRANCISCO • SINGAPORE • SYDNEY • TOKYO

William Andrew is an imprint of Elsevier

William Andrew is an imprint of Elsevier
The Boulevard, Langford Lane, Kidlington, Oxford, OX5 1GB, United Kingdom
50 Hampshire Street, 5th Floor, Cambridge, MA 02139, United States

British Library Cataloguing-in-Publication Data
A catalogue record for this book is available from the British Library

Library of Congress Cataloging-in-Publication Data
A catalog record for this book is available from the Library of Congress

ISBN: 978-0-323-40182-1

For Information on all William Andrew publications
visit our website at https://www.elsevier.com/

Publisher: Matthew Deans
Acquisition Editor: Simon Holt
Editorial Project Manager: Sabrina Webber
Production Project Manager: Kiruthika Govindaraju
Cover Designer: Greg Harris

Typeset by MPS Limited, Chennai, India

Contents

List of Contributors

O. Azzaroni Universidad Nacional de La Plata–CONICET, La Plata, Argentina

E. Fullwood University of Utah, Salt Lake City, UT, United States

E. Green University of Utah, Salt Lake City, UT, United States

W. Guo Chinese Academy of Sciences, Beijing, P.R. China

Q. Kang Chinese Academy of Sciences, Beijing, P.R. China

G. Pérez-Mitta Universidad Nacional de La Plata–CONICET, La Plata, Argentina

J. Selden University of Utah, Salt Lake City, UT, United States

I. Szleifer Northwestern University, Evanston, IL, United States

M. Tagliazucchi Universidad de Buenos Aires–CONICET, Buenos Aires, Argentina

M.E. Toimil-Molares GSI Helmholtzzentrum für Schwerionenforschung GmbH, Darmstadt, Germany

C. Trautmann GSI Helmholtzzentrum für Schwerionenforschung GmbH, Darmstadt, Germany

I. Zharov University of Utah, Salt Lake City, UT, United States

About the Editors

Mario Tagliazucchi

Mario Tagliazucchi was born in Buenos Aires, Argentina. He received his degree in Chemistry from the University of Buenos Aires (Argentina) in 2004 and obtained his PhD in Chemistry from the same university in 2009 under the supervision of Prof. Ernesto J. Calvo. From 2010 to 2015, he was a postdoctoral fellow at the Non-equilibrium Research Center at Northwestern University (United States), where he performed theoretical studies in the area of chemically modified nanopores and nanochannels working with Profs. Igal Szleifer and Emily Weiss. Since 2015, he is a Staff Scientist of CONICET (Science and Technology Council of Argentina). He has been awarded with CONICET-FUDETEC and Fulbright fellowships.

Igal Szleifer

Igal Szleifer was born in Buenos Aires, Argentina. He received his degree in Chemistry from the Hebrew University of Jerusalem (Israel) and his PhD in Chemistry from the same university, working with Prof. Avinoam Ben-Shaul. Igal Szleifer is currently the Christina Enroth-Cugell Professor of Biomedical Engineering at Northwestern University (United States) and he is a Professor of Chemistry, Chemical and Biological Engineering and Medicine. His work concentrates on the development of theoretical approaches to understand the coupling that exists between molecular organization and function. His studies on molecular modeling of complex multicomponent systems have applications in a range of systems ranging from responsive materials to biophysics. The understanding of the complexity of chemical reactions, physical interactions and their role in biological and biomimetic environments is at the center of his research interests.

Preface

The field of nanopores and nanochannels, a major area in modern nanotechnology, has emerged from the confluence of different research areas and technologies, which include (but are not limited to) membrane science, nanofluidics, nanopore (bio)analytics, nanoporous materials, and biological pores and channels. The confluence of such diverse areas required the participation of researchers from very different backgrounds and expertise, which makes this field a truly interdisciplinary enterprise. Historically, the first generations of nanopore membranes and single solid-state nanochannels and nanopores were used as prepared, without further chemical functionalization. However, researchers soon discovered the amazing potential of surface chemical modification to improve the transport properties of nanopores and nanochannels. Chemical functionalization allows controlling the degree of charge and hydrophobicity of the channels and it creates a handle to direct the transport of small ions, polymeric chains, and large particles. Chemical functionalization also has the ability to make pores and channels responsive to external stimuli, such as changes of pH, ionic strength, illumination, and the presence of analytes in solution. In summary, the area of chemically modified nanopores is driven by the promise of improved applications and fueled by the broad availability of tools for surface modification that chemists have been developing for decades to modify planar surfaces.

The link between chemical functionalization and transport properties is the key to design applications based on chemically modified nanochannels and nanopores. This book is built around this link, as it aims to equip the reader with the fundamental tools necessary to understand transport in these systems, as well as to examine the current state of the field by discussing examples in the recent literature. We have included two introduction chapters. The first one is focused on the experimental techniques used to prepare, chemically modify, and measure the transport properties of nanopores and nanochannels. The second introduction chapter discusses the theoretical basis necessary to understand transport, with special emphasis on the effects of surface charge. The following three chapters provide a closer look at nanopore and nanochannel systems of interest, both single channels and membranes with multiple pores. In particular, Chapters 3, Single Ion Track-Etched Nanochannels for Analytical Applications, and 4, Biomimetic Smart Nanopores and Nanochannels, discuss single nanochannels prepared by the ion-track-etch technique. Chemical modification of these channels is analyzed in terms of their applications in bioanalysis (chapter: Single Ion Track-Etched Nanochannels for Analytical Applications, by Omar Azzaroni and coworkers) and the ability to mimic the transport properties of biological systems (chapter: Biomimetic Smart Nanopores and Nanochannels, by Wei Guo and coworkers). Chapter 5, Chemically Modified Nanopores in Colloidal Crystals, by Ilya Zharov and coworkers, is focused on nanoporous membranes prepared by colloidal assembly and illustrates how chemical modification can tune the transport and separation behavior of these membranes. Finally, in Chapter 6, Advanced Modeling of Ion Transport in Polymer and Polyelectrolyte-Modified Nanochannels and Nanopores, we present new theoretical tools developed to model chemically modified nanochannels and nanopores. These tools are well suited to model these systems, as they explicitly consider the molecular properties of the chemical species attached to their inner walls. As a whole, this is the first book about nanopores and nanochannels that is strongly focused on chemical modification and their effects on transport, both from the theoretical and practical point of views. While the book has a chemical focus, a strong background in chemistry is not assumed, so we believe this work will be useful and enjoyable to researchers of diverse backgrounds, namely anyone interested in making, modifying, and using chemically modified nanopores and nanochannels in filtration membranes, sensors, or other emerging applications.

Acknowledgments

We thank our families for their constant support, patience, and encouragement. This book is dedicated to them.

We also thank our collaborators and coworkers in the field of nanopores and nanochannels for sharing their enthusiasm with us, bringing us interesting questions and problems, and making our work even more enjoyable than it is usually is. We are especially grateful to the authors of Chapters 3, 4, and 5—without their contribution this book would not exist.

Igal Szleifer insists on also dedicating this book to his coauthor, Mario Tagliazucchi. Despite Mario's reluctance to take additional credit, he took over the bulk of the work when Prof. Szleifer suffered a stroke in October 2015. If not for Mario's hard work and dedication, this book would never have become a reality. Igal will be forever grateful. Mario Tagliazucchi also wants to dedicate this book to the memory of his father, Jorge.

1

Introduction to Chemically Modified Nanochannels and Nanopores

M. Tagliazucchi

UNIVERSIDAD DE BUENOS AIRES–CONICET, BUENOS AIRES, ARGENTINA

CHAPTER OUTLINE

1.1 Types of Channels and Pores

Broadly speaking, the field of chemically modified nanochannels and nanopores embraces several different technologies and materials which share some key aspects. In this book, we will specifically focus on materials and devices that: (1) comprise a single or an array of nanochannels or nanopores with diameters in the range 1–1000 nm, (2) are chemically modified in order to confer functionality and responsiveness, and (3) serve as a medium for selective transport for diverse applications (filtration, sensing, separation, etc.). Fig. 1.1 shows microscopy images of solid-state nanopores and nanochannels that fulfil these criteria. We excluded from this list materials exhibiting porosity at the molecular scale, such as zeolites and metal-organic frameworks. Table 1.1 summarizes some properties of the systems in Fig. 1.1.

The systems in Fig. 1.1 are very diverse in their physical and chemical properties. Depending on the aspect ratio (i.e., L/R, where R and L are the radius and length, respectively) they can be classified either as pores (L/R of order unity) or channels (L/R much larger than one). It is also important to note that some of the systems in Fig. 1.1 are single-pore (or single-channel) systems and others contain multiple nanopores (or nanochannels). In the latter case, channels may be either well-ordered or randomly distributed. Regarding

FIGURE 1.1 Microscopy images of different types of nanopores and nanochannels: (A) top-view Scanning Electron Microscopy (SEM) image of nanochannels in track-etched poly(ethylene terephthalate) (PET) membrane; (B) top-view Transmission Electron Microscopy (TEM) image of single nanopore in an ultrathin silicon oxide membrane; (C) SEM image of a colloidal nanopore membrane (upper panel: top view, lower panel: lateral view, the inset in the upper panel shows that the nanopores are formed from the voids in the fcc lattice); (D) top-view SEM image of a mesoporous silica membrane; (E) optical (upper panel) and SEM (lower panel) image of a borosilicate nanopipette; (F) top-view SEM image of a microporous silicon membrane; (G) top-view SEM image of a nanochannel alumina membrane. Source: *Adapted from (A) Wen, L.; Tian, Y.; Ma, J.; Zhai, J.; Jiang, L., Construction of Biomimetic Smart Nanochannels with Polymer Membranes and Application in Energy Conversion Systems.* Phys. Chem. Chem. Phys. ***2012***, 14, *4027–4042, (B) Storm, A.J.; Chen, J.H.; Ling, X.S.; Zandbergen, H.W.; Dekker, C., Fabrication of Solid-State Nanopores with Single-Nanometre Precision.* Nat. Mater. ***2003***, 2, *537–540, (E) An, S.; Stambaugh, C.; Kim, G.; Lee, M.; Kim, Y.; Lee, K., et al., Low-Volume Liquid Delivery and Nanolithography Using a Nanopipette Combined with a Quartz Tuning Fork-Atomic Force Microscope.* Nanoscale ***2012***, 4, *6493–6500, (F) Velleman, L.; Shearer, C.J.; Ellis, A.V.; Losic, D.; Voelcker, N.H.; Shapter, J.G., Fabrication of Self-Supporting Porous Silicon Membranes and Tuning Transport Properties by Surface Functionalization.* Nanoscale ***2010***, 2, *1756–1761, and (G) Lee, W.; Ji, R.; Gosele, U.; Nielsch, K., Fast Fabrication of Long-Range Ordered Porous Alumina Membranes by Hard Anodization.* Nat. Mater. ***2006***, 5, *741–747; (D) courtesy of G. Soler-Illia's Lab.*

Table 1.1 Typical Dimensions and Membrane Materials for the Nanopore and Nanochannel Systems Shown in Fig. 1.1

System	Type	Typical Radius	Typical Length (Membrane Thickness)	Membrane Material
Track-etched polymer membranes	Single channel/ disordered array of channels	>10 nm (>5 nm with Au deposition)	~5–15 μm	PC, PET, PI
Pores prepared by focused electron or ion beam	Single pore	>2 nm	20–350 nm	SiO_2/Si_3N_4
Colloidal self-assembled membranes	Ordered 3D array of connected pores	5–20 nm	~5 μm	Silica
Nanopipettes	Single pore	>10–50 nm	>20 times the radius	Glass, quartz borosilicate
Templated mesoporous oxides	Ordered array of pores	3–9 nm	100–300 nm	MO_2 (e.g., SiO_2 TiO_2)
Porous silicon	Disordered pores	>2–6 nm	<100 μm	Silicon
Alumina membranes	Ordered array of nanochannels	>15 nm	<100 μm	Alumina (Al_2O_3)

their radial dimensions, porous materials are classified by IUPAC into micropores ($R < 2$ nm), mesopores (2 nm $< R <$ 50 nm), and macropores ($R > 50$ nm). These definitions are widely used for materials (e.g., porous silicon and mesoporous silica, Fig. 1.1D and F) but not for nanofabricated pores (e.g., e-beam drilled pores, Fig. 1.1B). When discussing pores in different systems is, therefore, important to consider the context: micropipettes have micrometer-sized openings, but pores in microporous silicon are smaller than 2 nm.

The systems shown in Fig. 1.1 are as follows:

1. Track-etched polymer membranes (Fig. 1.1A): They are prepared by irradiation of polymeric membranes with high-energy heavy ions (such as U and Bi). Due to their high energy, these ions are not stopped by the membrane but go through it, producing the degradation of the membrane in the form of a latent track. This latent track is then chemically etched in a highly alkaline solution order to produce very high aspect-ratio nanochannels.[1,2] The etching procedure can be controlled in order to produce different nanochannel shapes (cylindrical,[3] conical,[4] cigar-shaped,[5] etc.). Membranes containing multiple nanochannels are used for filtration and are commercially available, for example, polycarbonate (PC) membranes from Whatman (Nucleopore and Cyclopore) and PC, polyimide (PI), and polyester membranes from it4ip (ipPore). The ion-track technology has been extended in order to stop heavy-ion bombardment immediately after the impact of the first heavy ion,[6] which enables the fabrication of single ion-channel membranes. Due to their long aspect ratio and control over channel shape and size, single ion channels prepared by the track-etched technique are one of the most

versatile and popular platforms for the creation of nanochannel devices. Chemical modification of single ion channels prepared by this method and their applications are discussed in Chapter 3, Single Ion Track-Etched Nanochannels for Analytical Applications, and Chapter 4, Biomimetic Smart Nanopores and Nanochannels.

2. Electron- or ion-beam sculpted solid-state channels (Fig. 1.1B): This technique involves preparing a thin free-standing silicon oxide (SiO_2) or silicon nitride (Si_3N_4) membrane using microfabrication techniques. The membrane is then milled with a highly focused electron[7,8] or ion beam.[9] In a related preparation technique,[10] holes of a few tens of nanometer are prepared using electron-beam lithography and reactive-ion etching. Then, the as-drilled holes are shrunk by heating them with an electron beam.
3. Nanoporous colloidal films (Fig. 1.1C): These films are constructed by self-assembling monodisperse silica microparticles into a face-centered cubic (fcc) lattice.[11] The octahedral and tetrahedral voids of the lattice form an ordered three-dimensional (3D) network of interconnected pores. The pore size is given by the voids among spheres and, thus, it can be tuned by the size of the silica colloids (see inset in the top panel of Fig. 1.1C). The silanol groups on the surface of the particles enable chemical functionalization, as Zharov et al. and coworkers describe in Chapter 5, Chemically Modified Nanopores in Colloidal Crystals.
4. Mesoporous materials (Fig. 1.1D): These materials are prepared by combining a sol-gel process for the formation of the inorganic phase and self-assembled supramolecular aggregates as the removable template.[12–16] The sol-gel process is a wet-chemistry route to prepare metal oxides from small molecules in solution. Templating species (e.g., block copolymers or surfactants) are added to the precursor solution and form a self-assembled organic template, around which the metal oxide is formed and structured. The precursor solution is then deposited onto the substrate (silicon, glass, etc.) by dip- or spin-coating. Finally, the film is treated with a thermal ramp in order to remove the solvent, cross-link the inorganic material, and remove the organic template. The resulting mesoporous film consists of a network of well-ordered interconnected pores, whose geometry depends on the template molecule and the synthetic conditions.
5. Glass nanopipettes (Fig. 1.1E): Glass nanopipettes are glass capillaries terminated in a nanometric size hole. The fabrication of micropipettes (with openings in the micrometer range) using laser pulling of quartz or glass capillaries is a standard and well-established procedure. Nanopipettes are also prepared by laser pulling,[17] although the method may be technically challenging and, thus, some modified protocols have been developed to improve reproducibility. For example, in some cases, the tip of a micropipette is melted and resealed and then etched or polished until a nanopore of the desired size is formed.[18,19]
6. Porous silicon (Fig. 1.1F): Porous silicon membranes are prepared by electrochemical oxidation of silicon in HF solutions.[17,18] Microporous silicon has potential applications as a membrane material[20] and it is also widely studied due to its interesting optical and electronic properties.[21]
7. Anodized alumina (Fig. 1.1G): The electrochemical oxidation (anodization) of aluminum in acidic medium yields an aluminum oxide (alumina) layer with channels self-organized

into well-ordered hexagonal arrays.[22] The aluminum substrate supporting the porous alumina film can be etched after the anodization step in order to obtain free-standing nanochannel membranes. This method allows the fabrication of very well ordered arrays of nanochannels with very high aspect ratios.[22,23] Nanochannel alumina membranes with different channel radii are commercially available, for example, from Anopore (Anodisc).

1.2 Chemical Modification

Chemical modification is necessary in order to confer the native surface of the materials in Table 1.1 with desired surface properties (wetting behavior, surface charge), functionality (responsiveness, catalytic activity, biological affinity) or to reduce undesired interactions that may lead to fouling and clogging. An important exception is that of templated mesoporous oxide materials, for which the composition of the membrane can be tailored during synthesis using co-condensation reactions,[13,15] thus incorporating the desired functionalities. Chemical functionalization of nanochannels is, in many cases, a multistep synthetic process, where a first chemical reaction introduces chemical groups onto the inner surface of the channels that serve to attach the molecules of interest in a following step. In this section, we will discuss the methods available to modify native surfaces of nanochannels/nanopores for some of the systems in Table 1.1. These methods can be widely classified into covalent chemistry (chemical reactions leading to the formation of covalent bonds), non-covalent assembly, and co-condensation synthesis of mesoporous oxides. Chapters 3, Single Ion Track-Etched Nanochannels for Analytical Applications, 4, Biomimetic Smart Nanopores and Nanochannels, and 5, Chemically Modified Nanopores in Colloidal Crystals, will provide examples of more complex synthetic chemistries that can be performed on nanopore and nanochannel surfaces following the functionalization of the native surface.

1.2.1 Covalent Modification

1.2.1.1 Silica Surfaces

Many of the nanopore/nanochannel systems discussed above have silica surfaces: colloidal crystals based on self-assembled silica spheres, mesoporous silica, glass pipettes, and single pores drilled in SiO_2 ultrathin membranes. Silica surfaces also exists as a native oxide layer on SiN[26] and can be deposited as well on nanochannel membranes of other materials via atomic layer deposition.[27]

Silica surfaces are terminated in a combination of siloxane groups, Si−O−Si and silanol groups, Si−OH.[28,29] Silanols are classified into isolated silanols (whose Si maintains three bonds with the underlying silica structure), bridged or vicinal silanols (isolated silanols located on neighboring Si atoms, so they that can interact through hydrogen bonds), and germinal silanols (whose Si atoms have two hydroxyl groups and two bonds to the bulk silica). The surface density of each of these types of silanols is variable and depends on the type and degree of hydroxylation of the silica surface. For example, infrared studies indicate that a fully hydroxylated silica has a total surface density of 4.6 hydroxyls/nm^2, of which 25% are isolated, 62% bridged or vicinal, and 13% are germinal.[28,30] On the other hand,

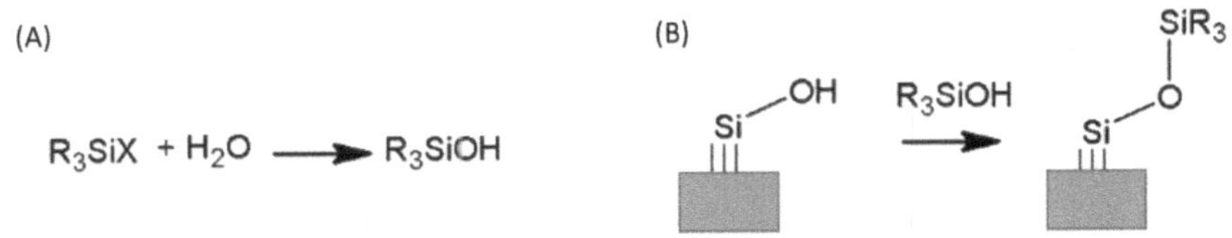

SCHEME 1.1 Chemical functionalization of a silica surface with an organosilane. The reaction involves the hydrolysis of the X group (A) and subsequent condensation of the resulting silanol with a surface-bound silanol (B).

acid–base titration of planar[31] and mesoporous[32] silica has suggested that 20% of the silanols are isolated and bridged and the remaining 80% are germinal.

Silanol groups can behave both as weak acids

$$\equiv \mathrm{SiOH} + \mathrm{H_2O} \rightleftarrows \equiv \mathrm{SiO^-} + \mathrm{H_3O^+} \tag{1.1}$$

and as weak bases.

$$\equiv \mathrm{SiOH} + \mathrm{H^+} \rightleftarrows \equiv \mathrm{SiOH_2}^+ \tag{1.2}$$

Despite being in principle amphoteric, silanols are very weak bases (the pK_a of $SiOH_2^+$ in Eq. (1.2) is -2.7[32]) and thus silica is generally negatively charged, except at very low pHs (the isoelectric point of silica is ~2–3[33]). The pK_a in Eq. (1.1) depends on the type of silanol: isolated and bridge silanols have been reported to have pK_as of 2–4.5, while the pK_a of germinal silanols is higher, around 8.5.[3,1,32] It is also important to note that the total surface density of silanol groups is very high and, therefore, the surface charge of silica is greatly affected by charge-regulation effects (see chapter: Theoretical Basis for Structure and Transport in Nanopores and Nanochannels).

Organofunctional silanes are arguably the most common reactants to modify silica surfaces. Silanes have general formulas $RSiX_3$, R_2SiX_2 or R_3SiX, where R is an organic group that confers functionality to the surface and X is a hydrolysable group that is lost during coupling of the silane to the silica surface (X = Cl, OEt, etc.). For example, 3-triethoxysilylpropanamine (APTES) has three hydrolysable ethoxy groups and an amine functional group, which remains unmodified upon binding to the silica surface. In the presence of water, the mechanism of reaction of organosilanes involves first the hydrolysis of the X group in solution (Scheme 1.1A) followed by reaction with the silica surface (Scheme 1.1B).

In general, dry organic solvents are preferred over water as the solvent for silanization, because uncontrollable hydrolysis in water leads to high concentrations of = Si(R)–OH, which polymerizes in solution. It is also important to note that the R group may participate in the silanization process, for example, the amino group in APTES forms hydrogen bonds with the silica surface and its basic and nucleophilic properties can accelerate the kinetics of condensation.[28]

1.2.1.2 Polymer Surfaces

Scheme 1.2 shows the chemical structure of typical polymeric materials used for track-etched nanochannel membranes: poly(ethylene terephtalate) (PET), polyimide (PI) and polycarbonate (PC). These polymers contain imide, carbonate, and ester groups, respectively,

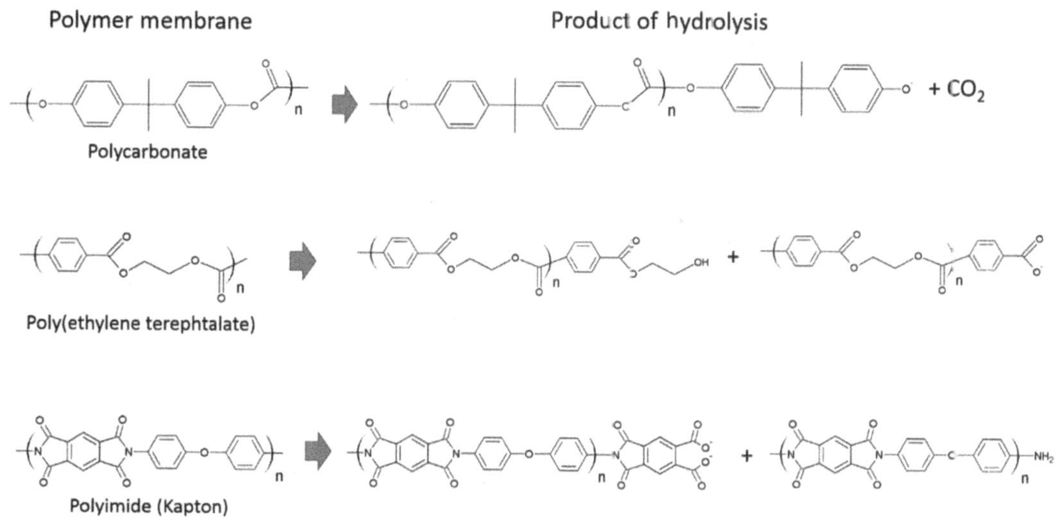

SCHEME 1.2 Chemical structure and surface-bound products of hydrolysis after etching for the different polymer materials used in track-etched membranes.

which become hydrolyzed during alkaline etching of the ion track (see chapter: Single Ion Track-Etched Nanochannels for Analytical Applications), leading to different surface functional groups. For example, hydrolyzed PET surfaces exhibit both alcohol and carboxylate groups,[34–36] PI membranes hydrolyze to amine and carboxylate groups[37,38] (the latter groups are in equilibrium with the corresponding carboxylic anhydrides[38]), and PC membranes hydrolyze to phenoxy groups with loss of CO_2.[34–36] The surface groups formed during the etching process determine the chemical reactivity and surface charge of the membrane. In the cases of PET and PI membranes, the carboxylate groups of the surface have a pK_a of ~ 5, so these membranes are negatively charged for $pH > 3$.[39] The maximum surface density of carboxylate groups in PET has been reported to be 1.5 group/nm^2.[40]

In the particular case of commercial PC membranes, a coating of poly(vinylpyrrolidone) is added during production in order to render these membranes hydrophilic. This coating gives the surface of the membrane ion-chelating and hydrogen-bonding properties.[41]

The chemical modification of PET and PI track-etched membranes, recently reviewed Jiang and collaborators,[42] in general involves functionalizing the carboxylate groups with an amine in solution using carbodiimide chemistry (Scheme 1.3). In a typical procedure,[43,44] the carboxyl groups are activated using an aqueous solution of 10 mM *N*-(3-dimethylaminopropyl)-*N*-ethylcarbodiimide hydrochloride (EDC) and 20 mM *N*-hydroxysulfosuccinimide (sulfo-NHS, 20 mM) for 30–60 min. This reaction yields a reactive sulfo-NHS-ester on the surface (Scheme 1.3C). Pentafluorophenol can be used as an alternative reactant to NHS to activate the surface carboxylates.[3,45] In a second step, the membrane is thoroughly washed and incubated overnight in a solution of an amine (1 μM–10 mM) in order to form the amide bond (Scheme 1.3D). Different amine-bearing molecules have been used in literature: diamines[46] (which allow a subsequent EDC/NHS reaction with a carboxylate in solution), enzymes,[47]

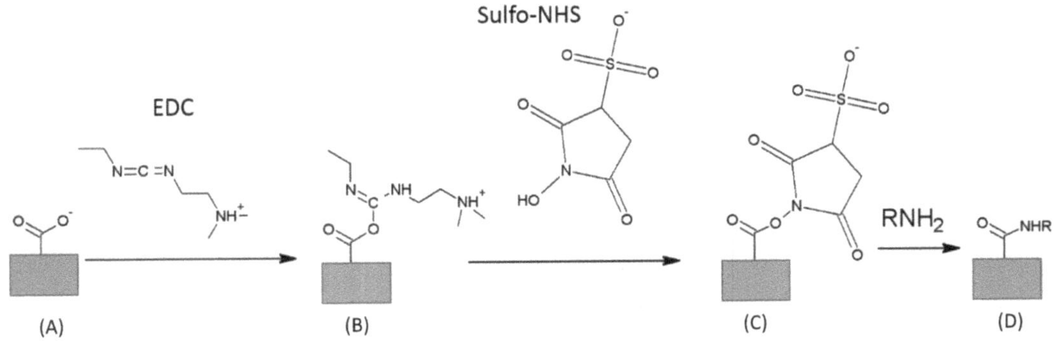

SCHEME 1.3 Chemical modification of surface carboxylates into amides with EDS/NHS. Surface modified by (A) carboxylate, (B) acylisourea active ester, (C) sulfo-NHS ester, (D) amide.

lysine (which reacts via the side-chain amino group),[5] polymerization initiators (which allow to grow a polymer brush on the inner channel walls),[3] amino-terminated DNA oligomers,[44,48] and small functional amines substituted with phosphoric acids,[45] terpyridine,[49] carbohydrates,[50] alkynes,[51] cyclodextrins,[52] etc.

Plasma modification is an alternative route to functionalize polymer membranes that is based on the generation of free radicals on the surface of the polymeric membrane by exposure to a plasma. Introducing monomer molecules into the plasma chamber leads to the formation of polymer brushes via radical polymerization.[48,53,54] This methodology can be used to asymmetrically modify single nanochannels in track-etched PET membranes by modifying different sides of the membrane with different polymers.[53,54]

1.2.1.3 Gold Deposition and Modification

Self-assembled monolayers (SAMs) of thiols on gold and other coinage metals are a very versatile route to introduce surface chemical functionalities.[55] While none of the systems in Fig. 1.1 has native gold surfaces, it is possible to coat the surfaces of nanochannels and nanopores with a thin gold layer and then form an SAM of thiols with the desired chemical functionalities.

Gold electroless method has been extensively used to coat the inner walls of nanochannels with a thin Au film. This method employs a mixture of a Au(I) complex and a strong chemical reductant (e.g., formaldehyde). This mixture is kinetically (but not thermodynamically) stable and therefore reduction of Au(I) does not occur in solution. However, small catalytic particles (typically Ag) immobilized on a surface catalyze the reduction of Au(I), which results in the formation of thin film of Au on that surface. In a typical electroless process (Fig. 1.2),[56] a track-etched PC membrane is exposed to a solution of Sn(II), which binds to the walls by complexation with surface chelating groups.[41] The membrane is rinsed and immersed into a Ag(I) solution, which results in the formation of Ag^0 deposits by localized reduction of Ag^+ ion with surface-immobilized Sn(II), which oxidizes to Sn(IV). Finally, the solution is rinsed again and immersed into the electroless solution to form the Au film. The thickness of the gold film can be controlled by adjusting the duration of the electroless deposition. The Au-coated membrane is finally modified by thiol self-assembly. The gold electroless methodology has been successfully applied to PET and PC track-etched

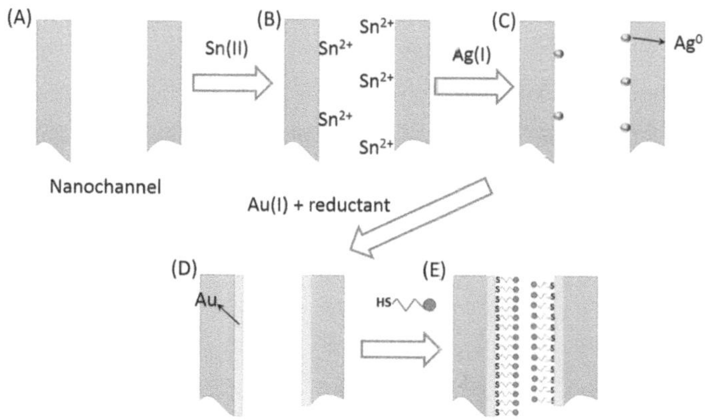

FIGURE 1.2 Scheme of chemical modification of a nanochannel by coating with electroless gold deposit on and subsequent formation of a thiol SAM. (A) bare nanochannel, (B) adsoption of Sn(II), (C) reduction of Ag(I) by Sn(II) to form Ag nanoparticles, (D) electroless deposition of Au catalized by the Ag nanoparticles, (E) formation of a thiol self-assembled monolayer on Au.

polymer membranes,[41,57–59] alumina membranes,[60,61] SiN nanopores,[62] and silica particles for colloidal membranes.[63] Note, however, that the modification of very narrow channels by electroless is challenging due to the fact that the gold films are typically composed of crystals of tens of nanometers, which can create rough surfaces at the nanoscale and may even occlude the channels. Muench et al. developed a protocol to reduce the size of the Au crystals below 5 nm by using 4-(dimethylamino)pyridine in the electroless solution.[56]

Nanochannels and nanopores can also be metallized with thermal evaporation or sputtering.[64–69] These techniques can coat only one end of the channel by depositing gold on only one side of the membrane, which allows the preparation of asymmetric nanochannels.[64–66,69] The good electrical conductivity of evaporated Au films enables to externally control their electrostatic potential against a reference electrode in solution using a potentiostat.[64,65,68] Control over the membrane potential can be used to create an nanofluidic diode[65] or to further functionalize the gold surface via electropolymerizarion.[64] Electrodeposition of gold inside nanochannels yields in general Au nanorods that completely fill the nanochannels, but some strategies have been devised to coat only the walls of the nanochannels,[70–72] thus obtaining gold-coated nanochannels.

Following the deposition of Au onto the inner surfaces of nanopores or nanochannels, standard Au-thiol chemistry can be used to form functional SAMs. This strategy has been used to modify nanopores with pH-responsive groups,[73–75] zwitterions,[75] thermoresponsive polymers,[68,76] ionophores,[77] hydrophobic alkyl chains,[74,78,79] antibodies,[80] DNA,[67,81] and PNA.[82]

1.2.2 Non-Covalent Modification

1.2.2.1 Electrostatic Self-Assembly: Polyelectrolyte Infiltration and Layer-by-Layer Deposition

Electrostatic adsorption of polyelectrolytes is a versatile alternative to covalent chemistry at the moment of chemically modifying nanochannels and nanopores with thin polymer

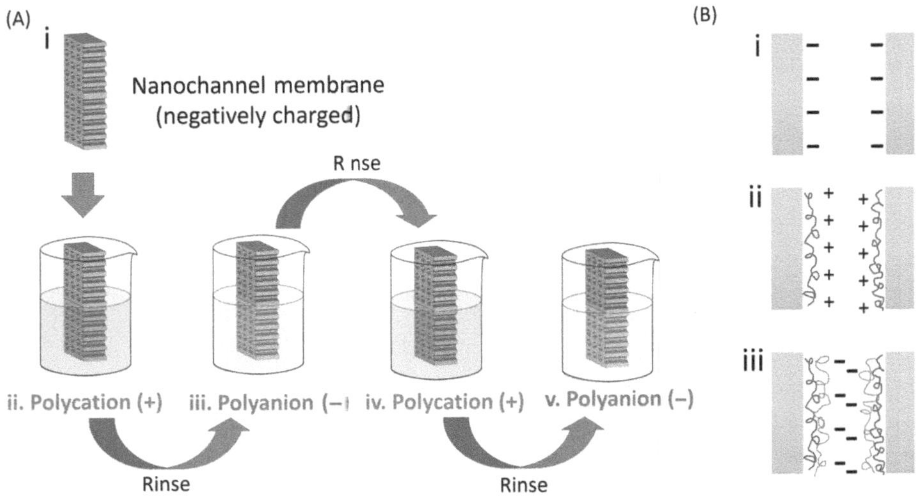

FIGURE 1.3 Electrostatic LbL deposition inside nanochannel membranes. The nanochannel membrane (i) is immersed into a solution of a polyelectrolyte having a surface charge opposite to that of the membrane (ii), a polycation in the present case. After polyelectrolyte adsorption, the membrane is rinsed and sequentially immersed into solutions of a polyanion and a polycation in order to build a LbL film on the inner walls of the channel (iii–v).

films. This methodology involves immersing the nanochannel or nanopore membrane into a solution of a polyelectrolyte with a charge opposite charge to that of the channel walls.

Polyelectrolyte deposition on flat surfaces is a well-studied process,[83] however adsorption in very narrow nanochannels and nanopores ($R \leq 10$ nm) is complicated by nanoconfinement effects and the need of a diffusion pathway enabling the polyelectrolyte to permeate into the channel. Therefore, successful modification of the inner surfaces of nanochannels and nanopores via electrostatic adsorption requires to carefully consider processing variables such as the diameter of the channel, the molecular weight of the polyelectrolyte, and the ionic strength and pH of the solution. Caruso and coworkers studied the adsorption of a single layer of positively charge poly(allylamine), PAH, onto the inner surfaces of negatively mesoporous silica colloids,[84] as well as the adsorption of a single layer of negatively charged poly(acrylic acid), PAA, onto the positive surfaces of mesoporous aminosilica particles.[85] These experiments showed a decrease in adsorption when increasing the molecular weight of the polyelectrolytes or decreasing nanopore size, which reveals limitations either to the diffusion of the polyelectrolyte into the channels or of its ability to adsorb in a nanoconfined environment. For PAH, the amount of adsorption was observed to increase with increasing ionic strength, which was ascribed to coiling of the polyelectrolyte with increasing salt concentration.[85] Brunsen and coworkers studied mesoporous silica films modified by an end-functionalized poly(vinylferrocene) on conductive Indium tin oxide (ITO) using cyclic voltammetry (CV). These experiments showed no redox signal from the ferrocene, which also suggests a hindrance for the polymer to diffuse within the mesoporous film and reach the underlying electrode.[86]

Following the adsorption of a first layer of polyelectrolyte, the net charge of the surface inverts and becomes the same as that of the adsorbed polyelectrolyte, a phenomenon known

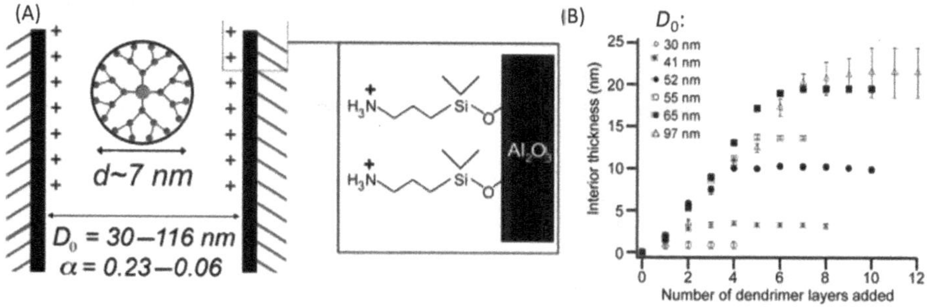

FIGURE 1.4 Example of electrostatic LbL adsorption as a method to chemically modify nanopores and nanochannels. Negatively and positively charged dendrimers were sequentially adsorbed within an aluminum oxide membrane premodified by a layer of positively charged aminopropyl-dimethylethoxysilane (A). The thickness of the film as a function of the number of adsorbed layers strongly depends on the radius of the nanochannel and number of adsorbed layers (B). Source: *Reproduced from Lazzara, T.D.; Lau, K.H.A.; Abou-Kandil, A.I.; Caminade, A.M.; Majoral, J.-P.; Knoll, W., Polyelectrolyte Layer-by-Layer Deposition in Cylindrical Nanopores.* ACS Nano ***2010**, 4, 3909–3920, with permission.*

as charge inversion.[87] Charge inversion enables to sequentially adsorb two polyelectrolytes of opposite charge in order to assembly a polyelectrolyte multilayer (Fig. 1.3). This process, known as layer-by-layer (LbL) self-assembly, has been introduced by Decher for planar surfaces.[87–89] The formation of LbL films within nanochannel and nanopore membranes have been also studied, see Refs. 90 and 91 for recent reviews. LbL assembly within nanoscale pores and channels is affected by the nanoconfinement effects described above for the adsorption of a single polyelectrolyte layer. Rubner and coworkers studied LbL in nanofluidic channels of 222 nm[92] and show that polyelectrolyte adsorption stops when the effective size of the channel is reduced below 11 nm due to the growth of the polyelectrolyte film. Lazzara studied the formation of LbL films of negatively and positively charged dendrimers with a diameter of approximately 7 nm in alumina membranes with diameters greater than 30 nm using optical waveguide spectroscopy (Fig. 1.4A). For the smallest diameters (30 nm), the thickness of the film on the inner surface of the channel saturates after the first layer.[93] On the other hand, for pores wider than 30 nm, the authors identified the cross-over from the linear growth typically observed for planar surfaces, where the thickness of the film increases linearly with the number of layers, to a nanoconfined regime, where the thickness reaches a *plateau* with the number of adsorptions steps (Fig. 1.4B). A related Molecular Dynamics study of LbL assembly of nanoparticles inside nanopores has shown that the main mechanism behind the saturation in the layer thickness is a hindrance for the particles to permeate into the channel due to repulsive electrostatic interactions with the particles adsorbed at the entrance of the pore.[94] Jonas and coworkers studied LbL of linear polyelectrolytes in track-etched nanochannels and also observed a linear and a saturated regime for film growth. In the saturated regime, the polyelectrolyte multilayer forms a gel that prevents further infiltration of polymeric chains into the channel.[95] Studies of LbL deposition with aluminum oxide membranes of different pore size have also shown that the thickness per bilayer of PAH/PSS decreases for decreasing pore size.[96] The saturation of LbL assembly in nanochannels is accompanied by a gradual disappearance of the charge overcompensation effect: Azzaroni and coworkers studied LbL in conical nanochannels with a tip diameter of approximately 36 nm and show that the degree of inversion of the surface charge upon polyelectrolyte adsorption damped with the number of layers until almost stopping for eight polyelectrolyte layers.[97]

1.2.2.2 Nanochannel Modification Through Non-Electrostatic Self-Assembly

Covalent chemistry and electrostatic self-assembly are arguably the most resorted tools to modify nanochannels, but non-electrostatic supramolecular self-assembly is arising a novel and powerful method to modify nanopore surfaces with stimuli responsive materials. For example, Li and coworkers used hydrogen-bonding LbL to prepare a multilayer film of PAA and poly(vinyl-pyridine) within PC track-etched membranes.[98] Hydrogen-bonded LbL films are appealing due to the possibility of disassembling in high pH solutions.[99] The first adsorbed layer in Ref. 98 was PAA, a hydrogen bond acceptor; therefore, it is possible that this layer is adsorbed on the poly(vinylpirrolidone) (an hydrogen bond acceptor) coating of the commercial PC membrane. In other example, charge transfer interactions have been used to bind a pyranine in solution to viologen units grafted on the inner surface of mesoporous silica films, which enables control over the surface charge and permselectivity of the material.[100]

1.2.3 Co-condensation

As we explained in Section 1.1, mesoporous silica films are prepared by the condensation of silanes in the presence of a templating molecule. In the co-condensation method,[13,15] a functional silane, i.e., a silane with a functional group and one or more reactive groups (such as −OR), is added to the precursor solution. The resulting film contains, therefore, the functional group both on the inner surfaces of the pores and inside the silica walls. For example, a precursor solution for co-condensation synthesis[101] comprises tetraethoxysilane as the inorganic oxide precursor, 3-aminopropyltrieoxysilane (APTES) as the organofunctional silane, and the F127 block copolymer as the structure-directing agent. The co-condensation method differs from all the other techniques described in this section in the sense that functionality is incorporated during the synthesis of the material itself, rather than introduced by post-functionalization. For this reason, co-condensation is regarded as a "one-pot" route to obtain functional nanoporous materials.[14] However, co-condensation and post-functionalization routes are not mutually exclusive but rather complementary, for example, the introduction of amino groups in mesoporous materials enables the subsequent post-functionalization with a carboxylic acid in solution through carbodiimide chemistry (Section 1.2.1.2).[102]

Organic groups introduced by co-condensation may be buried within the film and, therefore, not available for post-functionalization, for example, Soler-Illia and coworkers show that only 16% of the amino groups within an aminosilica mesoporous films are chemically accessible (compared with 37% of available amino groups for a mesoporous silica post-modified with APTES).[101] Moreover, the presence of the organofunctional molecule may affect the properties of the final material either by affecting the chemistry of the condensation reaction during the synthesis or due to its presence within the oxide walls of the material.

1.3 Transport Experiments

We can classify the methods to measure the transport properties of nanopores and nanochannels into three broad groups: those that measure ionic conductance (Fig. 1.5A), those that rely on the redox signal generated by electroactive probes (Fig. 1.5B), and those based on passive permeation against gradients of concentration (Fig. 1.5C).

1.3.1 Conductance Measurements

Fig. 1.2A schematizes a conductance experiment. A membrane containing one or multiple channels is mounted in a two-compartment cell made of a chemically inert material. Each side of the cell is filled by an identical solution (typically KCl solutions are used, since both K^+ and Cl^- have almost the same mobility) and an electrode (e.g., a reversible Ag/AgCl electrode) is inserted into each compartment. Upon applying an electric potential difference between the two electrodes, an ionic current flows across the system. For reversible Ag/AgCl electrodes, the ionic currents are transformed into electric currents by the following redox reaction.

$$AgCl(s) + e^- \rightleftarrows Ag(s) + Cl^-(aq) \tag{1.3}$$

In this reaction, the AgCl (which is a white/gray solid coating the Ag wire) is consumed in one electrode and produced in the other.

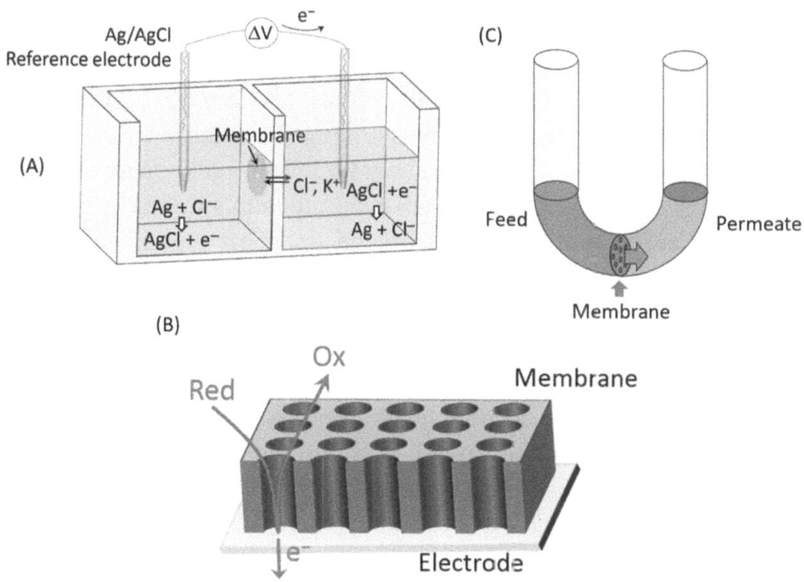

FIGURE 1.5 Schematic representations of experiments used to evaluate the transport properties of nanochannels and nanopores: (A) conductance measurement, (B) electroactive probe measurements, and (C) concentration-driven transport.

The electric current is measured as it flows from one electrode to the other through an external circuit. The instrumentation required to apply the voltage bias between electrodes and to measure the resulting current depends on whether the membrane contains one or multiple channels. Currents in single-channel systems are of the order of picoamperes and in general they can be measured with patch clamp amplifiers,[103] such as those used in electrophiosiology, or with pico/femtoammeters.[104] Single-pore recordings also requires electromagnetic shielding (e.g., a Faraday cage) to reduce noise. Membranes with arrays of nanochannel or nanopores exhibit currents in the order of the nano/microamperes, which can be measured with a potentiostat.[105]

Conductance measurements are useful to determine the transport properties of nanopores and nanochannels in self-standing membranes. These measurements are the basis of many proposed analytical applications, notably the DNA-translocation experiment.[106] Moreover, of the three experimental strategies discussed in this section, conductance measurements are the only ones that allow determining transport through single-pore membranes. This experiment also allows to study transport rectification, but, on the other hand, it does not provide straightforward information on ion selectivity (i.e., selective transport of anions or cations).

1.3.2 Electroactive Probe Measurements

Fig. 1.2B depicts the electroactive-probe experiment: one side of a nanochannel/nanopore membrane is in contact with the surface of a flat electrode and its other side is in contact with a solution of a redox-active species. The redox-active species diffuse through the membrane and are oxidized or reduced at the surface of the underlying electrode. The magnitude of the redox signal generated by this process is then related to the permeability of the membrane. This method is generally useful for nanopore membranes that can be deposited on electrode surfaces, such as the self-assembled colloidal films in Chapter 5, Chemically Modified Nanopores in Colloidal Crystals; however, free-standing membranes can also be measured by this method by coating one of its sides with a metal (typically gold using thermal evaporation or sputtering). The membrane/electrode system is mounted in a three-electrode cell.

The choice of the redox probe depends on the transport property under investigation. For example, electrostatic effects on permeation can be probed by comparing the redox signal generated by negative (e.g., $Fe(CN)_6^{4-/3-}$), positive (e.g., $Ru(NH_3)_6^{2+/3+}$), and neutral (e.g., ferrocene-methanol) probes. The use of these redox probes has some experimental caveats, for instance the probes $Fe(CN)_6^{4-/3-}$ and $Ru(NH_3)_6^{2+/3+}$ can strongly adsorb to different surfaces and give rise to nonideal behaviors. Moreover, both $Fe(CN)_6^{4-/3-}$ and ferrocene-methanol slowly decompose in aqueous solutions. In addition to the redox probe, the measuring solution contains added salt (supporting electrolyte), whose role is to provide ionic conductivity to the solution and to fix its ionic strength.

Among the electrochemical techniques that can be used to quantify the response of the probe at the electrode, CV is arguably the most used. In this technique, the electrode potential is scanned between a potential higher than the redox potential of the probe (i.e., a potential where the probe at the electrode surface is fully oxidized) and a potential lower

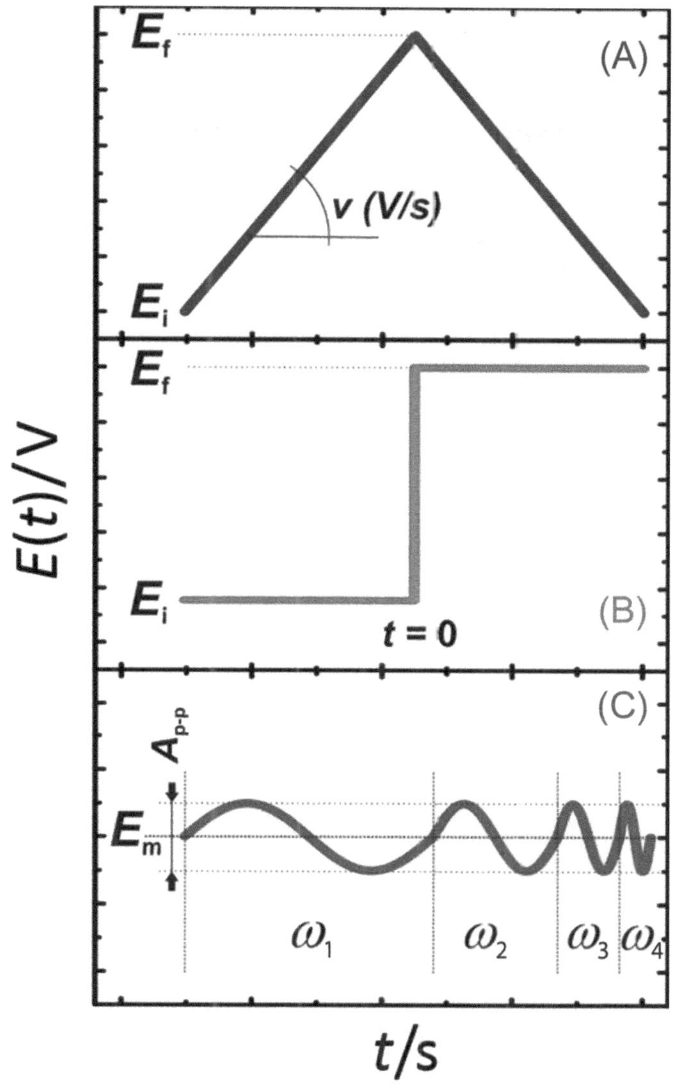

FIGURE 1.6 Potential-time programs for the most common electrochemical techniques: (A) CV, (B) chronoamperometry, and (C) electrochemical impedance spectroscopy. Some parameters of these techniques are: v = scan rate, E_i and E_f = initial and final potentials, E_m = amplitude of potential modulation, and ω = frequency of potential modulation.

than the redox potential of the probe (where the probe is fully reduced), or vice versa, see potential versus time program in Fig. 1.6A. The oxidation or reduction of the probe at the electrode surface gives rise to a signal in the current–potential plot, whose shape provides information on the diffusion of the redox probe through the film. For a detailed discussion of the implementation of the technique and its theoretical foundations, we refer the reader to chapters 6 and 14 of Bard and Faulkner's book.[107] We will discuss here the most salient features of current–potential waves that enable the interpretation of experimental results. We will also critically discuss how the shape of a cyclic voltammogram can be used to extract information about the permeability of nanopore/nanochannel membranes.

Fig. 1.7A shows a typical current–potential curve for a redox probe in a homogeneous solution reacting on a noncoated, macroscopic electrode. The curve shows a positive peak, which is assigned to the oxidation of the redox probe (anodic peak) and a negative peak, due to the reduction of the probe (cathodic peak). The positions of the peaks are shifted with respect to the redox potential of the redox probe due to diffusional limitations.[107] In this experiment, only the redox probes located within a diffusional distance to the electrode are able to electrochemically react at its surface. This characteristic distance is given by[108]:

$$l_d = \sqrt{\frac{2DRT}{F\upsilon}} \tag{1.4}$$

where υ is the scan rate (see Fig. 1.6A), D is the diffusion coefficient, T is the temperature, and F and R are Faraday's and the ideal gas constants, respectively. Note the dependence l_d on the scan rate (which dictates the timescale of the experiment): the faster the scan rate, the shorter the diffusional distance. In the diffusional-controlled CV response, the peak current (i_p) is given by the Randles–Sevcik equation[107]:

$$i_p = 0.446 nFAc_{probe}\left(\frac{nF\upsilon D}{RT}\right)^{1/2} \tag{1.5}$$

where c_{probe} is the concentration of the redox molecule and A is the area of the electrode. The Randles–Sevcik equation indicates that the peak current in the cyclic-voltammetry experiment depends both on the diffusion constant and the concentration of the redox probe inside the film. In other words, while i_p characterizes the permeability of the nanopore or nanochannel membrane, it cannot separate concentration and mobility effects.

The ideal voltammetric response depicted in Fig. 1.6A is strictly valid for a noncoated substrate, but it is typically observed for electrodes coated by porous nanochannel/nanopore membranes as well (e.g., see Ref. 102). The scaling law between peak current and scan rate predicted by the Randles–Sevcik equation (Eq. (1.5)), $i_p \propto \upsilon^{1/2}$, is also observed in these experiments. Note, however, that if the film thickness (h) is smaller than the diffusional timescale l_d, then the redox signal has contributions both from probe molecules within the film and probe molecules free in solution, and therefore estimating c_{probe} or D using Eq. (1.5) lacks of physical meaning. In the opposite case, $h \ll l_d$, the diffusional wave is restricted to the interior of the film and thus Eq. (1.5) can be safely used to extract quantitative information of the concentration and/or mobility of the probe in the film.

In many cases, the concentration of the redox probes is greatly enhanced within the film with respect to solution due to attractive interactions between the probe and the walls of the

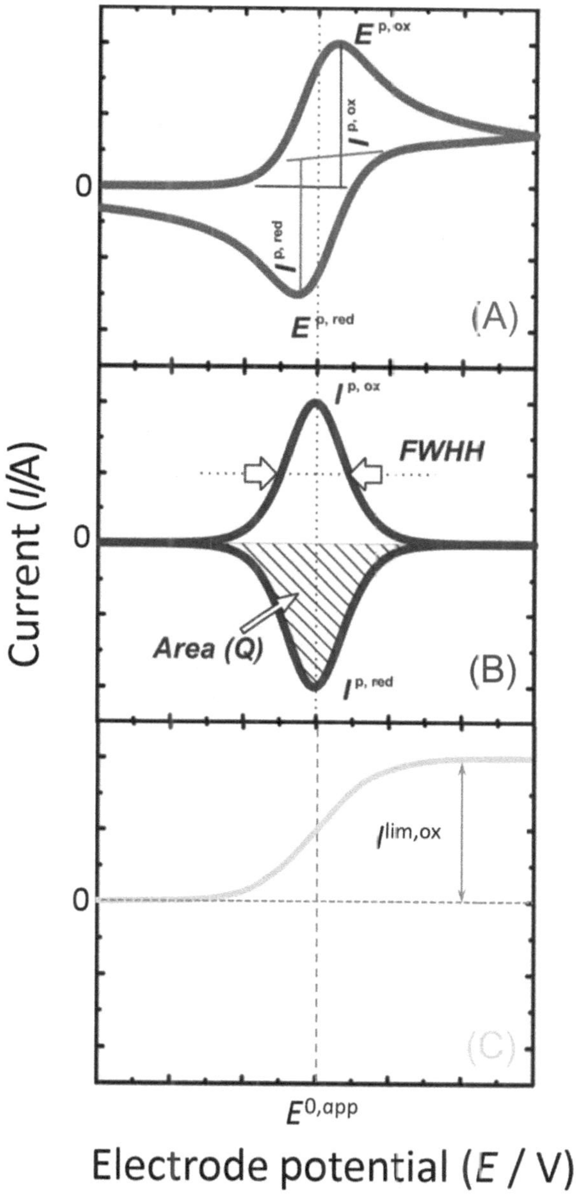

FIGURE 1.7 Ideal current–potential curves for redox probes in different systems: (A) diffusion-controlled voltammogram for the reduction of a redox probe in solution on a macroscopic electrode, (B) fully reversible response for a redox probe confined in a thin membrane, and (C) diffusion-controlled response for the reduction of a redox probe in solution on a UME. The vertical dashed lines show the apparent redox potential of the redox probe in solution, $I^{p,\ red}$ and $I^{p,\ ox}$ are the peak currents for the reduction and oxidation peaks in (A) and (B), $I^{lim,\ ox}$ is the limiting current for oxidation in (C), and FWHH is the full width at half height of the redox peak.

nanochannels/nanopores. In this case, diffusional transport of charge, either by physical diffusion of the redox species inside the film or by electron hopping between immobilized species,[109] is limited to the interior of the film. For slow scan rates (when the diffusional wave spans the whole film) the CV response is that of a surface-confined species,[110] shown in Fig. 1.7B. In the cyclic voltammograms of Fig. 1.7B, both the anodic and cathodic peaks appear at the same potential, which is equal to the apparent redox potential of the probe. The concentration of redox molecules within the film (or at least those that are electrically connected to the electrode) can be determined from the charge of the voltammetric peak, Q (i.e., the area under the peak, see Fig. 1.7B) as:

$$c_{\text{probe}} = \frac{Q}{FAh} \tag{1.6}$$

The current–potential waves in Fig. 1.7A and B correspond to macroscopic planar electrodes. We will discuss now the typical CV response for a diffusion-limited redox probe on an ultramicroelectrode (UME: an electrode with radius $<25\ \mu\text{m}$), see Fig. 1.7C. This signal is observed for the disk microelectrodes coated by colloidal-crystal nanoporous films discussed in Chapter 5, Chemically Modified Nanopores in Colloidal Crystals (see also Ref. 11). For a planar electrode, the redox molecules experience semiplanar diffusion, which creates diffusional limitations and give rise to a decay in the current with time, thus producing the redox peaks in the current–potential plot of Fig. 1.7A. On the other hand, diffusional access to a microelectrode is hemispherical and can sustain a steady-state (nondecaying) current.[107] Therefore, the current in a microelectrode reaches a steady-state value, known as the limiting current, and it does not develop a peak in the CV (see also the discussion for the conduction of single vs multiple nanochannels in chapter: Theoretical Basis for Structure and Transport in Nanopores and Nanochannels). In a bare disk UME, the limiting current due to diffusional mass transfer is[107]:

$$i_{\text{bare}} = 4\pi r_0 F D_{\text{bulk}} c_{\text{probe, bulk}} \tag{1.7}$$

where D_{bulk} and $c_{\text{probe,bulk}}$ are the bulk diffusion coefficient and concentration of the probe and r_0 is the radius of the UME. On the other hand, the maximum current through a film (assuming no mass-transport limitations in the solution) is:

$$i_{\text{film}} = \frac{A}{h} F D_{\text{film}} \Delta c_{\text{probe}} = \frac{\pi r_0^2}{h} F D_{\text{film}} \Delta c_{\text{probe}} \tag{1.8}$$

where Δc_{probe} is the difference of the probe concentration between the electrode surface ($z = 0$) and the film/solution interface ($z = h$). The maximum current occurs when $c_{\text{probe}}(z = 0) \sim 0$ (fast reaction on the electrode) and thus $\Delta c_{\text{probe}} = c_{\text{probe}}(z = h)$. The parameters D_{film} and Δc_{probe} can be further related to aspects of the nanoporous/nanochannel film such as tortuosity, porosity, permselectivity, etc. (see chapter: Chemically Modified Nanopores in Colloidal Crystals).

The total limiting current approximately is,

$$i_{lim}^{-1} = i_{bare}^{-1} + i_{film}^{-1} \tag{1.9}$$

Thus, the total current is dominated by the smallest contribution. Since $i_{bare} \propto r_0$ and $i_{film} \propto r_0^2/h$, the limiting current can provide information on the permeability of the film for small UMEs or thick films since in these cases $i_{lim} \sim i_{film}$.

In addition to CV, there are other electrochemical methods that can be used to probe nanochannel/nanopore membranes, such as are chronoamperometry (Fig. 1.6B), where the current is measured in time following a jump of the electrode potential, and electrochemical impedance spectroscopy (Fig. 1.6C), in which the current is measured for low-amplitude sinusoidal perturbations of the electrode potential of varying frequency, which allows to calculate the complex impedance as a function of the perturbation frequency (impedance spectrum).

1.3.3 Concentration-Driven Transport

This experiment evaluates the permeation rate of molecules through a nanochannel or nanopore membrane. The membrane is mounted into a two-compartment cell (e.g., a U-tube cell, see Fig. 1.2C). One side of the cell (the feed side) is filled with a solution containing a dye, while the other side (the permeate side) is filled with a dye-free solution (the ionic strength and pH of the solutions on both sides of the membranes are usually the same).[57,111] The dye flows from the feed side to the permeate side through the membrane driven by its concentration gradient. The permeation rate is quantified by monitoring the change in dye concentration in the permeate side as a function of time. For short permeation times, the concentration of the dye in the permeate cell increases linearly with time,[111] which allows to calculate the flux through the membrane.

Comparison between different dyes allows to study the selectivity of the membrane, for example, size selectivity can be inferred by comparing dyes of similar charge but different size (e.g., $Ru(bpy)_3^{2+}$ and methylviologen^{2+})[111] and charge selectivity can be measured by comparing dyes of similar size and different charge (e.g., $Ru(bpy)_3^{2+}$ and 2,6-naphthalenedisulfonate^{2-}).[112] Note also that this method cannot be straightforwardly applied to single nanochannel membranes: measurable changes of concentration can only be obtained with arrays of nanopores or channels.

References

1. Hulteen, J. C.; Martin, C. R. A General Template-Based Method for the Preparation of Nanomaterials. *J. Mater. Chem.* **1997,** *7,* 1075–1087.
2. Spohr, R. *Ion Tracks and Microtechnology;* Vieweg Verlag, 1990.
3. Yameen, B.; Ali, M.; Neumann, R.; Ensinger, W.; Knoll, W.; Azzaroni, O. Synthetic Proton-Gated Ion Channels Via Single Solid-State Nanochannels Modified with Responsive Polymer Brushes. *Nano Lett.* **2009,** *9,* 2788–2793.

4. Yameen, B.; Ali, M.; Neumann, R.; Ensinger, W.; Knoll, W.; Azzaroni, O. Proton-Regulated Rectified Ionic Transport Through Solid-State Conical Nanopores Modified with Phosphate-Bearing Polymer Brushes. *Chem. Commun.* **2010,** *46,* 1908–1910.
5. Ali, M.; Ramirez, P.; Nguyen, H. Q.; Nasir, S.; Cervera, J.; Mafe, S., et al. Single Cigar-Shaped Nanopores Functionalized with Amphoteric Amino Acid Chains: Experimental and Theoretical Characterization. *ACS Nano* **2012,** *6,* 3631–3640.
6. Chtanko, N.; Toimil Molares, M. E.; Cornelius, T.; Dobrev, A.; Neumann, R. Etched Single-Ion-Track Templates for Single Nanowire Synthesis. *J. Phys. Chem. B* **2004,** *108,* 9950–9954.
7. Krapf, D.; Wu, M. Y.; Smeets, R. M. M.; Zandbergen, H. W.; Dekker, C.; Lemay, S. G. Fabrication and Characterization of Nanopore-Based Electrodes with Radii Down to 2 nm. *Nano Lett.* **2006,** *6,* 105–109.
8. Xander, J. A. J.; Magnus, P. J.; Calin, P.; Gautam, V. S.; Cees, D.; Nynke, H. D. Rapid Manufacturing of Low-Noise Membranes for Nanopore Sensors by Trans -Chip Illumination Lithography. *Nanotechnology* **2012,** *23,* 475302.
9. Li, J.; Stein, D.; McMullan, C.; Branton, D.; Aziz, M. J.; Golovchenko, J. A. Ion-Beam Sculpting at Nanometre Length Scales. *Nature* **2001,** *412,* 166–169.
10. Storm, A. J.; Chen, J. H.; Ling, X. S.; Zandbergen, H. W.; Dekker, C. Fabrication of Solid-State Nanopores with Single-Nanometre Precision. *Nat. Mater.* **2003,** *2,* 537–540.
11. Zharov, I.; Khabibullin, A. Surface-Modified Silica Colloidal Crystals: Nanoporous Films and Membranes with Controlled Ionic and Molecular Transport. *Acc. Chem. Res.* **2014,** *47,* 440–449.
12. Wan, Y.; Zhao, D. On the Controllable Soft-Templating Approach to Mesoporous Silicates. *Chem. Rev.* **2007,** *107,* 2821–2860.
13. Soler-Illia, G. J.; Azzaroni, O. Multifunctional Hybrids by Combining Ordered Mesoporous Materials and Macromolecular Building Blocks. *Chem. Soc. Rev.* **2011,** *40,* 1107–1150.
14. Soler-Illia, G. J.; Sanchez, C.; Lebeau, B.; Patarin, J. Chemical Strategies to Design Textured Materials: From Microporous and Mesoporous Oxides to Nanonetworks and Hierarchical Structures. *Chem. Rev.* **2002,** *102,* 4093–4138.
15. Mehdi, A.; Reye, C.; Corriu, R. From Molecular Chemistry to Hybrid Nanomaterials. Design and Functionalization. *Chem. Soc. Rev.* **2011,** *40,* 563–574.
16. Sanchez, C.; Boissière, C.; Grosso, D.; Laberty, C.; Nicole, L. Design, Synthesis, and Properties of Inorganic and Hybrid Thin Films Having Periodically Organized Nanoporosity. *Chem. Mater.* **2008,** *20,* 682–737.
17. Karhanek, M.; Kemp, J. T.; Pourmand, N.; Davis, R. W.; Webb, C. D. Single DNA Molecule Detection Using Nanopipettes and Nanoparticles. *Nano Lett.* **2005,** *5,* 403–407.
18. Zhang, B.; Zhang, Y.; White, H. S. The Nanopore Electrode. *Anal. Chem.* **2004,** *76,* 6229–6238.
19. Gao, C.; Ding, S.; Tan, Q.; Gu, L. Q. Method of Creating a Nanopore-Terminated Probe for Single-Molecule Enantiomer Discrimination. *Anal. Chem.* **2009,** *81,* 80–86.
20. Velleman, L.; Shearer, C. J.; Ellis, A. V.; Losic, D.; Voelcker, N. H.; Shapter, J. G. Fabrication of Self-Supporting Porous Silicon Membranes and Tuning Transport Properties by Surface Functionalization. *Nanoscale* **2010,** *2,* 1756–1761.
21. Parkhutik, V. Porous Silicon—Mechanisms of Growth and Applications. *Solid · State Electron.* **1999,** *43,* 1121–1141.
22. Thompson, G.; Wood, G. Porous Anodic Film Formation on Aluminium. *Nature* **1981,** *290,* 230–232.
23. Lee, W.; Ji, R.; Gosele, U.; Nielsch, K. Fast Fabrication of Long-Range Ordered Porous Alumina Membranes by Hard Anodization. *Nat. Mater.* **2006,** *5,* 741–747.
24. Wen, L.; Tian, Y.; Ma, J.; Zhai, J.; Jiang, L. Construction of Biomimetic Smart Nanochannels with Polymer Membranes and Application in Energy Conversion Systems. *Phys. Chem. Chem. Phys* **2012,** *14,* 4027–4042.

25. An, S.; Stambaugh, C.; Kim, G.; Lee, M.; Kim, Y.; Lee, K., et al. Low-Volume Liquid Delivery and Nanolithography Using a Nanopipette Combined with a Quartz Tuning Fork-Atomic Force Microscope. *Nanoscale* **2012,** *4,* 6493–6500.
26. Wanunu, M.; Meller, A. Chemically Modified Solid-State Nanopores. *Nano Lett.* **2007,** *7,* 1580–1585.
27. Velleman, L.; Triani, G.; Evans, P. J.; Shapter, J. G.; Losic, D. Structural and Chemical Modification of Porous Alumina Membranes. *Microporous Mesoporous Mater.* **2009,** *126,* 87–94.
28. Vansant, E. F.; Van Der Voort, P.; Vrancken, K. C. *Characterization and Chemical Modification of the Silica Surface;* Elsevier, 1995.
29. Papirer, E. *Adsorption on Silica Surfaces,* vol. 90. CRC Press, 2000.
30. Zhuravlev, L. T. The Surface Chemistry of Amorphous Silica. Zhuravlev Model. *Colloids Surf. A* **2000,** *173,* 1–38.
31. Ong, S.; Zhao, X.; Eisenthal, K. B. Polarization of Water Molecules at a Charged Interface: Second Harmonic Studies of the Silica/Water Interface. *Chem. Phys. Lett.* **1992,** *191,* 327–335.
32. Rosenholm, J. M.; Czuryszkiewicz, T.; Kleitz, F.; Rosenholm, J. B.; Linden, M. On the Nature of the Bronsted Acidic Groups on Native and Functionalized Mesoporous Siliceous SBA-15 as Studied by Benzylamine Adsorption from Solution. *Langmuir* **2007,** *23,* 4315–4323.
33. Iler, R. K. *The Chemistry of Silica: Solubility, Polymerization, Colloid and Surface Properties, and Biochemistry;* Wiley, 1979.
34. Marchard-Brynaerd, J. Polymer Membranes. In *Encyclopedia of Surface and Colloid Science;* Somasundaran, P., Ed.; Vol. 7; CRC Press, 2006.
35. Dauginet, L.; Duwez, A. S.; Legras, R.; Demoustier-Champagne, S. Surface Modification of Polycarbonate and Poly(Ethylene Terephthalate) Films and Membranes by Polyelectrolyte Deposition. *Langmuir* **2001,** *17,* 3952–3957.
36. Marchard-Brynaert, J. Surface Functionalization of Polymer Membranes. In *Surface Chemistry and Electrochemistry of Membranes;* Sorenson, T. S., Ed.; Vol. 79; CRC Press, 1999.
37. Siwy, Z.; Dobrev, D.; Neumann, R.; Trautmann, C.; Voss, K. Electro-Responsive Asymmetric Nanopores in Polyimide with Stable Ion-Current Signal. *Appl. Phys. A* **2003,** *76,* 781–785.
38. Perrot, C.; Gonon, L.; Marestin, C.; Gebel, G. Hydrolytic Degradation of Sulfonated Polyimide Membranes for Fuel Cells. *J. Membr. Sci.* **2011,** *379,* 207–214.
39. Siwy, Z.; Apel, P.; Baur, D.; Dobrev, D. D.; Korchev, Y. E.; Neumann, R., et al. Preparation of Synthetic Nanopores with Transport Properties Analogous to Biological Channels. *Surf. Sci.* **2003,** *532,* 1061–1066.
40. Siwy, Z.; Kosińska, I. D.; Fuliński, A.; Martin, C. R. Asymmetric Diffusion Through Synthetic Nanopores. *Phys. Rev. Lett.* **2005,** *94.*
41. Menon, V. P.; Martin, C. R. Fabrication and Evaluation of Nanoelectrode Ensembles. *Anal. Chem.* **1995,** *67,* 1920–1928.
42. Hou, X.; Zhang, H. C.; Jiang, L. Building Bio-Inspired Artificial Functional Nanochannels: From Symmetric to Asymmetric Modification. *Angew. Chem. Int. Ed.* **2012,** *51,* 5296–5307.
43. Ali, M.; Neumann, R.; Ensinger, W. Sequence-Specific Recognition of DNA Oligomer Using Peptide Nucleic Acid (PNA)-Modified Synthetic Ion Channels: PNA/DNA Hybridization in Nanoconfined Environment. *ACS Nano* **2010,** *4,* 7267–7274.
44. Hou, X.; Guo, W.; Xia, F.; Nie, F. Q.; Dong, H.; Tian, Y., et al. A Biomimetic Potassium Responsive Nanochannel: G-Quadruplex DNA Conformational Switching in a Synthetic Nanopore. *J. Am. Chem. Soc.* **2009,** *131,* 7800–7805.
45. Ali, M.; Mafe, S.; Ramirez, P.; Neumann, R.; Ensinger, W. Logic Gates Using Nanofluidic Diodes Based on Conical Nanopores Functionalized with Polyprotic Acid Chains. *Langmuir* **2009,** *25,* 11993–11997.

46. Yameen, B.; Ali, M.; Neumann, R.; Ensinger, W.; Knoll, W.; Azzaroni, O. Ionic Transport through Single Solid-State Nanopores Controlled with Thermally Nanoactuated Macromolecular Gates. *Small* **2009,** *5,* 1287–1291.

47. Ali, M.; Ramirez, P.; Tahir, M. N.; Mafe, S.; Siwy, Z.; Neumann, R., et al. Biomolecular Conjugation Inside Synthetic Polymer Nanopores Via Glycoprotein-Lectin Interactions. *Nanoscale* **2011,** *3,* 1894–1903.

48. Hou, X.; Liu, Y. J.; Dong, H.; Yang, F.; Li, L.; Jiang, L., et al. Transport Nanodevice: Asymmetric Chemical Modification of Single Nanochannels. *Adv. Mater.* **2010,** *22,* 2440.

49. Ali, M.; Nasir, S.; Nguyen, Q. H.; Sahoo, J. K.; Tahir, M. N.; Tremel, W., et al. Metal Ion Affinity-Based Biomolecular Recognition and Conjugation Inside Synthetic Polymer Nanopores Modified with Iron-Terpyridine Complexes. *J. Am. Chem. Soc.* **2011,** *133,* 17307–17314.

50. Ali, M.; Nasir, S.; Ramirez, P.; Cervera, J.; Mafe, S.; Ensinger, W. Carbohydrate-Mediated Biomolecular Recognition and Gating of Synthetic Ion Channels. *J. Phys. Chem. C* **2013,** *117,* 18234–18242.

51. Sun, Z.; Han, C.; Song, M.; Wen, L.; Tian, D.; Li, H., et al. Fabrication of Cysteine-Responsive Biomimetic Single Nanochannels by a Thiol-yne Reaction Strategy and Their Application for Sensing in Urine Samples. *Adv. Mater.* **2014,** *26,* 455–460.

52. Han, C.; Hou, X.; Zhang, H.; Guo, W.; Li, H.; Jiang, L. Enantioselective Recognition in Biomimetic Single Artificial Nanochannels. *J. Am. Chem. Soc.* **2011,** *133,* 7644–7647.

53. Hou, X.; Yang, F.; Li, L.; Song, Y. L.; Jiang, L.; Zhu, D. B., et al. Responsive Single Nanochannel. *J. Am. Chem. Soc.* **2010,** *132,* 11736–11742.

54. Zhang, H.; Hou, X.; Zeng, L.; Yang, F.; Li, L.; Yan, D., et al. Bioinspired Artificial Single Ion Pump. *J. Am. Chem. Soc.* **2013,** *135,* 16102–16110.

55. Love, J. C.; Estroff, L. A.; Kriebel, J. K.; Nuzzo, R. G.; Whitesides, G. M. Self-Assembled Monolayers of Thiolates on Metals as a Form of Nanotechnology. *Chem. Rev.* **2005,** *105,* 1103–1169.

56. Muench, F.; Kunz, U.; Neetzel, C.; Lauterbach, S.; Kleebe, H. J.; Ensinger, W. 4-(Dimethylamino)Pyridine as a Powerful Auxiliary Reagent in the Electroless Synthesis of Gold Nanotubes. *Langmuir* **2011,** *27,* 430–435.

57. Nishizawa, M.; Menon, V. P.; Martin, C. R. Metal Nanotubule Membranes with Electrochemically Switchable Ion-Transport Selectivity. *Science* **1995,** *268,* 700–705.

58. Gao, P.; Martin, C. R. Voltage Charging Enhances Ionic Conductivity in Gold Nanotube Membranes. *ACS Nano* **2014,** *8,* 8266–8272.

59. De Leo, M.; Pereira, F. C.; Moretto, L. M.; Scopece, P.; Polizzi, S.; Ugo, P. Towards a Better Understanding of Gold Electroless Deposition in Track-Etched Templates. *Chem. Mater.* **2007,** *19,* 5955–5964.

60. Kohli, P.; Wharton, J. E.; Braide, O.; Martin, C. R. Template Synthesis of Gold Nanotubes in an Anodic Alumina Membrane. *J. Nanosci. Nanotechnol.* **2004,** *4,* 605–610.

61. Yu, Y.; Kant, K.; Shapter, J. G.; Addai-Mensah, J.; Losic, D. Gold Nanotube Membranes Have Catalytic Properties. *Microporous Mesoporous Mater.* **2012,** *153,* 131–136.

62. Whelan, J. C.; Karawdeniya, B. I.; Bandara, Y. M.; Velleco, B. D.; Masterson, C. M.; Dwyer, J. R. Electroless Plating of Thin Gold Films Directly onto Silicon Nitride Thin Films and into Micropores. *ACS Appl. Mater. Interfaces* **2014,** *6,* 10952–10957.

63. Ignacio-de Leon, P. A.; Zharov, I. SiO2@Au Core–Shell Nanospheres Self-Assemble to Form Colloidal Crystals That Can Be Sintered and Surface Modified to Produce pH-Controlled Membranes. *Langmuir* **2013,** *29,* 3749–3756.

64. Pérez-Mitta, G.; Marmisollé, W. A.; Trautmann, C.; Toimil-Molares, M. E.; Azzaroni, O. Nanofluidic Diodes with Dynamic Rectification Properties Stemming from Reversible Electrochemical Conversions in Conducting Polymers. *J. Am. Chem. Soc.* **2015,** *137,* 15382–15385.

65. Kalman, E. B.; Sudre, O.; Vlassiouk, I.; Siwy, Z. S. Control of Ionic Transport Through Gated Single Conical Nanopores. *Anal. Bioanal. Chem.* **2009,** *394,* 413–419.

66. Macias, G.; Hernández-Eguía, L. P.; Ferré-Borrull, J.; Pallares, J.; Marsal, L. F. Gold-Coated Ordered Nanoporous Anodic Alumina Bilayers for Future Label-Free Interferometric Biosensors. *ACS Appl. Mater. Interfaces* **2013,** *5*, 8093–8098.

67. Tian, Y.; Zhang, Z.; Wen, L.; Ma, J.; Zhang, Y.; Liu, W., et al. Mercury(II)-Gated Single Nanochannel. *Chem. Commun.* **2013,** *49*, 10679–10681.

68. Guo, W.; Xia, H.; Xia, F.; Hou, X.; Cao, L.; Wang, L., et al. Current Rectification in Temperature-Responsive Single Nanopores. *ChemPhysChem* **2010,** *11*, 859–864.

69. Tian, Y.; Hou, X.; Jiang, L. Biomimetic Ionic Rectifier Systems: Asymmetric Modification of Single Nanochannels by Ion Sputtering Technology. *J. Electroanal. Chem.* **2011,** *656*, 231–236.

70. McPhillips, J.; Murphy, A.; Jonsson, M. P.; Hendren, W. R.; Atkinson, R.; Höök, F., et al. High-Performance Biosensing Using Arrays of Plasmonic Nanotubes. *ACS Nano* **2010,** *4*, 2210–2216.

71. Cheng, C. L.; Lin, J. S.; Chen, Y. F. Fabrication and Growth Mechanism of Metal (Zn, Sn) Nanotube Arrays and Metal (Cu, Ag) Nanotube/Nanowire Junction Arrays. *Mater. Lett.* **2008,** *62*, 1666–1669.

72. Cao, G.; Liu, D. Template-Based Synthesis of Nanorod, Nanowire, and Nanotube Arrays. *Adv. Colloid Interface Sci.* **2008,** *136*, 45–64.

73. Chun, K. Y.; Stroeve, P. Protein Transport in Nanoporous Membranes Modified with Self-Assembled Monolayers of Functionalized Thiols. *Langmuir* **2002,** *18*, 4653–4658

74. Lee, S. B.; Martin, C. R. Controlling the Transport Properties of Gold Nanotubule Membranes Using Chemisorbed Thiols. *Chem. Mater.* **2001,** *13*, 3236–3244.

75. Lee, S. B.; Martin, C. R. pH-Switchable, Ion-Permselective Gold Nanotubule Membrane Based on Chemisorbed Cysteine. *Anal. Chem.* **2001,** *73*, 768–775.

76. Nasir, S.; Ali, M.; Ramirez, P.; Gómez, V.; Oschmann, B.; Muench, F., et al. Fabrication of Single Cylindrical Au-Coated Nanopores with Non-Homogeneous Fixed Charge Distribution Exhibiting High Current Rectifications. *ACS Appl. Mater. Interfaces* **2014,** *6*, 12486–12494.

77. Jágerszki, G.; Takács, Á.; Bitter, I.; Gyurcsányi, R. E. Solid-State Ion Channels for Potentiometric Sensing. *Angew. Chem. Int. Ed.* **2011,** *50*, 1656–1659.

78. Hulteen, J. C.; Jirage, K. B.; Martin, C. R. Introducing Chemical Transport Selectivity into Gold Nanotubule Membranes. *J. Am. Chem. Soc.* **1998,** *120*, 6603–6604.

79. Jirage, K. B.; Hulteen, J. C.; Martin, C. R. Effect of Thiol Chemisorption on the Transport Properties of Gold Nanotubule Membranes. *Anal. Chem.* **1999,** *71*, 4913–4918.

80. Kim, B. Y.; Swearingen, C. B.; Ho, J. A.; Romanova, E. V.; Bohn, P. W.; Sweedler, J. V. Direct Immobilization of Fab' in Nanocapillaries for Manipulating Mass-Limited Samples. *J. Am. Chem. Soc.* **2007,** *129*, 7620–7626.

81. Harrell, C. C.; Kohli, P.; Siwy, Z.; Martin, C. R. DNA-Nanotube Artificial Ion Channels. *J. Am. Chem. Soc.* **2004,** *126*, 15646–15647.

82. Jágerszki, G.; Gyurcsányi, R. E.; Höfler, L.; Pretsch, E. Hybridization-Modulated Ion Fluxes through Peptide-Nucleic-Acid-Functionalized Gold Nanotubes. A New Approach to Quantitative Label-Free DNA Analysis. *Nano Lett.* **2007,** *7*, 1609–1612.

83. Netz, R. R.; Andelman, D. Neutral and Charged Polymers at Interfaces. *Phys. Rep.* **2003,** *380*, 1–95.

84. Wang, Y.; Bansal, V.; Zelikin, A. N.; Caruso, F. Templated Synthesis of Single-Component Polymer Capsules and Their Application in Drug Delivery. *Nano Lett.* **2008,** *8*, 1741–1745.

85. Wang, Y.; Angelatos, A. S.; Dunstan, D. E.; Caruso, F. Infiltration of Macromolecules into Nanoporous Silica Particles. *Macromolecules* **2007,** *40*, 7594–7600.

86. Elbert, J.; Krohm, F.; Rüttiger, C.; Kienle, S.; Didzoleit, H.; Balzer, B. N , et al. Polymer-Modified Mesoporous Silica Thin Films for Redox-Mediated Selective Membrane Gating. *Adv. Funct. Mater.* **2014,** *24*, 1591–1601.

87. Decher, G. Fuzzy Nanoassemblies: Toward Layered Polymeric Multicomposites. *Science* **1997,** *277,* 1232–1237.

88. Decher, G.; Schlenoff, B. J. *Multilayer Thin Films;* Wiley-VCH: Weinheim, 2003.

89. Decher, G.; Hong, J. D.; Schmitt, J. Buildup of Ultrathin Multilayer Films by a Self-Assembly Process: III. Consecutively Alternating Adsorption of Anionic and Cationic Polyelectrolytes on Charged Surfaces. *Thin Solid Films* **1992,** *210–211,* 831–835.

90. Ariga, K.; Ji, Q.; Hill, J. P.; Vinu, A. Coupling of Soft Technology (Layer-by-Layer Assembly) with Hard Materials (Mesoporous Solids) to Give Hierarchic Functional Structures. *Soft Matter* **2009,** *5,* 3562–3571.

91. Azzaroni, O.; Lau, K. H. A. Layer-by-Layer Assemblies in Nanoporous Templates: Nano-Organized Design and Applications of Soft Nanotechnology. *Soft Matter* **2011,** *7,* 8709–8724.

92. DeRocher, J. P.; Mao, P.; Han, J.; Rubner, M. F.; Cohen, R. E. Layer-by-Layer Assembly of Polyelectrolytes in Nanofluidic Devices. *Macromolecules* **2010,** *43,* 2430–2437.

93. Lazzara, T. D.; Lau, K. H. A.; Abou-Kandil, A. I.; Caminade, A. M.; Majoral, J.-P.; Knoll, W. Polyelectrolyte Layer-by-Layer Deposition in Cylindrical Nanopores. *ACS Nano* **2010,** *4,* 3909–3920.

94. Carrillo, J. M. Y.; Dobrynin, A. V. Layer-by-Layer Assembly of Charged Nanoparticles on Porous Substrates: Molecular Dynamics Simulations. *ACS Nano* **2011,** *5,* 3010–3019.

95. Roy, C. J.; Dupont-Gillain, C.; Demoustier-Champagne, S.; Jonas, A. M.; Landoulsi, J. Growth Mechanism of Confined Polyelectrolyte Multilayers in Nanoporous Templates. *Langmuir* **2010,** *26,* 3350–3355.

96. Cho, Y.; Lee, C.; Hong, J. Pore Size Effect on the Formation of Polymer Nanotubular Structures within Nanoporous Templates. *Colloids Surf. A* **2014,** *443,* 195–200.

97. Ali, M.; Yameen, B.; Cervera, J.; Ramirez, P.; Neumann, R.; Ensinger, W., et al. Layer-by-Layer Assembly of Polyelectrolytes into Ionic Current Rectifying Solid-State Nanopores: Insights from Theory and Experiment. *J. Am. Chem. Soc.* **2010,** *132,* 8338–8348.

98. Tian, Y.; He, Q.; Cui, Y.; Tao, C.; Li, J. Assembly of Nanotubes of Poly(4-Vinylpyridine) and Poly (Acrylic Acid) Through Hydrogen Bonding. *Chemistry* **2006,** *12,* 4808–4812.

99. Sukhishvili, S. A.; Granick, S. Layered, Erasable Polymer Multilayers Formed by Hydrogen-Bonded Sequential Self-Assembly. *Macromolecules* **2002,** *35,* 301–310.

100. Kumar, B.; Rao, K. V.; Sampath, S.; George, S. J.; Eswaramoorthy, M. Supramolecular Gating of Ion Transport in Nanochannels. *Angew. Chem. Int. Ed.* **2014,** *126,* 13289–13293.

101. Calvo, A.; Joselevich, M.; Soler-Illia, G. J. A. A.; Williams, F. J. Chemical Reactivity of Amino-Functionalized Mesoporous Silica Thin Films Obtained by Co-Condensation and Post-Grafting Routes. *Microporous Mesoporous Mater.* **2009,** *121,* 67–72.

102. Andrieu-Brunsen, A.; Micoureau, S.; Tagliazucchi, M.; Szleifer, I.; Azzaroni, O.; Soler-Illia, G. J. A. A. Mesoporous Hybrid Thin Film Membranes with Pmetac@Silica Architectures: Controlling Ionic Gating Through the Tuning of Polyelectrolyte Density. *Chem. Mater.* **2015,** *27,* 808–821.

103. Siwy, Z.; Trofin, L.; Kohli, P.; Baker, L. A.; Trautmann, C.; Martin, C. R. Protein Biosensors Based on Biofunctionalized Conical Gold Nanotubes. *J. Am. Chem. Soc.* **2005,** *127,* 5000–5001.

104. Karnik, R.; Castelino, K.; Fan, R.; Yang, P.; Majumdar, A. Effects of Biological Reactions and Modifications on Conductance of Nanofluidic Channels. *Nano Lett.* **2005,** *5,* 1638–1642.

105. Harrell, C. C.; Lee, S. B.; Martin, C. R. Synthetic Single-Nanopore and Nanotube Membranes. *Anal. Chem.* **2003,** *75,* 6861–6867.

106. Edel, J.; Albretch, T. *Engineered Nanopores for Bioanalytical Applications;* Elsevier, 2013.

107. Bard, A. J.; Faulkner, L. R. *Electrochemical Methods,* 2nd ed.; John Wiley & Sons: New York, 2001.

108. Tagliazucchi, M.; Calvo, E. J. Charge Transport in Redox Polyelectrolyte Multilayer Films: The Dramatic Effects of Outmost Layer and Solution Ionic Strength. *ChemPhysChem* **2010,** *11*, 2957–2968.
109. Blauch, D. N.; Saveant, J. M. Dynamics of Electron Hopping in Assemblies of Redox Centers. Percolation and Diffusion. *J. Am. Chem. Soc.* **1992,** *114*, 3323–3332.
110. Laviron, E. General Expression of the Linear Potential Sweep Voltammogram in the Case of Diffusionless Electrochemical Systems. *J. Electroanal. Chem.* **1979,** *101*, 19–28.
111. Martin, C. R.; Nishizawa, M.; Jirage, K.; Kang, M. Investigations of the Transport Properties of Gold Nanotubule Membranes. *J. Phys. Chem. B* **2001,** *105*, 1925–1934.
112. Kang, M. S.; Martin, C. R. Investigations of Potential-Dependent Fluxes of Ionic Permeates in Gold Nanotubule Membranes Prepared Via the Template Method. *Langmuir* **2001,** *17*, 2753–2759.

2

Theoretical Basis for Structure and Transport in Nanopores and Nanochannels

M. Tagliazucchi[1], I. Szleifer[2]

[1]UNIVERSIDAD DE BUENOS AIRES–CONICET, BUENOS AIRES, ARGENTINA

[2]NORTHWESTERN UNIVERSITY, EVANSTON, IL, UNITED STATES

CHAPTER OUTLINE

Chemically Modified Nanopores and Nanochannels.

2.1 Fundamental Concepts of Thermodynamics and Transport

2.1.1 The Chemical Potential

Chemically modified nanopores and nanochannels are highly inhomogeneous and chemically complex systems. In order to provide a solid foundation for the description of their equilibrium and transport properties, we will introduce in this section important fundamental concepts in thermodynamics and transport. The tendency of a chemical species to flow between different parts of the system is given by its chemical potential. In equilibrium, the chemical potentials of all species are constant across the system and no net flow of molecules occurs. On the other hand, if heterogeneities of chemical potential exist within the system, then the molecules will flow in order to decrease these differences. Note that the difference of chemical potential of one component in the system may still increase, provided that this increase is accompanied by an even larger decrease in the differences of the other components, see for example Ref. 1.

Formally, the chemical potential of a species i is the derivative of the free energy of the system with respect to the number of moles of i at constant temperature (T), number of moles of the other species n_j, and pressure, p, (for the Gibbs free energy, G) or volume, V (for the Helmholtz free energy, H), namely:

$$\mu_i = \left.\frac{\partial G}{\partial n_i}\right)_{n_{j(j\neq i)},p,T}, \quad \mu_i = \left.\frac{\partial H}{\partial n_i}\right)_{n_{j(j\neq i)},V,T} \tag{2.1}$$

Based on this definition, we show below that the chemical potential describes the tendency of molecules to flow from one region of the system to other. Let us consider a small change in the Gibbs free energy of a system:

$$dG = -SdT + Vdp + \sum_i \mu_i dn_i \tag{2.2}$$

(where S is the entropy of the system). We will now partition this system into two homogeneous regions, α and β, and consider the exchange of species i between them, while keeping p, T, and n_j constant. The change in free energy of this process is:

$$dG = \mu_i^{\alpha} dn_i^{\alpha} + \mu_i^{\beta} dn_i^{\beta} = -(\mu_i^{\alpha} - \mu_i^{\beta}) dn_i^{\alpha\to\beta} \tag{2.3}$$

where dn_i^{α} and dn_i^{β} are the change in the number of moles of i in regions α and β, respectively. In the second equality of Eq. (2.3) we defined $dn_i^{\alpha\to\beta}$ as the number of moles of region α transferred into region β and used the fact that $dn_i^{\alpha\to\beta} = dn_i^{\beta} = -dn_i^{\alpha}$. The free energy G always decreases for spontaneous processes at constant T and p, so dG in Eq. (2.3) must be negative, which tells us that if $\mu_i^{\alpha} > \mu_i^{\beta}$, then $dn_i^{\alpha\to\beta}$ must be positive and molecules of i will flow from region α into region β. In other words, molecules flow from regions of high chemical potential into regions of low chemical potential. Equilibrium is achieved for $dG = 0$, which, in terms of Eq. (2.3), requires $\mu_i^\alpha = \mu_i^\beta$.

The fact that molecules flow from regions of high chemical potential to regions of low chemical potential is also valid when the chemical potential continuously vary across the system. In such inhomogeneous systems, the flow of molecules occurs against the gradient of chemical potential, $\nabla\mu_i$ (where ∇ is the gradient operator, e.g., in one dimension, $\nabla\mu_i = d\mu_i/dx$). If the gradient of chemical potential is not very large and the influence of species $j \neq i$ on the diffusion of species i can be neglected (i.e., we neglect off-diagonal elements in the Onsager coefficient matrix[2]), then the flux of species i will be proportional to its gradient of chemical potential:

$$\mathbf{J}_i(\mathbf{r}) = -D_i c_i(\mathbf{r}) \nabla \beta \mu_i(\mathbf{r}) \tag{2.4}$$

where D_i and $c_i(\mathbf{r})$ are the diffusion coefficient and concentration of species i, respectively, and $\mathbf{r}$ is the position within the system. Note that Eq. (2.4) uses an equilibrium concept (the chemical potential) in nonequilibrium conditions. This approximation is known as the local equilibrium hypothesis[3,4] and assumes that there exists a length scale where equilibrium can be achieved locally and that, within that length scale, thermodynamic variables have the same properties and relationships among them as in equilibrium.

It is useful to separate the chemical potential into different contributions:

$$\mu_i(\mathbf{r}) = \mu_i^0 + k_B T \ln\left(\frac{c_i(\mathbf{r})}{c^*}\right) + \mu_i^{ex}(\mathbf{r}) \tag{2.5}$$

where μ_i^0 is the standard chemical potential, which describes the chemical potential of the system in an ideal reference state (in general, the infinite dilution reference state is used for solutes), c_i is the concentration of species i, c^* is a standard concentration, μ_i^{ex} is the excess chemical potential, and k_B is Boltzmann's constant. Alternatively, the excess chemical potential can be thought of as an effective mean-field potential acting on the particle due to the presence of other particles and external forces. This potential is known as the potential of mean force, U^{mf}, and it is equal to the excess chemical potential up to a constant value:[5]

$$\mu_i(\mathbf{r}) = K + k_B T \ln\left(\frac{c_i(\mathbf{r})}{c^*}\right) + U_i^{mf}(\mathbf{r}) \tag{2.6}$$

The first two terms in the right side of Eq. (2.5) are the ideal contributions to the chemical potential. More specifically, they represent the chemical potential of the system in the absence of interactions. This contribution contains only translational entropy, given by the second term in Eq. (2.5). The third term in Eq. (2.5) accounts for the presence of intramolecular interactions and external fields. For example, if the effect of the electrostatic potential ψ is considered (and other interactions are neglected), $\mu_i^{ex}(\mathbf{r}) = q_i\psi(\mathbf{r})$ (where q_i is the charge of one i molecule) and thus,

$$\mu_i(\mathbf{r}) = \mu_i^0 + k_B T \ln\left(\frac{c_i(\mathbf{r})}{c^*}\right) + q_i\psi(\mathbf{r}) \tag{2.7}$$

In some research fields (mainly electrochemistry and solid-state physics), the chemical potential including electrostatic interactions, such as that in Eq. (2.7), is known as "electrochemical potential." In this book, we will not make a distinction between chemical and electrochemical potential. Finally, there are, of course, many other possible contributions to μ_i^{ex} besides the electrostatics, such as intermolecular interactions, gravitational fields, and magnetic fields.[6] Since these contributions have little general relevance for ion transport in nanochannels and pores, we will not address them in detail.

2.1.2 Poisson–Boltzmann Equation

More often than not, the walls of solid-state nanochannels and nanopores are charged. This charge can be either the native charge of the material or be introduced by chemical modification (see chapter: Introduction to Chemically Modified Nanochannels and Nanopores). Adsorption or grafting of polyelectrolytes also results in large amounts of charge introduced into the system.

Charged surfaces immersed in electrolyte solutions generate an electrical double layer (EDL), where the ions of charge opposite to the surface (counterions) are enriched with respect to the bulk solution and the ions of the same charge as the surface (co-ions) are depleted. The cornerstone of all continuum models for the EDL is the Poisson equation which exactly describes the relationship between the electrostatic potential, ψ, and the number density of charges, c_Q:

$$\nabla(\varepsilon(\mathbf{r})\varepsilon_0 \nabla\psi(\mathbf{r})) = -c_Q(\mathbf{r}) \tag{2.8}$$

where $\mathbf{r}$ denotes the position in the system and $\varepsilon(\mathbf{r})$ and ε_0 are the relative dielectric function at $\mathbf{r}$ and the vacuum dielectric permittivity, respectively. In general, c_Q is approximated by its ensemble-averaged value, $\langle c_Q \rangle$ If the dielectric function is constant across the system, Eq. (2.8) reduces to:

$$\varepsilon\varepsilon_0 \nabla^2\psi(\mathbf{r}) = -\langle c_Q(\mathbf{r})\rangle \tag{2.9}$$

The average density of charges is given by the sum of the contributions of all ionic species,

$$\langle c_Q(\mathbf{r})\rangle = \sum_i q_i c_i(\mathbf{r}) \tag{2.10}$$

where q_i and c_i are the charge and ensemble-averaged number density of ion i, respectively, and the sum runs over all ionic species in solution. In equilibrium, $c_i(\mathbf{r})$ can be expressed as a function of its excess chemical potential as:

$$c_i(\mathbf{r}) = c_i^{bulk} \exp\left(-\frac{\mu_i^{ex}(\mathbf{r}) - \mu_i^{ex,bulk}}{k_B T}\right) \tag{2.11}$$

where we used Eq. (2.5) to relate the excess chemical potential and concentration at **r** with those in the bulk solution (i.e., the homogeneous solution far from the surface). At this point, we will make an approximation and consider only electrostatic interactions, $\mu_i^{ex} = q_i\psi$ (i.e., the same approximation used for Eq. (2.7)). Combination of Eqs. (2.10) and (2.11) results in the well-known Poisson–Boltzmann (PB) equation:

$$\varepsilon\varepsilon_0 \nabla^2 \psi(\mathbf{r}) = -\sum_i q_i c_i^{bulk} \exp\left(-\frac{q_i\psi(\mathbf{r})}{k_B T}\right) \tag{2.12}$$

It is worthwhile to stress that the assumption $\mu_i^{ex} = q_i\psi$ is the simplest one that captures the right physics of the problem, but more sophisticated approximations can be used. For example, the molecular theory described in Chapter 6, Advanced Modeling of Ion Transport in Polymer and Polyelectrolyte-Modified Nanochannels and Nanopores, considers additional contributions to the excess chemical potential in order to deal with the presence of polyelectrolytes and osmotic forces. On the other hand, modified PB theories including steric correlations and solvation forces are required to study very narrow pores ($R < 1-2$ nm), where layering of ions is observed.[7]

2.1.3 Solutions of the Poisson–Boltzmann Equation

The PB equation is a nonlinear equation in $\psi(\mathbf{r})$. In the case of a planar surface with a fixed surface charge density σ_q and assuming constant ε, the analytical solution of the PB equation is known as the Gouy–Chapman model:

$$\psi_0 = \text{asin h}\left(\frac{\sigma_q}{4\lambda_D c_{salt}|e|}\right)\frac{2k_B T}{|e|} \tag{2.13}$$

where ψ_0 is the potential at the surface (surface potential) and $\lambda_D = \sqrt{\varepsilon\varepsilon_0 k_B T/(2c_{salt}|e|^2)}$ is a characteristic distance known as the Debye length of the solution. As shown below, the Debye length dictates the decay length scale of electrostatic interactions in an electrolyte solution.

The potential profile across the interface, $\psi(z)$, is given by the transcendental equation[8]:

$$\tan\text{h}\left(\frac{|e|\psi(z)}{4k_B T}\right) = \tan\text{h}\left(\frac{|e|\psi_0}{4k_B T}\right)\exp(-z/\lambda_D) \tag{2.14}$$

where z is the distance from the surface. In the limit of small ψ_0, Eq. (2.14) can be approximately solved as (see Justification Box 2.1):

$$\psi(z) = \psi_0 \exp(-z/\lambda_D) \tag{2.15}$$

In other words, the thickness of the EDL (i.e., the region where $\psi(z)$ decays) is of the order of the Debye length, λ_D.

JUSTIFICATION BOX 2.1 LINEARIZED POISSON-BOLTZMANN EQUATION

In many cases, the nonlinear nature of the PB equation, Eq. (2.12), prevents finding analytical expressions. For a 1:1 electrolyte in the limit when $|q_i\psi(\mathbf{r})/k_BT| \ll 1$ (i.e., when the electrostatic energy is much smaller than the thermal energy), we can use the series expansion $e^x \sim 1 + x$ to approximate the PB equation as:

$$\varepsilon\varepsilon_0\nabla^2\psi(\mathbf{r}) = |e|c_{\text{salt}}^{\text{bulk}}\frac{2\psi(\mathbf{r})}{k_BT}$$

which is now a linear equation in ψ. This equation is the basis of many important theories, such as the Debye Hückel theory of electrolyte solutions. Note that $|q_i\psi(\mathbf{r})| < |q_i\psi_0|$, where ψ_0 is the surface potential, thus the condition for linearization can be more conveniently expressed as $|q_i\psi_0/k_BT| \ll 1$, which implies $|\psi_0| \ll \frac{k_BT}{|e|} = 25$ mV.

In the case of Eq. (2.14), linearization is performed using $\tan\text{h}(x) = (e^x - e^{-x})/(e^x + e^{-x}) \sim x$ for $|x| \ll 1$, so

$$\tan\text{h}\left(\frac{e\psi_0}{4k_BT}\right) \sim \frac{|e|\psi_0}{4k_BT}$$

Note that the potential–charge relationship given by Eq. (2.13) is the exact solution of the PB equation for planar surface. It is also an approximate solution for the EDL for cylindrical nanochannels or nanopores if the radius of the pore R is much larger than λ_D. For narrow pores, there is no analytical solution for Eq. (2.13), but some approximate solutions exist.

Table 2.1 compiles solutions to the PB equation for some systems of interest in nanopore science.

2.1.4 Surface Charge Dependent on pH

So far, we discussed the electrostatic potential next to surfaces with a fixed charge density. In many cases, however, surfaces are coated by chemical species that are either acidic or basic and whose state of charge depends on the solution pH. For example, the colloidal particles making the films discussed in Chapter 5, Chemically Modified Nanopores in Colloidal Crystals, are made of silica and, thus, have charged silanol groups on their surface. These silanol groups are weak acids and may exist either in a neutral state (SiOH) or in a negatively charged state (SiO^-), which are related by an acid–base chemical equilibrium:

$$\text{HAc} \overset{K_a}{\rightleftharpoons} \text{Ac}^- + \text{H}^+ \tag{2.16}$$

where HAc is the acid form (SiOH) and Ac^- is the deprotonated species (SiO^-). In other examples in literature, chemical functionalization of nanopore surfaces introduces weakly basic amino functionalities:

$$\text{B} + \text{H}_2\text{O} \overset{K_b}{\rightleftharpoons} \text{BH}^+ + \text{OH}^- \tag{2.17}$$

where B is the base form (e.g., NH_2) and BH^+ is the protonated species (e.g., NH_3^+).

Table 2.1 Analytical Solutions of the PB Equation for Electrostatic Potential Profile as a Function of the Surface Charge σ_q for Systems of Interest

Geometry	Approximations	Equation	Reference
Planar surface	Exact solution	$\tanh\left(\frac{\lvert e\rvert\psi(z)}{4k_BT}\right) = \tanh\left(\frac{\lvert e\rvert\psi_0}{4k_BT}\right)\exp(-z/\lambda_D)$	8
Infinite channel	$R \gg \lambda_D$		
Planar surface	Linear PB ($\psi_0 \ll 4k_BT/\lvert e\rvert$)	$\psi(z) = \psi_0 \exp(-z/\lambda_D)$	8
Infinite channel	Linear PB ($\psi_0 \ll 4k_BT/\lvert e\rvert$)	$\psi(r) = \frac{\sigma_q\lambda_D}{\varepsilon\varepsilon_0}\frac{I_0(r/\lambda_D)}{I_1(R/\lambda_D)}$	9,10
Infinite channel	Matching asymptotic expansions of the PB equation	Complex solutions but larger range of validity than linear PB	11–13
Spherical pore	Linear PB, $r/\lambda_D \ll 2$	$\psi(r) = \frac{k_BT}{\lvert e\rvert}(f(r/\lambda_D) - f(\xi/\lambda_D))$ for $f(x) = 1 - \sinh(x)/x$ and $\left.\frac{df}{dx}\right\rvert_{x=\xi} = -\frac{4\pi\lvert e\rvert^2}{\varepsilon\varepsilon_0 k_BT}\frac{R}{\sigma_q\xi}$	14
Spherical pore	Series expansions	Complex solutions but larger range of validity than linear PB	13,14

In the equations, I_0 and I_1 are the zeroth and first modified Bessel functions of the first kind.

The effective charge density σ_q of surfaces coated by weak acids and weak bases is equal to the product of the total number of acid or basic sites per unit area (σ_{HAc/Ac^-}) and the fractions of these sites that are charged, f_{Ac^-} or f_{BH^+} for acid and basic groups, respectively. For example, for an acid, $\sigma_q = \sigma_{HAc/Ac^-}f_{Ac^-}$. The fractions f_{Ac^-} or f_{BH^+} for weak acid or basic groups on a surface may be very different from those of the same chemical group in the bulk solution. Let us consider, for example, a planar surface modified by a monolayer of a weak acid. The deprotonation of HAc species will be more difficult on the surface than in bulk solution due to the negative electrostatic potential at the plane of the surface, ψ_0, generated by the Ac^- species on the surface. In other words, the negative electrostatic potential at the surface inhibits the formation of an additional negative charge by deprotonation of HAc. Hence, for a given pH, the fraction of the Ac^- species on the surface will be smaller than that corresponding to the same chemical species free in solution. This effect, by which the state of charge of a weak acid–base group in an heterogeneous environment is shifted with respect to its bulk value in order to decrease electrostatic repulsions, is known as charge regulation.[15]

In order to formulate a quantitative model for the charge-regulation effect, let us write the expression for the acid–base equilibrium on the surface:

$$K_a = \frac{c_{H^+}c_{Ac^-}}{c_{HAc}} = \frac{c_{H^+}(z=0)f_{Ac^-}}{1-f_{Ac^-}} \tag{2.18}$$

where f_{Ac^-} is the fraction of the acid species on the surface. The concentration of protons on the surface follows from Eq. (2.11) (assuming $\mu_i^{ex} = q_i\psi$),

$$c_{H^+}(z=0) = c_{H^+}^{bulk} \exp\left(-\frac{q_{H^+}\psi_0}{k_B T}\right) \quad (2.19)$$

Combining Eqs. (2.19) with (2.18) and using $\sigma_q = \sigma_{HAc/Ac-} f_{Ac-}$, we get:

$$\sigma_q = \frac{K_a \sigma_{HAc/Ac-}}{c_{H^+}^{bulk} \exp\left(-\frac{q_{H^+}\psi_0}{k_B T}\right) + K_a} \quad (2.20)$$

which relates the charge and the potential of the surface. In order to find the electrostatic potential at a flat surface modified by a weak acid, we need to solve the set of two Eqs. (2.13) and (2.20) with two unknowns (σ_q and ψ_0). Unfortunately, these equations are highly nonlinear and defy analytical solution, therefore the treatment of problems with charge-regulating surfaces should be solved either by numerical methods or using a parametric solution (i.e., parametrization in f_{Ac-}, see Ref. 16).

2.1.5 Polyelectrolyte-Modified Nanopores and Nanochannels

In many of the examples in this book, pores and channels are modified by a polyelectrolyte layer instead of a surface charge. In the limit where the thickness of this polyelectrolyte layer is much smaller than the Debye length of the solution, one can of course approximate this layer as a plane of charge and use the models described above for charged surfaces. On the other hand, when the thickness of the polyelectrolyte layer is larger than λ_D, it is possible to propose a two-phase model, known as the Donnan model (Fig. 2.1A).[17] In this model, there are two well-defined phases: the polyelectrolyte phase, which has a concentration of charged segments c_p due to the polyelectrolyte chains (negative charges in the example of Fig. 2.1A), and the solution phase. In each phase, the electroneutrality condition must be fulfilled, therefore the concentration of counterions (cations in the current example) is higher in the polyelectrolyte phase than in the solution phase. Thus, in order to sustain this imbalance in counterion concentration, an interfacial potential $\Delta\psi_D$ develops (see Fig. 2.1), which is known as the Donnan potential. It can be shown that for a 1:1 electrolyte, $\Delta\psi_D$ is given by,[17,18]

$$\Delta\psi_D = \frac{k_B T}{|e|} \ln\left[\frac{c_p^{pol} q_p + \sqrt{\left(c_p^{pol} q_p\right)^2 + \left(2c_{salt}^{sol}|e|\right)^2}}{2c_{salt}^{sol}|e|}\right] \quad (2.21)$$

where c_{salt}^{bulk} is the salt density in the solution phase and c_p^{pol} and q_p are the concentration and charge of the charged groups of the polymer in the polymer phase. The Donnan potential ($\Delta\psi_D$) and the surface potential (ψ_0) described in the previous sections are two limiting cases of the same phenomena, valid for film thicknesses much smaller and much larger than the Debye length, respectively (see Fig. 2.1B). The most general case

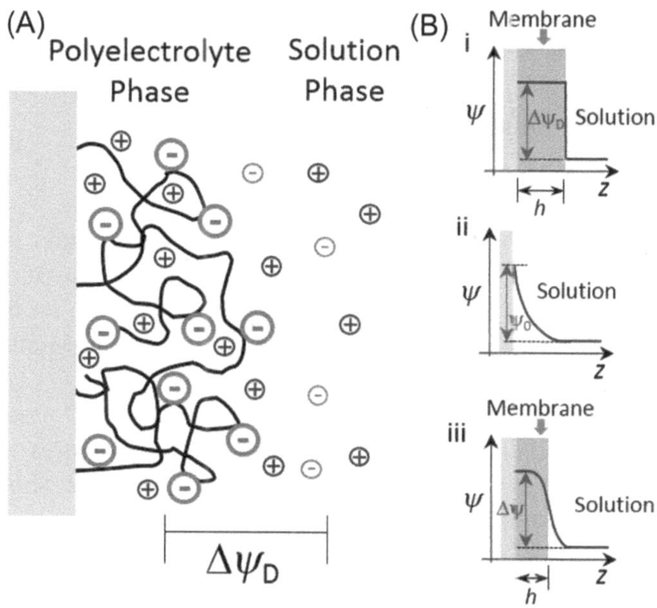

FIGURE 2.1 (A) Scheme showing the interphase between a polyelectrolyte film and an electrolyte solution. Due to the presence of fixed negative charges in the film, there is an interfacial potential (Donnan potential, $\Delta\psi_D$) between the two regions. (B) Scheme showing a comparison between the electrostatic potential profile (red (dark gray in print version) curve) for (i) a two-phase system (Donnan potential, valid when $h \gg \lambda_D$), (ii) a surface-confined system (surface potential, valid for $h \rightarrow 0$), and (iii) a general case (important for $h \sim \lambda_D$, i.e., films with thicknesses of a few nanometers).

where charges are distributed over a region of arbitrary thickness was studied by Ohshima and Ohki.[19]

2.1.6 Nernst–Planck Equation

So far we discussed the structure and electrostatic potential distribution in equilibrium. We now focus our attention back to transport properties. Replacing the definition of chemical potential, Eqs. (2.5) and (2.6) into Eq. (2.4) yields, (for the sake of readability, we omit hereafter the explicit dependence of all functions with **r**):

$$\mathbf{J}_i = -D_i \nabla c_i - D_i \nabla \beta \mu_i^{ex} = -D_i \nabla c_i - D_i \nabla \beta U_i^{mf} \tag{2.22}$$

This equation shows that the driving force of the ion fluxes has two contributions: an ideal entropic contribution due to the gradient of concentration and a nonideal one due to the change of the dependence of mean force with the position within the system. Taking only the ideal part of the chemical potential (i.e., neglecting the excess chemical potential in Eq. (2.5)) and inserting it into Eq. (2.4) leads to Fick's first law of diffusion:

$$\mathbf{J}_i = -D_i \nabla c_i \tag{2.23}$$

Fick's law can describe only concentration-driven flow, as it neglects all interactions and external forces, including the electrostatic potential. Adding the electrostatic contribution to the excess chemical potential (Eq. (2.7), also known as the electrochemical potential) leads to the Nernst–Planck (NP) diffusion equation:

$$\mathbf{J}_i = -D_i(\nabla c_i + c_i q_i \beta \nabla \psi) \tag{2.24}$$

As we discussed above, the excess chemical potential may contain other contributions besides electrostatics (such as osmotic gradients, van de Waals interactions, etc.), which will lead to diffusion equations that are more general than Eq. (2.24), see for example Chapter 6, Advanced Modeling of Ion Transport in Polymer and Polyelectrolyte-Modified Nanochannels and Nanopores.

The set of Eqs. (2.26) (NP) and (2.8) (Poisson) are known as the Poisson–Nernst–Planck (PNP) equations and are widely used to model transport in nanopores when solvent flows can be neglected. When solvent flow cannot be neglected, one should resort to the Navier–Stokes equations, as we discuss next.

2.1.7 Solvent Flow

In the case where there is a net flow of solution, we should add a convection term to Eq. (2.4), which results in:

$$\mathbf{J}_i = -D_i c_i \nabla \beta \mu_i + c_i \mathbf{u} \tag{2.25}$$

where $\mathbf{u}$ is the velocity of the solution. Once again, using the common approximation that involves considering only the electrostatic contributions to the chemical potential, yields:

$$\mathbf{J}_i = -D_i(\nabla c_i + c_i q_i \beta \nabla \psi) + c_i \mathbf{u} \tag{2.26}$$

Eqs. (2.25) or (2.26) should be coupled with the Navier–Stokes equation describing the motion of the fluid:

$$\frac{\partial \mathbf{u}}{\partial t} = -\frac{1}{\rho}\nabla p - (\mathbf{u} \cdot \nabla)\mathbf{u} + \nu \nabla^2 \mathbf{u} + \frac{1}{\rho}\mathbf{F} \tag{2.27}$$

where υ is the kinematic viscosity of the fluid (viscosity divided by density), ρ is the fluid density, and $\mathbf{F}$ is a body force, i.e., a force acting on a volume element of the fluid due to external fields, which in the presence of an inhomogeneous electrostatic potential is:

$$\mathbf{F} = -\nabla \psi \sum_i q_i c_i \tag{2.28}$$

According to Eq. (2.28), gradients of electrostatic potential in the system can act on a volume element of the fluid, accelerating it. The resulting electrically driven flow is known as electroosmotic flow.

The terms in the Navier–Stokes equation account for: mechanical forces due to pressure gradients (first term), inertial forces (second term), viscous forces (third term), and external body forces (fourth term). The inertial forces can be in general neglected in nanopores (see Justification Box 2.2), which results in the Stokes flow (also called creeping flow):

$$\frac{\partial \mathbf{u}}{\partial t} = -\frac{1}{\rho}\nabla p + \nu\nabla^2\mathbf{u} + \frac{1}{\rho}\mathbf{F} \tag{2.29}$$

JUSTIFICATION BOX 2.2 VISCOUS VERSUS INERTIAL FORCES IN THE NAVIER–STOKES EQUATION.

Let us define the following dimensionless variables: $\mathbf{r}' = \mathbf{r}/X$, $\mathbf{u}' = \mathbf{u}/U$, $p' = p/(\rho U^2)$, $\nabla' = X\nabla$ (gradient operator), $\partial/\partial t' = \partial/\partial t(L/U)$, $\mathbf{F}' = \mathbf{F}/(\rho U^2/X)$, where U and X are characteristic velocity and length scales. Using these definitions, we can write Eq. (2.27) as

$$\frac{\partial \mathbf{u}'}{\partial t'}\left(\frac{U^2}{X}\right) = -\nabla' p'\left(\frac{U^2}{X}\right) - (\mathbf{u}'\cdot\nabla')\mathbf{u}'\left(\frac{U^2}{X}\right) + \nu\nabla'^2\mathbf{u}'\left(\frac{U}{X^2}\right) + \frac{1}{\rho}\mathbf{F}'\left(\frac{\rho U^2}{X}\right)$$

Diving this equation by U^2/X leads to

$$\frac{\partial \mathbf{u}'}{\partial t'} = -\nabla' p' - (\mathbf{u}'\cdot\nabla')\mathbf{u}' + Re^{-1}\nabla'^2\mathbf{u}' + \mathbf{F}'$$

where Re is the Reynolds' number, $Re = UX/\nu$. For nanochannels, typical values for X are very small (in the few nanometer range), so Re is also very small. In the equation above, the pressure force and the body force are externally fixed, so Re provides a balance between the relative magnitudes of the viscous and inertial forces in the system. For $Re \to 0$ ($Re^{-1} \to \infty$), the inertial term $(\mathbf{u}'\cdot\nabla')\mathbf{u}'$ becomes negligible in comparison with the viscous term $\nabla'^2\mathbf{u}'$, leading to Eq. (2.29).

The set of Eqs. (2.26) (NP), (2.27) (Navier–Stokes), and (2.8) (Poisson) are known as the Poisson–Nernst–Planck/Navier–Stokes (PNP-NS) equations. It is interesting to note that the convection term was included in Eq. (2.25) based on a phenomenological argument, rather than a first-principle derivation. While this is a good approximation that has proven its value in a variety of situations, a first-principle derivation would be interesting because it will open the door to advanced models coupling hydrodynamic flows with non-electrostatic interactions, presence of polyelectrolyte chains, etc. The reader interested in this topic is referenced to some recent work in the literature.[2,20–23]

2.1.8 Steady-State Conditions

So far, we discussed fluxes of ions and solvent that can change in time. If one is interested in the behavior of the system at long times, but not on how the system reaches such behavior (i.e., transients) then it is reasonable to assume that all these flows and the overall molecular organization of the system are independent of time. Assuming such steady-state conditions allows to greatly simplify the numerical problem of finding a solution for the transport equations. The change in the local concentration of a species i is given by the continuity condition:

$$\frac{\partial c_i}{\partial t} = -\nabla \cdot \mathbf{J}_i \tag{2.30}$$

where $\nabla \cdot \mathbf{J}_i$ is the divergence of $\mathbf{J}_i$.

In the steady state, $\nabla \cdot \mathbf{J}_i = 0$, which combined with the PNP equation (2.26) yields,

$$-\nabla(D_i(\nabla c_i + c_i q_i \beta \nabla \psi)) + \nabla(c_i \mathbf{u}) = 0 \tag{2.31}$$

Moreover, the Stokes flow equation in the steady state (Eq. 2.29) becomes:

$$-\frac{1}{\rho}\nabla p + \nu \nabla^2 \mathbf{u} + \frac{1}{\rho}\mathbf{F} = 0 \tag{2.32}$$

2.2 Solution of Transport Equations

Solving a transport problem in nanopore science requires to solve either the PNP (Eqs. 2.8 and 2.24) or the PNP-NS (Eqs. 2.8, 2.26, and 2.27) set of equations. These equations rarely have an exact analytical solution and, thus, one must rely on approximated analytical solutions, numerical methods, or, usually, a combination of both. For instance, numerically solving the transport equations in three dimensions is highly demanding even for current computers, thus a reduction of the dimensionality of the equations is usually necessary. Cylindrical nanopores and nanochannels are symmetric with respect to the rotation around the channel axis and, thus, the original three-dimensional problem can be exactly reformulated in two dimensions using cylindrical coordinates, where the solution depends on the radial coordinate (r) and the axial coordinate (z), but not on the angular coordinate (θ). In some situations, such two-dimensional (2D) problem can be further approximated into a one-dimensional (1D) problem (e.g., a solution that depends only on z or r), which greatly simplifies the solution but introduces further approximations that must be validated. In Table 2.2, we have summarized work in literature that provides approximate solutions of the PNP and PNP-NS equations for transport through nanopores and nanochannels of different shape and surface charge. We have classified these approximate solutions depending on the geometry and charge distribution of the system and whether the solution is a fully analytical equation or a 1D model in the channel radius (1D-radial) or axis (1D-axial) that needs to be numerically solved.

Numerical methods are used when approximated analytical solutions are unavailable or the approximations involved are undesirable. Current computer hardware allows numerically solving the PNP and PNP-NS equations in two dimensions for relatively small pores. The Finite Element Method (FEM) is by far the most used method for this task, in Appendix 1 we

Table 2.2 Summary of Some Approximated Analytical Solutions for the PNP and PNP-NS Equations

Geometry	Charge	Type of Solution	Reference
Cylindrical long channel	Homogeneously distributed	Analytical	10,24–26
	Inhomogeneous distributed	Analytical	27
	Inhomogeneous distributed	1D-axial	27
	Bipolar	1D-axial	28
	Bipolar	Analytical	29,30
	Bipolar (outer walls)	Analytical	31
	Modified by polyelectrolytes	Analytical	32
	Modified by polyelectrolytes	1D-radial (low bias regime)	32
	pH-regulated	Analytical	8,33,34
	pH-regulated	1D-axial	24,25
	pH-regulated	Analytical	34–36
Conical long channel	Homogeneously distributed	1D-axial	37,37–42
	Homogeneously distributed	Analytical	42,43
Cigar-shaped channel	Amphoteric amino acids	Analytical	44
Long channels with noncylindrical geometries	Inhomogeneous distributed	Analytical (low bias regime)	45
	Inhomogeneous distributed	1D-axial	40,45,46
	Modified by polyelectrolytes	1D-radial (low bias regime)	47
Short cylindrical pore	Modified by polyelectrolytes	Analytical (low bias regime)	48

The type of solution may be fully analytical (Analytical) or a 1D approximation that should be solved numerically as a function of the position along the channel axis (1D-axial) or radius (1D-radial).

have included a tutorial for the calculation of the current through a nanochannel using the FEM software Comsol.

In the rest of this section, we will focus on some relevant physical results for ion and solvent transport in nanochannels. We will also discuss the main difference between transport in nanochannels and microchannels, which is originated in two important differences: (i) nanoscale channels exhibit a much larger area/volume ratio than microchannels and (ii) their characteristic radii ($\sim 5-100$ nm) is commensurable with the Debye length of the solution ($\lambda_D = 1$ nm).

2.2.1 Unipolar Transport

The thickness of the EDL near a charged surfaced is of the order of the Debye length of the solution. As the channel radius is narrowed down to this characteristic length scale, the channel becomes highly permselective and its inner solution is rich in ions with charge opposite to that of the inner wall (counterions) and depleted in ions of the same charge (co-ions), see panels A and B in Fig. 2.2. The solution inside the channel is thus usually referred as an unipolar solution.[49] The imbalance between the concentration of counterions within the channel and in the bulk solution creates a potential difference between the bulk solution and the interior of the pore (see Fig. 2.2B),[24,50] which is analog to the Donnan potential discussed in Section 2.1.5 for polyelectrolyte films. In fact, assuming that the electrostatic

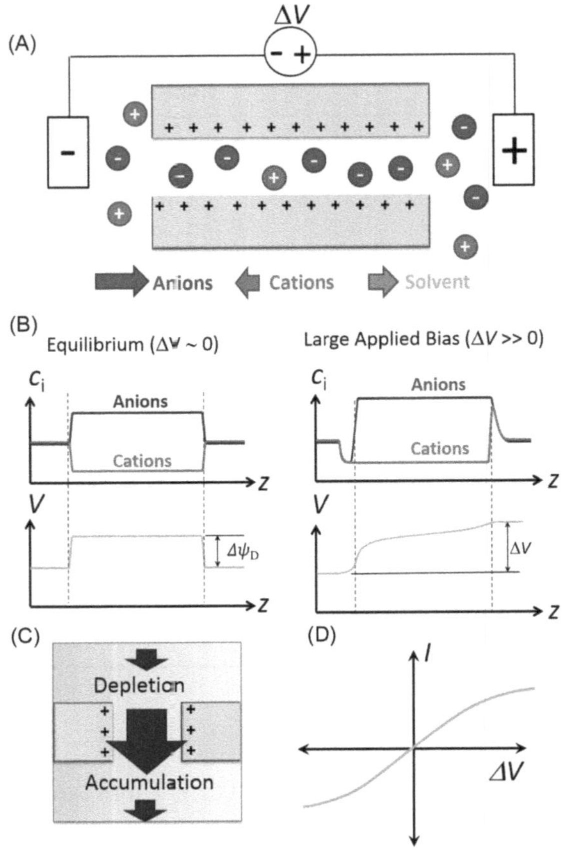

FIGURE 2.2 (A) Scheme of a nanochannel with positively charged inner walls between two reservoirs. Due to the charges on the inner walls, the anions (counterions) are enriched inside the channel and the cations (co-ions) are depleted. Upon applying a potential ΔV to reversible electrodes located at the reservoirs, the ions and the solvent flow according to the direction of the arrows. (B) Concentration of ions (upper plot) and profile of electrostatic potential (lower plot) as a function of the position along the axis of the channel (z). The vertical dashed lines show the positions of the reservoirs. Schematic plots are shown for equilibrium (left plots) and nonequilibrium (right plots) conditions: $\Delta\psi_D$ is a Donnan-type potential between the nanochannel and the solution. (C) Scheme showing the origin of the concentration polarization effect. (D) Current–potential curve for the system in (A).

potential within the channel is homogeneous in the r and θ directions (i.e., a 1D approximation)[10,24] allows to use Eq. (2.21) to calculate this potential difference:

$$\Delta\psi_D = \frac{k_B T}{|e|} \ln\left[\frac{c_f^{\text{channel}}|e| + \sqrt{\left(c_f^{\text{channel}}|e|\right)^2 + \left(2c_{\text{salt}}^{\text{sol}}|e|\right)^2}}{2c_{\text{salt}}^{\text{sol}}|e|}\right] \tag{2.33}$$

where $c_{\text{salt}}^{\text{sol}}$ is the salt concentration in solution and c_f^{channel} is the r- and θ-averaged concentration of fixed charges in the channel, which can be estimated as the ratio between the total number of fixed charges on the walls and the channel volume, which for a cylindrical geometry is:

$$c_{\mathrm{f}}^{\mathrm{channel}} = \frac{2\pi R\sigma_{\mathrm{q}}L}{\pi R^2 L|e|} = 2\frac{\sigma_{\mathrm{q}}}{R|e|} \tag{2.34}$$

where R and σ_{q} are the radius of the channel and the surface charge of the inner wall. Combining Eqs. (2.33) and (2.34) leads to,[51]

$$\Delta\psi_{\mathrm{D}} = \frac{k_{\mathrm{B}}T}{|e|}\ln\left[Du + \sqrt{Du^2+1}\right] \tag{2.35}$$

where $Du = \sigma_{\mathrm{q}} \cdot (c_{\mathrm{salt}}^{\mathrm{sol}} R)^{-1}$ is a dimensionless quantity, known as the Dukhin number,[52] that characterizes the ratio between surface and bulk conductivities in a variety of phenomena.

The unipolar solution within the nanochannel leads to a very high selectivity toward one type of ion, i.e., the channel selectively transports counterions and blocks co-ions. If the motilities of counterions and co-ions are very similar (as it is the case for KCl), the permselectivity ratio, f_{perm}, which is the fraction of the total current transported by the counterions, is equal to the ratio of the number of counterions over the total number of ions within the channel. The effective counterion concentration is $\sim c_{\mathrm{f}}^{\mathrm{channel}}$ because we need counterions to compensate for the fixed charges of the inner walls. The total concentration of co-ions is smaller than in solution, $c_{\mathrm{salt}}^{\mathrm{sol}}$. These arguments provide a lower bound of the permselectivity of the channel:

$$f_{\mathrm{perm}} \sim \frac{c_{\mathrm{f}}^{\mathrm{channel}}}{c_{\mathrm{f}}^{\mathrm{channel}} + c_{\mathrm{salt}}^{\mathrm{sol}}} > 1 + 2\frac{\sigma}{Rc_{\mathrm{salt}}^{\mathrm{sol}}} = 1 + 2Du \tag{2.36}$$

Many works in literature have proposed that ion selectivity in nanochannels depends on the dimensionless number λ_{D}/R (the ratio of the Debye length to the nanochannel radii). However, Eqs. (2.35) and (2.36) (as well as more quantitative work in literature[31,52,53]) show that the best quantity to describe channel permselectivity and unipolar composition and transport is the Dukhin number.

2.2.2 Concentration Polarization

The distribution of the ions in a charged nanochannel in the steady state can be different from that in equilibrium. This effect results from the reorganization of the ions in response to the applied potential and gives rise important effects in nanofluidics, such as the concentration polarization effect described in this section and the current rectification effect described in Section 2.2.3. In the concentration polarization effect, the concentrations of both the counterions and the co-ions are depleted with respect to their equilibrium values at the side of the membrane where the counterions enter the channel and they are enhanced on the other side (see Fig. 2.2B). Concentration polarization arises because the concentration of counterions inside the channel is much larger than at the entrances, therefore, the interior of the channel can sustain larger currents than the entrances. As a consequence of this imbalance of the conductance of the pore and the reservoirs, the counterions deplete at the entrance to the channel and accumulate at its exit (see Fig. 2.2C). The concentration of the co-ions at the entrances is similar to that of the counterions in order to avoid large deviations from electroneutrality. Concentration polarization impacts on the current–potential response, which levels off for large applied potentials instead of showing an ohmic linear relationship (Fig. 2.2D).

2.2.3 Current Rectification

Current rectification arises when the magnitude of the current through a nanopore or nanochannel depends on the sign of the applied bias, see for example the asymmetrical current–potential curve in Fig. 2.3. Current rectification requires breaking the symmetry of the system with respect to the plane of the membrane. The nature of this requirement becomes clear if we review the polarization concentration effect shown by the symmetric system in Fig. 2.2 For $\Delta V > 0$, a depletion region develops in the left reservoir and an accumulation region forms in the right one (Fig. 2.2B). This ion redistribution leads to a non-ohmic I–ΔV curve (Fig. 2.2D). However, given the symmetry of the system, for $\Delta V < 0$, an exactly inverse situation arises: the depletion region forms in the right reservoir and the accumulation one develops in the left one. The resulting I–ΔV curve is thus symmetric with respect to inversion of the sign of both I and ΔV. In order to break this symmetry, the reservoirs cannot be equivalent anymore. The symmetry can be broken either by using asymmetrical channel shapes (e.g., conical nanopores/channels[37,42,54–62]), by having symmetrical channels but with an asymmetrical charge distribution, e.g., a cylindrical nanopore with inner walls that are charged positively in one half of the pore and negatively in the other half,[27,29–31,63–68] or by using solutions of different compositions in the reservoirs.[44,62,69] Depending on the sign of ΔV, current-rectifying nanochannels operate either in an open state (also referred as *on* state or forward bias, in the case of Fig. 2.3, $\Delta V > 0$), where the current is large or in a closed state (also referred as *off* state or reverse bias, in the case of Fig. 2.3, $\Delta V < 0$), where the current is small. The efficiency of the rectification is in general evaluated in terms of the rectification ratio, which is equal to $\alpha(|\Delta V|) = -I^{open}(|\Delta V|)/I^{closed}(|\Delta V|)$, where I^{open} and I^{closed} are the currents in the open and closed states for a applied bias of magnitude $|\Delta V|$.

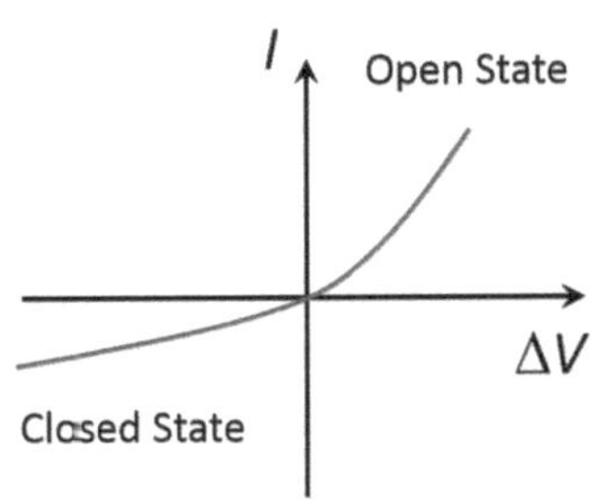

FIGURE 2.3 Current–potential curve showing current rectification.

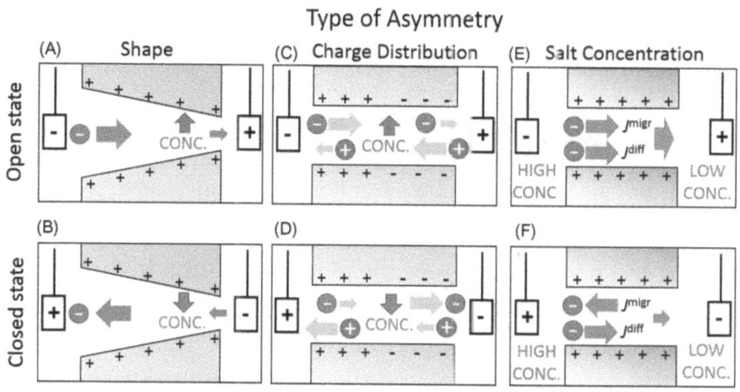

FIGURE 2.4 Schemes showing the different types of asymmetry that can give rise to current rectification.

Fig. 2.4 explains the origin of current rectification for the different types of asymmetry mentioned in the previous paragraph. Panels A and B in Fig. 2.4 show a shape-asymmetric (conical) nanochannel. To understand the behavior of this system, we should note that the conductance at the tip is much smaller than at the base due to the reduction of the cross-sectional area. The device operates in its open state when the electrode that faces the base has a polarity opposite to that of the inner walls. In these conditions, the counterions (whose current is much larger than that of the co-ions, as explained above) flow toward the tip, as shown in Fig. 2.4A. Since the tip region sustains a smaller current than the base region, the concentration of counterions accumulates at the tip. The concentration of co-ions also accumulates at the tip in order to guarantee electroneutrality. This increase of electrolyte concentration at the tip results in an overall increase in the conductance and, therefore, the channel operates in its open state. When the sign of the potential bias is reversed, there is a depletion of ion concentration at the cone tip (Fig. 2.4B), which results in the close state.

Fig. 2.4C and D shows a system where the upper half of the channel has a charge that is opposite to the lower half and, therefore, the asymmetry is imposed by charge distribution rather than the geometry. This system operates in the open state when the positive electrode faces the negatively charged side of the membrane, see Fig. 2.4C. In these conditions, the cations enter the channel through the negatively charged side of the membrane, but they are blocked inside the channel due to the difficulty to exit via the positively charged side. Therefore, the cation concentration is enhanced at the center of the channel. The anions flow in a direction opposite to that of the cations, but they also experience an opposite interaction with the walls of the channel. Thus, the concentration of anions is also boosted at the center of the channel. The increase in ion concentration inside the channel leads to a decrease in the resistance of the channel, hence the device operates in its open state. Upon inverting the sign of the applied bias, the directions of the ions fluxes are reversed so the ions can easily leave the channel but face a large resistance to enter it. Therefore, the concentration at the center of the channel is smaller than in equilibrium conditions and the nanochannel operates in its closed state. This mechanism also applies, although the details differ, to systems where the charges are located on the outer walls rather than the inner walls of the channel, see Chapter 6, Advanced Modeling of Ion Transport in Polymer and Polyelectrolyte-Modified Nanochannels and Nanopores.

Finally, in Fig. 2.4E and F we show a system where the symmetry is broken by an imbalance of the salt concentration in the reservoirs. To understand this device, it is useful to decompose the total ionic flux as the sum of a diffusional flux, J^{diff}, driven by the difference of salt concentration in the reservoirs and a migrational flux, J^{migr}, due to the applied bias. The device operates in the open state when the electrode in the high-concentration reservoir has a polarity opposite to that of the inner wall of the channel.[69] Under these conditions, the counterions move from the high-concentration reservoir to the low-concentration one (see Fig. 2.4E). Therefore, the potential-driven and the concentration-driven fluxes of counterions have the same direction, which results in a high conductance. The closed state is observed when the polarity of the electrode is reversed, so that the potential-driven and concentration-driven fluxes have opposite directions. Note that the current–potential curves in this case do not have $I = 0$ for $\Delta V = 0$ due to the presence of the diffusional flux even in the absence of the applied bias.[69]

Current rectification has been also observed in systems where both reservoirs have the same salt concentration but different pH and the inner charge of the channel is pH-dependent (i.e., chemical modification by weak acid or basic species).[44,68] In these systems, the pH difference between the reservoirs leads to different state of protonation for the weak groups at different positions within the channel, which gives rise to an inhomogeneous distribution of surface charges. Therefore, the general mechanism for current rectification is similar to that described for nanochannels with an asymmetric charge distribution in Fig. 2.4C and D.

2.2.4 Access Resistance versus Channel Resistance

Single nanochannels differ from single microchannels in the relative magnitudes of the access and channel resistances. The access resistance is the resistance of the solution located between the channel entrance and the electrodes at the reservoirs. Let us estimate the channel and access resistances for the simplest case of a single channel with neutral walls near equilibrium (we will neglect electroosmosis in these calculations). Inside the nanochannel, the ion fluxes occur in the direction parallel to the axis of the channel and their magnitude is independent of the radial coordinate (see Fig. 2.5). Thus, we can write the total current through the system as,

$$I = |e|A(J_{\mathrm{C}} - J_{\mathrm{A}}) \tag{2.37}$$

where $A = \pi R^2$ is the cross-sectional area of the channel (R is the channel radius) and J_{i} is the flux of species i (i = A or C for anions or cations, respectively) in the axial direction. These fluxes are given by the Nernst–Planck equation, Eq. (2.24), to yield:

$$J_{\mathrm{i}}(z) = -D_{\mathrm{i}} c_{\mathrm{i}} q_{\mathrm{i}} \beta \frac{\partial \psi}{\partial z} \tag{2.38}$$

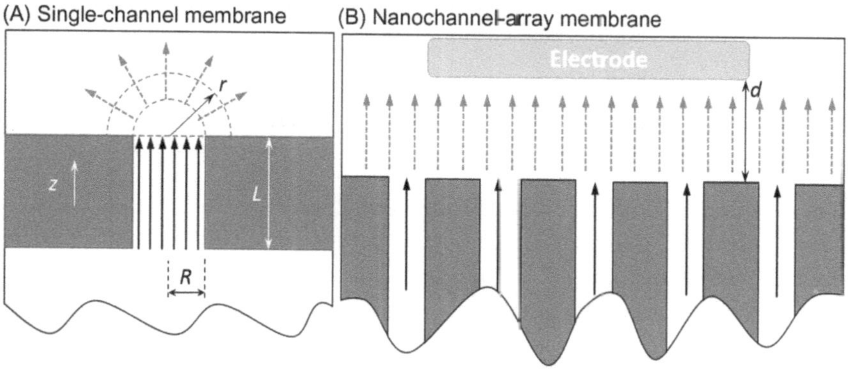

FIGURE 2.5 Schemes of a single-nanochannel (A) and nanochannel-array (B) membranes. The arrows indicate the direction of the ion fluxes involved in the nanochannel resistance (solid arrows) and the access resistance (dashed arrows).

(note that the diffusion term has not been included in Eq. (2.38), since the concentration is constant across the system). For a long nanochannel, the electric field does not depend on the *z*-direction, so integration of Eq. (2.38) results in,

$$J_i = -D_i c_i q_i \beta \frac{\Delta V}{L} \tag{2.39}$$

where ΔV is the potential difference between both entrances of the channel and L is the length of the channel (membrane thickness). Combining Eqs. (2.37) and (2.39) finally yields,

$$\Omega^{\text{channel, single}} = \frac{\Delta V}{I} = \frac{L}{R^2 \pi c_{\text{salt}} \beta |e|^2 (D_C + D_A)} \tag{2.40}$$

where $\Omega^{\text{channel, single}}$ is the resistance of the channel and c_{salt} is the salt concentration. Let us now estimate the resistance of the reservoirs. In the reservoirs, the ion fluxes have hemispherical symmetry (see Fig. 2.5A). The flux of ions along the radial direction (r) therefore is:

$$J_i(r) = -D_i c_i q_i \beta \frac{\partial \psi}{\partial r} \tag{2.41}$$

and the total current is:

$$I = |e| A(r)(J_C(r) - J_A(r)) \tag{2.42}$$

where $A(r)$ is the area of a hemispherical surface at position r, $A(r) = 2\pi r^2$. Combining Eqs. (2.41) and (2.42) yields,

$$I = 2\pi r^2 (D_C + D_A) c_{\text{salt}} \beta |e|^2 \frac{\partial \psi}{\partial r} \tag{2.43}$$

Let us now rearrange and integrate Eq. (2.43) between $r = R$ (i.e., the entrance of the channel) and $r \to \infty$,

$$\int_R^\infty \frac{I}{r^2} \mathrm{d}r = 2\pi(D_\mathrm{C} + D_\mathrm{A})c_\mathrm{salt}\beta|e|^2 \int_R^\infty \mathrm{d}\psi \tag{2.44}$$

Note that we take the limit $r \to \infty$ since the contribution of the hemispherical surface at r to the access resistance decreases progressively as we get away from the entrance of the channel, thus the exact location of the reference electrode is irrelevant. Defining ΔV as the potential difference between the reference electrode and the channel entrance, $\Delta V = V(r = \infty) - V(r = R)$, allows rewriting Eq. (2.44) as,

$$\Omega^{\text{access, single}} = \frac{\Delta V}{I} = \frac{1}{2\pi R(D_\mathrm{C} + D_\mathrm{A})c_\mathrm{salt}\beta|e|^2} \tag{2.45}$$

where $\Omega^{\text{access, single}}$ is the access resistance. The ratio between the access resistance for a single nanochannel (Eq. (2.45)) and the channel resistance (Eq. (2.40)) is:

$$\frac{\Omega^{\text{access, single}}}{\Omega^{\text{channel, single}}} = \frac{R}{2L} \tag{2.46}$$

Therefore, for a typical membrane thickness of $L \sim 10\ \mu\mathrm{m}$, the access resistance can be completely neglected with respect to the resistance of a single nanochannel. However, in the case of a single microchannel with $R \sim 1-10\ \mu\mathrm{m}$, the access resistance and the channel resistance become comparable. On the other hand, Eq. (2.46) also shows that in single nanopores prepared in ultrathin membranes with $L < 100$ nm (see chapter: Introduction to Chemically Modified Nanochannels and Nanopores), the access resistance is commensurable with the channel resistance.

The relative magnitudes of the access and membrane resistances are important for applications. For example, for separation technologies, the membrane resistance should be as small as possible. On the other hand, applications in sensors are based on measuring changes in the conductance of the membrane, which can be either transient changes (resistive pulses) upon the translocation of a biomolecule[70–73] or steady-state changes as the nanochannel gets blocked by binding of the analyte of interest.[56,59,74,75] In both cases, in order to easily resolve changes in the membrane resistance, it should be larger than the access resistance.

Eq. (2.46) is valid for single channels. For nanochannel arrays, the calculation of the access resistance changes because the current lines cannot be longer assumed to have hemispherical symmetry. In the case of a membrane with a high density of homogeneously distributed channels, we can assume that the current lines are normal to the membrane and the electrostatic potential is constant on planes parallel to it (see Fig. 2.5B). We can use Eq. (2.38) and integrate it between $z = 0$ (surface of the membrane) and $z = d$ (location of the electrode), which yields:

$$J_\mathrm{i} = -D_\mathrm{i}c_\mathrm{i}q_\mathrm{i}\beta\frac{\Delta V}{d} \tag{2.47}$$

The total current is

$$I = |e|A^{\text{membrane}}(J_C - J_A) \tag{2.48}$$

where A^{membrane} is the area of the membrane exposed to the electrode. Combination of Eqs. (2.47) and (2.48) yields,

$$\Omega^{\text{access, array}} = \frac{\Delta V}{I} = \frac{d}{c_{\text{salt}}\beta|e|^2 A^{\text{membrane}}(D_C + D_A)} \tag{2.49}$$

On the other hand, the resistance of the channels is given by the parallel combination of the resistances of all nanochannels in the membrane:

$$\Omega^{\text{channel, array}} = \frac{\Omega^{\text{channel, single}}}{N} = \frac{L}{NR^2\pi c_{\text{salt}}\beta|e|^2(D_C + D_i)} \tag{2.50}$$

where N is the total number of channels in the membrane,

$$N = \frac{A^{\text{membrane}}}{\pi R^2} m_p \tag{2.51}$$

In Eq. (2.51), m_p is the porosity, i.e., the fraction of the membrane area occupied by the channel. We can now estimate the ratio of access to membrane resistance for the nanochannel array using Eqs. (2.49) and (2.50):

$$\frac{\Omega^{\text{access, array}}}{\Omega^{\text{channel, array}}} = \frac{d \cdot m_p}{L} \tag{2.52}$$

For a track-etched membrane, with typical values of $L \sim 10\ \mu\text{m}$, $d \sim 1\ \text{mm}$, and $m_p \sim 0.05-0.2$,[76] equation shows that the access resistance can be 5–20 times larger than the channel resistance. For thinner and more porous membranes (such as alumina or mesoporous silica), the access resistance will represent an even larger contribution to the total resistance.

2.2.5 Electroosmotic Flow

The analysis of ion transport in the previous sections neglected electroosmotic flow. In order to include this effect, we should consider the flow of solvent through nanochannels as dictated by the Stokes flow equation, Eq. (2.32). Rice and Whitehead theoretically studied the electrokinetic flow in nanocapillaries.[9] For very long nanochannels, they proposed that the flow of solvent occurs only in the axial direction, z, and its magnitude is independent of z and θ (the angular coordinate) due to translational and rotational symmetries, respectively. Thus $\mathbf{u}(\mathbf{r}) = u_z(r)\hat{z}$ and Eq. (2.32) becomes,

$$-\frac{1}{\rho}\frac{\partial p}{\partial z}+\nu\frac{1}{r}\frac{\partial}{\partial r}\left(r\frac{\partial u_z(r)}{\partial r}\right)-\frac{1}{\rho}\frac{\partial \psi}{\partial z}\sum_{i=A,C} q_i c_i(r)=0 \tag{2.53}$$

where the z-component of the external force has been replaced by the electrostatic force according to Eq. (2.28). The boundary condition for Eq. (2.53) at the channel wall is $u_z(R)=0$ (nonslip boundary condition). We will also assume that the pressure p depends on z but not on r. In order to solve Eq. (2.53), we need the r-dependent concentration of the ions. We can obtain such expression in the low potential limit by writing down the linearized PB equation for the ions when $|q_i\psi(R)/k_BT| \ll 1$ (see Section 2.1.2 and Justification Box 2.1),

$$c_i(r)=c_i^{\text{bulk}}\exp\left(-\frac{q_i\psi(r)}{k_BT}\right)\sim c_i^{\text{bulk}}\left(1-\frac{q_i\psi(r)}{k_BT}\right) \tag{2.54}$$

Isolating $\psi(r)$ from this equation and using the solution for the EDL of an infinite nanochannel (Table 2.1):

$$\sum_{i=A,C} q_i c_i(r) \simeq -\frac{\sigma_q}{\lambda_D}\frac{I_0(r/\lambda_D)}{I_1(R/\lambda_D)} \tag{2.55}$$

Replacing Eq. (2.55) for i = A, C into Eq. (2.53) and solving for $u_z(r)$ finally yields:

$$u_z(r)=-\frac{\partial p}{\partial z}\frac{1}{4\nu\rho}(R^2-r^2)+\frac{\partial\psi}{\partial z}\frac{\sigma_q\lambda_D}{\nu\rho}\left(1-\frac{I_0(r/\lambda_D)}{I_1(R/\lambda_D)}\right) \tag{2.56}$$

Eq. (2.56) shows that the velocity of the fluid within the channel has two contributions: the hydrodynamic flow due to a gradient of pressure (first term) and the electroosmotic flow due to the presence of an electric field, $E_z=-\partial\psi/\partial z$. The magnitude of the hydrodynamic term scales as R^2, therefore its contribution to u_z in nanochannels will be small unless very high gradients of pressure exist. The electroosmotic term, on the other hand, can efficiently induce solvent flow providing that λ_D is not much smaller than R. The hydrodynamic flow is always in a direction opposite to the gradient of pressure (the solvent flows from regions of large pressure to regions of low pressure). On the other hand, the direction of the electro-osmotic flow depends both on the direction of the applied potential bias and the sign of the surface charge, σ_q. For example, Eq. (2.56) shows that the solvent flows in the direction of increasing potential (i.e., to the positive reservoir) if $\sigma_q > 0$ (see Fig. 2.2A). This effect has a simple explanation: for $\sigma_q > 0$, the inner solution of the channel is enriched in anions, which drag the solvent with them as they move toward the positive reservoir.

Hydrodynamic flows can create a net flow of ions even in the absence of an applied electrical field or a concentration gradient. Such hydrodynamic flow will carry ions according the combination of Eqs. (2.25) (for $\nabla\mu_i=0$) and (2.56) (for $\partial\psi/\partial z=0$),

$$J_i(r)=-c_i(r)\frac{\partial p}{\partial z}\frac{1}{4\nu\rho}(R^2-r^2) \tag{2.57}$$

In the example of Fig. 2.2A, $c_A(r) \gg c_C(r)$, therefore the hydrodynamic flow will create a net electric current through the channel, known as streaming current (the total current can be evaluated by numerically integrating the ion fluxes over the cross-sectional area of the channel). This current can be harvested by placing reversible electrodes in the reservoirs connected by the channel, as a method to generate electrical energy from a pressure difference.[77,78] If the streaming current is not harvested, a potential difference (known as streaming potential) will build up between the reservoirs. The streaming potential creates an electric current that opposes that created by the hydrodynamic force, so that in the steady state there is no net electric current flowing through the channel.

Appendix I Calculation of the Conductance of a Nanopore by Finite Elements

This tutorial shows an example of how to calculate the steady-state ionic conductance of a short nanopore of $R = 5$ nm and $L = 50$ nm with charged inner walls. We will use the Finite Element Software Comsol Multiphysics (version 4.3a has been used in this tutorial). The mathematical problem involves the simultaneous solution of the Nernst-Planck equation (2.31) for both anions and cations, the Stokes flow equation (Eq. (2.32)) and the Poisson equation of electrostatics (Eq. (2.9)).

In order to set up this type of study, we follow the steps:

1. In the Model Wizard, select 2D axisymmetric in Select Space Dimension and press ⇨ in the upper bar. This will create a model using cylindrical coordinates and assuming homogeneity in the angular dimension (i.e., the solution will depend on r and z only).
2. In Add Physics, under Chemical Species Transport select Transport of Diluted Species (chds) and click + to add the equations to the Selected physics window (requires the Chemical Reaction Engineering Module).
3. Click on Transport of Diluted Species (chds) in the Selected physics and scroll down. Change the Number of species to 2 and rename the variables under Concentration to "cc" and "ca" (these variables will be the cation and anion concentrations).
4. Under AC/DC select Electrostatics (es) and add it to the Selected physics window.
5. Under Fluid Flow, select Single-Phase Flow, then select Laminar Flow (spf) and add it to the Selected physics window. Press ⇨ in the upper bar.
6. Now under Select Study Type, select Stationary and click on the checkered-flag button in the upper bar.
   We are now in conditions to build the geometry of our system.
7. In the Model Builder (left panel), select Geometry 1. In the middle panel, go to the Units folder and under Length unit select "nm" (this will select the most useful length units for our system).
8. Now right click on Global Definitions on the Model Builder and select Parameters from the list. This will create a list of parameters, which are useful to define important variables of our problem (see Fig. 2.6). Enter the parameters of the system as shown in Fig. 2.6.

Name	Expression	Value	Description
c0	0.1 [mol/dm^3]	100.00 mol/m³	Bulk salt concentration
R	5 [nm]	5.0000E-9 m	Pore radius
L	50 [nm]	5.0000E-8 m	Pore length
sigma	0.1 [nm^-2]	1.0000E17 1/m²	Charge density of pore walls
Rdim	1 [um]	1.0000E-6 m	Reservoir size
DV	0.5 [V]	0.50000 V	Applied potential

FIGURE 2.6 Definitions of the parameters of the model.

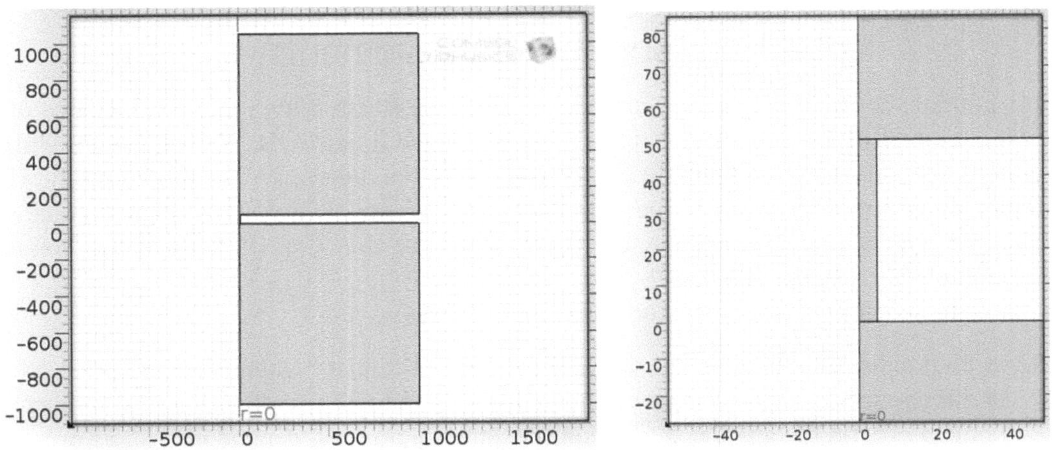

FIGURE 2.7 Geometry for a short nanopore between two reservoirs. The *y*-axis shows the axial coordinate (*z*) and the *x*-axis shows the radial coordinate (*r*). In both cases, the scales are in nanometers.

9. In the Model Builder right click on Geometry 1 and pick Rectangle from the list to add a rectangle to the model. In the middle panel, enter "R" in Width and "L" in Height. This rectangle will be the inner part of the pore.
10. In the Model Builder right click on Geometry 1 and pick Rectangle from the list again. Enter "Rdim" both in Width and Height. Enter "L" in z under the Position tab. This rectangle will be the upper reservoir. Create another rectangle and enter "Rdim" in Width and Height and "-Rdim" in z. This rectangle will be the lower reservoir.
11. Press Build All on the top bar to generate the model. Pressing on the Graphic window on the right will show the entire system (left panel in Fig. 2.7), the nanochannel region can be viewed using the zoom button, (right panel in Fig. 2.7).
12. In the Model Builder right click on Materials and select Open Material Browser. In the Material Browser, go to Liquids and Gases, then Liquids and pick Water. Click the on the top bar to add Water to the list of materials. Go to Water (mat 1) under materials in the Model Builder and enter a Relative permittivity of 78.5.

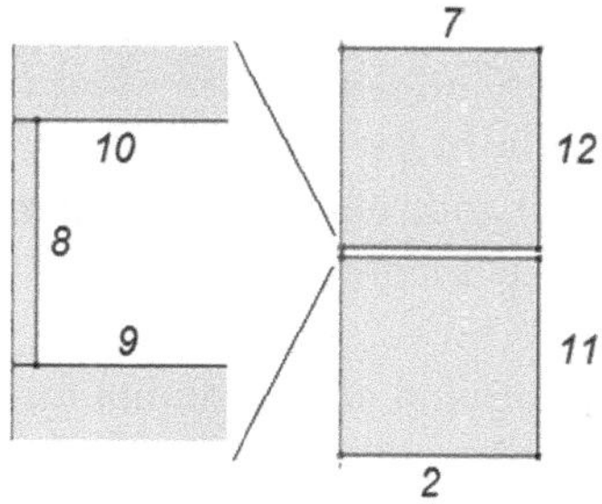

FIGURE 2.8 Numbering for the boundaries used in this tutorial.

At this point, we can define the equations that we will use to solve the problem.

13. In the Model Builder click on Transport of Diluted Species (chds). In the central panel, make sure that both Convection and Migration in electric field are checked.

14. Under Transport of Diluted Species (chds) click on Convection, Diffusion and Migration. Under Velocity field, pick Velocity field (spf/fp1) in the scroll down menu. Under Electric potential pick Electric potential (es). Scroll down to Diffusion and enter "2e-9 [m^2/s]" for both Dcc and Dca (these are the diffusion coefficients of K^+ and Cl^-). Move down and set the charge of cc to "1" and the charge of ca to " − 1".

15. Right click on Transport of Diluted Species (chds) and select Concentration. Under Concentration 1 enable Species ca and Species cc and enter "c0" in both. In the graph on the right, pick boundaries 2, 7, 11, and 12 (click on the boundary in the graph to select it, see boundaries in Fig. 2.8).

16. Just below the Model Builder tab (at the left top of the screen), press the 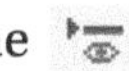 and select Advanced Physics Options. A Constraint Settings tab shows up in Concentration 1, unfold it and check the Use weak constraints option. This will enable the use of Lagrange multipliers for the concentration at the boundaries, which are required for an accurate calculation of the ionic fluxes.

17. Under Transport of Diluted Species (chds), No flux 1, make sure boundaries 8, 9, and 10 are selected.

18. Under Transport of Diluted Species (chds), Initial Values 1, set the initial values of both cc and ca to "c0".

19. Right click on Electrostatics (es) and select Space Charge Density. Select the three rectangles (domains 1, 2, and 3) and under Space charge density in the middle window enter "F_const*(cc-ca)". This is the total charge density determined from the concentrations of the ions to be used in the Poisson equation (F_const is Faraday's constant, i.e., the total charge of a mol of elemental charges).

20. Right click on Electrostatics (es) and select Ground. Select boundaries 2 and 11. These boundaries have an electrostatic potential of zero.

21. Right click on Electrostatics (es) and select Electric Potential. Select boundaries 7 and 12. Enter "DV" under Electric potential. These boundaries have an electrostatic potential of DV.

22. Right click on Electrostatics (es) and select Surface Charge Density. Select boundary 8. Enter "sigma*F_const/N_A_const" under Surface charge density. This is the surface charge density on the inner walls of the pore, where N_A_const is Avogadro's number.
23. Right click on Electrostatics (es) and select Zero Charge. Select boundaries 9 and 10.
24. Under Laminar Flow (spf) select "Incompressible flow" from the Compressibility scroll down list.
25. Right click on Laminar Flow (spf) and add a volume force. Select the three rectangles (domains 1, 2, and 3). Under Volume Force enter "(cc-ca)*F_const*es.Er" for the r component of the force and "(cc-ca)*F_const*es.Ez" for the z component. The variables es.Er and es.Ez are the r and z components of the electric field.
26. Right click on Laminar Flow (spf) and select Outlet. Select boundaries 2, 7, 11, 12, keep the pressure at zero.
27. Under Laminar Flow (spf), Wall 1, make sure the No slip boundary condition is selected for boundaries 8, 9, and 10.

We have now finished to set up the equations for the system. In all modules, we can check the equations in the model by unfolding the ► Equations tab.

Before solving the model, we need to mesh the system. For the problem of interest, we should take special care with the discretization next to the charged wall. The EDL in the 0.1 M solution used here has a thickness of ~ 1 nm (see Section 2.1.2), thus we should use a discretization length scale much smaller than 1 nm in order to accurately describe the interface.

28. Under Mesh 1 pull down the Element size menu and select Finer.
29. Right click on Mesh 1 and select Free Triangular. Right click on the newly created Free Triangular 1 and select Size. Go to the newly created Size 1 element, pull down the Geometric entity level menu and select Boundary. This will allow to create a finer mesh close to a boundary. Select the boundary 8. Under Element Size, select Custom. Unfold the Element Size Parameters, check Maximum size element and enter 0.1 nm. This will ensure that the discretization close to the inner surface of the pore is fine enough to capture the details of the EDL.
30. Click on Build All on the top bar to build the Mesh (see Fig. 2.9).

We are now in conditions to solve the model. Right click on Study 1 and select Compute or press F8. Solving this particular model will demand more or less time depending on the hardware. Once the system is successfully solved, we can examine the solution under the results group.

31. Right click on Results and select 2D plot group. Right click on the newly created 2D plot group and select Surface. In Surface 1, enter "V" under expression and click on the Plot button in the upper bar. Fig. 2.10 shows the resulting plot for the electrostatic potential, we observe that the potential in the upper boundary is DV = 0.5 V, while that at the lower one is zero.
32. Now enter "ca" under Expression and click on the Plot button in the upper bar again. This plots the density of the anions in the system. Zoom on the upper entrance of the pore. Fig. 2.11 shows that the anions are enriched near the positively charged surface.

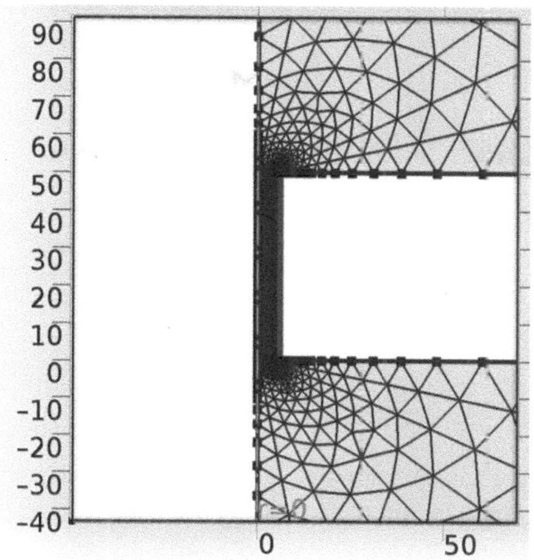

FIGURE 2.9 Meshing of the nanopore.

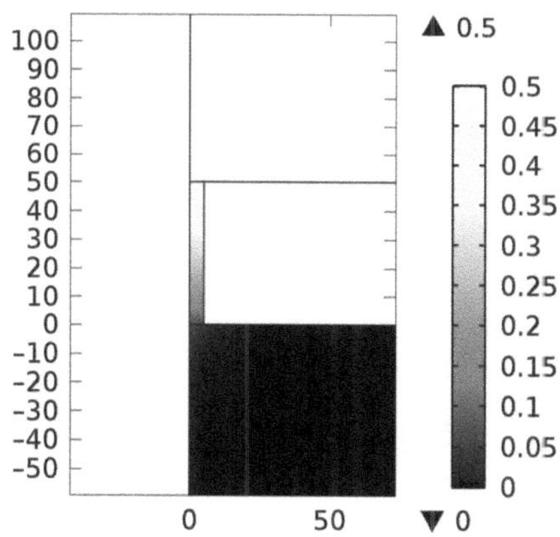

FIGURE 2.10 Color map showing the electrostatic potential (the scale bar is in units of V, the *r* and *z* axis are in nanometers). (The colors of this plot has been modified from Comsol default colors in order to facilitate reproduction in black and white.)

We also appreciate that the thickness of the EDL is approximately 1 nm in this condition.

33. Plotting the concentration of "ca" and "cc" at a larger scale reveals a depletion region at the entrance of the lower reservoir and an enhancement region at the entrance of the upper one (see Fig. 2.12). This is a manifestation of the polarization concentration effect discussed in Section 2.2.2.

34. Right click on Results and select 2D plot group again. Right click on the newly created 2D plot group and select Arrow Surface. Enter "u" in the r component and "w" in the z one (these are the components of the solution velocity field). In Arrow Positioning,

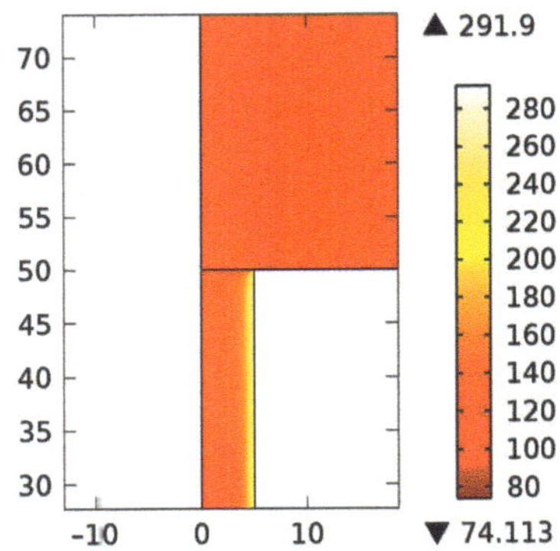

FIGURE 2.11 Color map showing the concentration of anions. The scale bar is in units of mol/m^3 (100 mol/m^3 = 0.1 M), the *r* and *z* are in nanometers.

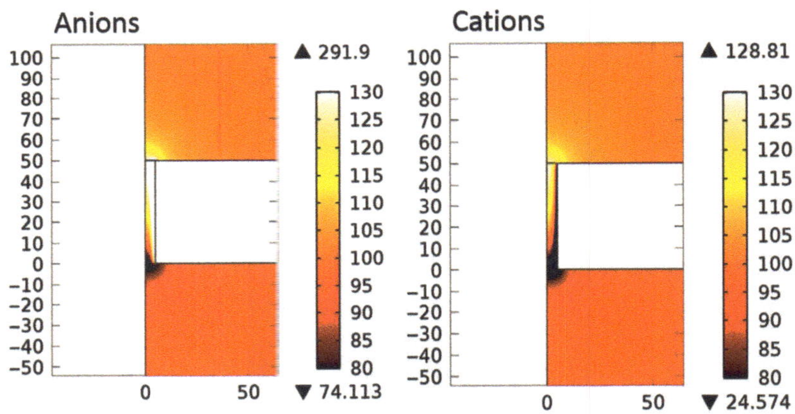

FIGURE 2.12 Color map showing the concentration of anions and cations. The scale bars are in units of mol/m^3 (100 mol/m^3 = 0.1 M), the *r* and *z* are in nanometers.

enter 800 in r grid points and 300 in z grid points. In Coloring and Style, check Scale factor and enter a scale factor of 50. Press Plot. The plot (Fig. 2.13) shows the velocity field of the electroosmotic flow The inner solution of the pore is mainly composed of anions due to the positive charge on the inner surface (see Fig. 2.11). As the anions move upward toward the positively charged reservoir (Fig. 2.10), they drag the solvent toward the upper reservoir.

We can now use the model in order to predict the *I*–*V* curve for the system. To do that, we should scan the variable DV.

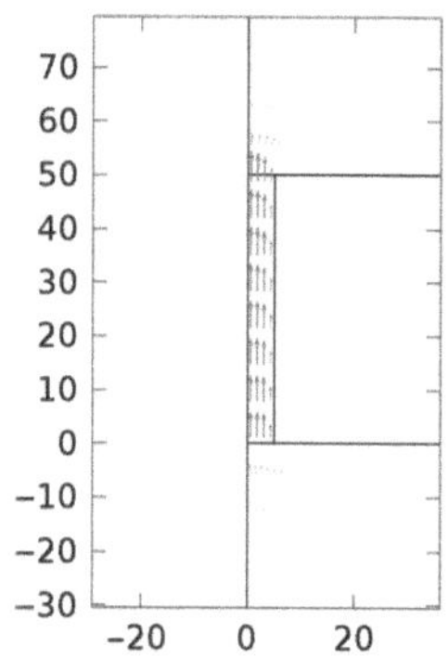

FIGURE 2.13 Arrow plot showing the velocity profile of the solution. The scale bar of the *r* and *z* are in nanometers.

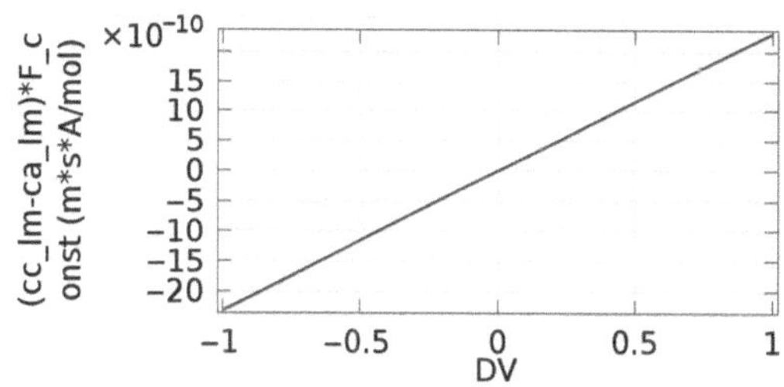

FIGURE 2.14 Current–potential curve for the short nanochannel studied in this section. The *x*-axis shows the applied potential in V and the *y*-axis shows the current in A.

35. Right click on Study 1 and select Parametric Sweep. Under the newly created Parametric Sweep element, press the + in Study Settings to add a parameter name, select DV. Under the Parameter value list enter "range(-1,0.1,1)". This will scan the value of the applied potential between −1 V and +1 V in steps of 0.1 V. Right click on Study 1 and select Compute or press F8.

36. After the solving procedure is finished, go to Results, right click on Derived Values and select Integration, Line Integration. Select boundaries 7 and 12. In Expression enter "(cc_lm-ca_lm)*F_const". The variables cc_lm and ca_lm are Lagrange multipliers for the concentrations, which allow an accurate calculation of the fluxes. Press the Evaluate button on the top bar. The integrated total current (in units of Ampere) appears in the right bottom window. Press the button to plot the results. Fig. 2.14 shows the calculated *I*–*V* curve.

Appendix II Symbols Used in This Chapter

Symbol	*Definition*	*Symbol*	*Definition*
A	Area	$\mathbf{u}$	Velocity of the solvent
c_i	Concentration of species i	V	Volume
c_Q	Total density of charge	z	Distance from a planar surface; axial coordinate of a nanochannel
c_{salt}	Salt concentration in solution	β	Inverse temperature, $1/k_BT$
Du	Dukhin number	ΔV	Potential bias applied between electrodes at both sides of a nanochannel
d	Distance between the nanochannel membrane and the counter electrode	$\Delta\psi_D$	Donnan potential
$\mathbf{E}$	Electric field	ε	Relative dielectric function
$\lvert e\rvert$	Elemental charge	ε_0	Vacuum dielectric function
$\mathbf{F}$	Body force	ψ	Electrostatic potential
f_i	Fraction of species in protonation state i	ψ_0	Surface potential
f_{perm}	Permselectivity of a nanochannel	λ_D	Debye length
G	Gibbs free energy	μ_i	Chemical potential of species i
H	Helmholtz free energy	μ_i^{ex}	Excess chemical potential of species i
h	Film thickness	υ	Kinematic viscosity of the fluid
I	Current	Ω	Resistance
I_n	Modified Bessel function of the first kind of order n	ρ	Density of the fluid
$\mathbf{J}_i$	Flux of species i	σ	Surface density of species i
k_B	Boltzmann's constant	σ_q	Surface charge
K_a	Acid equilibrium constant	θ	Angular coordinate
L	Length of the nanochannel		
q_i	Charge of species i		
m_p	Membrane porosity		
n_i	Number of moles of species i		
p	Pressure		
R	Nanochannel or nanopore radius		
Re	Reynolds' number		
r	Radial coordinate, distance from the axis of the nanochannel		
S	Entropy		
T	Temperature		

References

1. Cussler, E. L.; Evans, D. F.; Matesich, S. M. A. Theoretical and Experimental Basis for a Specific Countertransport System in Membranes. *Science* **1971,** *172,* 377–379.
2. Prigogine, I. *Introduction to Thermodynamics of Irreversible Processes;* Interscience Publishers: New York, 1955.

3. Lebon, G.; Jou, D.; Casas-Vásquez, J. *Understanding Non-Equilibrium Thermodynamics;* Springer-Verlag: Berlin, 2008.
4. De Groot, S. R.; Mazur, P. *Non-Equilibrium Thermodynamics;* Courier Corporation, 2013.
5. Berendsen, H. J. *Simulating the Physical World: Hierarchical Modeling from Quantum Mechanics to Fluid Dynamics;* Cambridge University Press, 2007.
6. Job, G.; Herrmann, F. Chemical Potential—a Quantity in Search of Recognition. *Eur. J. Phys.* **2006,** *27,* 353–371.
7. Kalcher, I.; Schulz, J. C. F.; Dzubiella, J. Electrolytes in a Nanometer Slab-Confinement: Ion-Specific Structure and Solvation Forces. *J. Chem. Phys.* **2010,** 133.
8. Bard, A. J.; Faulkner, L. R. *Electrochemical Methods,* 2nd ed.; John Wiley & Sons: New York, 2001.
9. Rice, C. L.; Whitehead, R. Electrokinetic Flow in a Narrow Cylindrical Capillary. *J. Phys. Chem.* **1965,** *69,* 4017.
10. Vlassiouk, I.; Smirnov, S.; Siwy, Z. Ionic Selectivity of Single Nanochannels. *Nano Lett.* **2008,** *8,* 1978–1985.
11. Petsev, D. N.; Lopez, G. P. Electrostatic Potential and Electroosmotic Flow in a Cylindrical Capillary Filled with Symmetric Electrolyte: Analytic Solutions in Thin Double Layer Approximation. *J. Colloid Interface Sci.* **2006,** *294,* 492–498.
12. Yaroshchuk, A. E. Transport Properties of Long Straight Nano-Channels in Electrolyte Solutions: A Systematic Approach. *Adv. Colloid Interface Sci.* **2011,** *168,* 278–291.
13. Šamaj, L.; Trizac, E. Poisson–Boltzmann Thermodynamics of Counterions Confined by Curved Hard Walls. *Phys. Rev. E.* **2016,** *93,* 012601.
14. Akoum, F.; Parodi, O. Electrostatic Interactions inside the Aqueous Core of Spherical Reversed Micelles (L//2 Phase). *Journal de physique Paris* **1985,** *46,* 1675–1681.
15. Ninham, B. W.; Parsegian, A. Electrostatic Potential between Surfaces Bearing Ionizable Groups in Ionic Equilibrium with Physiologic Saline Solution. *J. Theor. Biol.* **1971,** *31,* 405–428.
16. Ricci, A. M.; Tagliazucchi, M.; Calvo, E. J. Charge Regulation in Redox Active Monolayers Embedded in Proton Exchanger Surfaces. *Phys. Chem. Chem. Phys.* **2012,** *14,* 9988–9995.
17. Doblhofer, K.; Vorotyntsev, M. In *Electroactive Polymer Electrochemistry;* Lyons, M. E. G., Ed.; Plenum Press: New York, 1994.
18. Tagliazucchi, M.; Williams, F. J.; Calvo, E. J. Effect of Acid-Base Equilibria on the Donnan Potential of Layer-by-Layer Redox Polyelectrolyte Multilayers. *J. Phys. Chem. B.* **2007,** *111,* 8105–8113.
19. Ohshima, H.; Ohki, S. Donnan Potential and Surface Potential of a Charged Membrane. *Biophys. J.* **1985,** *47,* 673–678.
20. Archer, A. J. Dynamical Density Functional Theory for Molecular and Colloidal Fluids: A Microscopic Approach to Fluid Mechanics. *J. Chem. Phys.* **2009,** 130.
21. Goddard, B. D.; Nold, A.; Kalliadasis, S. Multi-Species Dynamical Density Functional Theory. *J. Chem. Phys.* **2013,** *138.*
22. Goddard, B. D.; Nold, A.; Savva, N.; Yatsyshin, P.; Kalliadasis, S. Unification of Dynamic Density Functional Theory for Colloidal Fluids to Include Inertia and Hydrodynamic Interactions: Derivation and Numerical Experiments. *J. Phys. Condens. Matter* **2013,** 25.
23. Marini Bettolo Marconi, U.; Melchionna, S. Charge Transport in Nanochannels: A Molecular Theory. *Langmuir* **2012,** *28,* 13727–13740.
24. Ramírez, P.; Mafé, S.; Aguilella, V. M.; Alcaraz, A. Synthetic Nanopores with Fixed Charges: An Electrodiffusion Model for Ionic Transport. *Phys. Rev. E.* **2003,** *68,* 119101–119108.

25. Ramírez, P.; Mafé, S.; Alcaraz, A.; Cervera, J. Modeling of pH-Switchable Ion Transport and Selectivity in Nanopore Membranes with Fixed Charges. *J. Phys. Chem. B.* **2003,** *107*, 13178–13187.

26. Cervera, J.; Ramírez, P.; Manzanares, J. A.; Mafé, S. Incorporating Ionic Size in the Transport Equations for Charged Nanopores. *Microfluid Nanofluid* **2010,** *9*, 41–53.

27. Ramírez, P.; Gómez, V.; Cervera, J.; Schiedt, B.; Mafé, S. Ion Transport and Selectivity in Nanopores with Spatially Inhomogeneous Fixed Charge Distributions. *J. Chem. Phys.* **2007,** *126*, 194703.

28. Szymczyk, A.; Zhu, H.; Balannec, B. Ion Rejection Properties of Nanopores with Bipolar Fixed Charge Distributions. *J. Phys. Chem. B.* **2010,** *114*, 10143–10150.

29. Vlassiouk, I.; Smimov, S.; Siwy, Z. Nanofluidic Ionic Diodes. Comparison of Analytical and Numerical Solutions. *ACS Nano* **2008,** *2*, 1589–1502.

30. Daiguji, H.; Oka, Y.; Shirono, K. Nanofluidic Diode and Bipolar Transistor. *Nano Lett.* **2005,** *5*, 2274–2280.

31. Tagliazucchi, M.; Rabin, Y.; Szleifer, I. Transport Rectification in Nanopores with Outer Membranes Modified with Surface Charges and Polyelectrolytes. *ACS Nano* **2013,** *7*, 9085–9097.

32. Tagliazucchi, M.; Azzaroni, O.; Szleifer, I. Responsive Polymers End-Tethered in Solid-State Nanochannels: When Nanoconfinement Really Matters. *J. Am. Chem. Soc.* **2010,** *132*, 12404–12411.

33. Ma, Y.; Xue, S.; Hsu, S.-C.; Yeh, L.-H.; Qian, S.; Tan, H. Programmable Ionic Conductance in a pH-Regulated Gated Nanochannel. *Phys. Chem. Chem. Phys.* **2014,** *16*, 20138–20146.

34. Huang, M.-J.; Mei, L.; Yeh, L.-H.; Qian, S. pH-Regulated Nanopore Conductance with Overlapped Electric Double Layers. *Electrochem. Commun.* **2015,** *55*, 60–63.

35. Ma, Y.; Yeh, L. H.; Qian, S. pH-Regulated Ionic Conductance in a Nanopore. *Electrochem. Commun.* **2014,** *43*, 91–94.

36. Xue, S.; Yeh, L. H.; Ma, Y.; Qian, S. Tunable Streaming Current in a pH-Regulated Nanochannel by a Field Effect Transistor. *J. Phys. Chem. C.* **2014,** *118*, 6090–6099.

37. Cervera, J.; Schiedt, B.; Ramírez, P. A Poisson/Nernst–Planck Model for Ionic Transport through Synthetic Conical Nanopores. *EPL (Europhys. Lett.)* **2005,** *71*, 35–41.

38. Cervera, J.; Schiedt, B.; Neumann, R.; Mafé, S.; Ramírez, P. Ionic Conduction, Rectification, and Selectivity in Single Conical Nanopores. *J. Chem. Phys.* **2006,** *124*, 104706.

39. Constantin, D.; Siwy, Z. S. Poisson–Nernst–Planck Model of Ion Current Rectification through a Nanofluidic Diode. *Phys. Rev. E.* **2007,** *76*, 041202.

40. Patricio, R.; Pavel, Yu, A.; Javier, C.; Salvador, M. Pore Structure and Function of Synthetic Nanopores with Fixed Charges: Tip Shape and Rectification Properties. *Nanotechnology* **2008,** *19*, 315707.

41. Yan, Y.; Wang, L.; Xue, J.; Chang, H.-C. Ion Current Rectification Inversion in Conic Nanopores: Nonequilibrium Ion Transport Biased by Ion Selectivity and Spatial Asymmetry. *J. Chem. Phys.* **2013,** *138*, 044706.

42. Kosińska, I. D.; Goychuk, I.; Kostur, M.; Schmid, G.; Hänggi, P. Rectification in Synthetic Conical Nanopores: A One-Dimensional Poisson–Nernst–Planck Model. *Phys. Rev. E.* **2008,** 77.

43. Momotenko, D.; Cortés-Salazar, F.; Josserand, J.; Liu, S.; Shao, Y.; Girault, H. H. Ion Current Rectification and Rectification Inversion in Conical Nanopores: A Perm-Selective View. *Phys. Chem. Chem. Phys.* **2011,** *13*, 5430–5440.

44. Ali, M.; Ramirez, P.; Nguyen, H. Q.; Nasir, S.; Cervera, J.; Mafe, S., et al. Single Cigar-Shaped Nanopores Functionalized with Amphoteric Amino Acid Chains: Experimental and Theoretical Characterization. *ACS Nano* **2012,** *6*, 3631–3640.

45. Frament, C. M.; Dwyer, J. R. Conductance-Based Determination of Solid-State Nanopore Size and Shape: An Exploration of Performance Limits. *J. Phys. Chem. C.* **2012,** *116*, 23315–23321.

46. Marconi, U. M. B.; Melchionna, S.; Pagonabarraga, I. Effective Electrodiffusion Equation for Non-Uniform Nanochannels. *J. Chem. Phys.* **2013,** *138,* 244107.
47. Gilles, F. M.; Tagliazucchi, M.; Azzaroni, O.; Szleifer, I. Ionic Conductance of Polyelectrolyte-Modified Nanochannels: Nanoconfinement Effects on the Coupled Protonation Equilibria of Polyprotic Brushes. *J. Phys. Chem. C.* **2016,** *120,* 4789–4798.
48. Tagliazucchi, M.; Rabin, Y.; Szleifer, I. Ion Transport and Molecular Organization Are Coupled in Polyelectrolyte Modified Nanopores. *J. Am. Chem. Soc.* **2011,** *133,* 17753–17763.
49. Karnik, R.; Fan, R.; Yue, M.; Li, D.; Yang, P.; Majumdar, A. Electrostatic Control of Ions and Molecules in Nanofluidic Transistors. *Nano Lett.* **2005,** *5,* 943–948.
50. Schoch, R. B.; Han, J.; Renaud, P. Transport Phenomena in Nanofluidics. *Rev. Mod. Phys.* **2008,** *80,* 839–883.
51. Cheng, L.-J.; Guo, L. J. Nanofluidic Diodes. *Chem. Soc. Rev.* **2010,** *39,* 923–938.
52. Zangle, T. A.; Mani, A.; Santiago, J. G. Theory and Experiments of Concentration Polarization and Ion Focusing at Microchannel and Nanochannel Interfaces. *Chem. Soc. Rev.* **2010,** *39,* 1014–1035.
53. Tagliazucchi, M.; Szleifer, I. Salt Pumping by Voltage-Gated Nanochannels. *J. Phys. Chem. Lett.* **2015,** *6,* 3534–3539.
54. Cervera, J.; Schiedt, B.; Neumann, R.; Mafá, S.; Ramírez, P. Ionic Conduction, Rectification, and Selectivity in Single Conical Nanopores. *J. Chem. Phys.* **2006,** 124.
55. Yameen, B.; Ali, M.; Neumann, R.; Ensinger, W.; Knoll, W.; Azzaroni, O. Single Conical Nanopores Displaying pH-Tunable Rectifying Characteristics. Manipulating Ionic Transport with Zwitterionic Polymer Brushes. *J. Am. Chem. Soc.* **2009,** *131,* 2070.
56. Ali, M.; Yameen, B.; Neumann, R.; Ensinger, W.; Knoll, W.; Azzaroni, O. Biosensing and Supramolecular Bioconjugation in Single Conical Polymer Nanochannels. Facile Incorporation of Biorecognition Elements into Nanoconfined Geometries. *J. Am. Chem. Soc.* **2008,** *130,* 16351–16357.
57. Sexton, L. T.; Horne, L. P.; Sherrill, S. A.; Bishop, G. W.; Baker, L. A.; Martin, C. R. Resistive-Pulse Studies of Proteins and Protein/Antibody Complexes Using a Conical Nanotube Sensor. *J. Am. Chem. Soc.* **2007,** *129,* 13144–13152.
58. Yameen, B.; Ali, M.; Neumann, R.; Ensinger, W.; Knoll, W.; Azzaroni, O. Proton-Regulated Rectified Ionic Transport through Solid-State Conical Nanopores Modified with Phosphate-Bearing Polymer Brushes. *Chem. Commun.* **2010,** *46,* 1908–1910.
59. Siwy, Z.; Trofin, L.; Kohli, P.; Baker, L. A.; Trautmann, C.; Martin, C. R. Protein Biosensors Based on Biofunctionalized Conical Gold Nanotubes. *J. Am. Chem. Soc.* **2005,** *127,* 5000–5001.
60. Sexton, L. T.; Horne, L. P.; Martin, C. R. Developing Synthetic Conical Nanopores for Biosensing Applications. *Mol. BioSyst.* **2007,** *3,* 667–685.
61. Ali, M.; Mafe, S.; Ramirez, P.; Neumann, R.; Ensinger, W. Logic Gates Using Nanofluidic Diodes Based on Conical Nanopores Functionalized with Polyprotic Acid Chains. *Langmuir* **2009,** *25,* 11993–11997.
62. Siwy, Z.; Kosińska, I. D.; Fuliński, A.; Martin, C. R. Asymmetric Diffusion through Synthetic Nanopores. *Phys. Rev. Lett.* **2005,** *94.*
63. Vlassiouk, I.; Siwy, Z. S. Nanofluidic Diode. *Nano Lett.* **2007,** *7,* 552–556.
64. Cheng, L.-J.; Guo, L. J. Ionic Current Rectification, Breakdown, and Switching in Heterogeneous Oxide Nanofluidic Devices. *ACS Nano* **2009,** *3,* 575–584.
65. Yan, R.; Liang, W.; Fan, R.; Yang, P. Nanofluidic Diodes Based on Nanotube Heterojunctions. *Nano Lett.* **2009,** *9,* 3820–3825.
66. Kalman, E. B.; Vlassiouk, I.; Siwy, Z. S. Nanofluidic Bipolar Transistors. *Adv. Mater.* **2008,** *20,* 293–297.
67. Daiguji, H. Ion Transport in Nanofluidic Channels. *Chem. Soc. Rev.* **2010,** *39,* 901–911.

68. Zhang, H.; Hou, X.; Zeng, L.; Yang, F.; Li, L.; Yan, D., et al. Bioinspired Artificial Single Ion Pump. *J. Am. Chem. Soc.* **2013,** *135,* 16102–16110.

69. Cao, L.; Guo, W.; Wang, Y.; Jiang, L. Concentration-Gradient-Dependent Ion Current Rectification in Charged Conical Nanopores. *Langmuir* **2012,** *28,* 2194–2199.

70. Bezrukov, S. M.; Vodyanoy, I.; Parsegian, V. A. Counting Polymers Moving through a Single Ion Channel. *Nature* **1994,** *370,* 279–281.

71. Dekker, C. Solid-State Nanopores. *Nat. Nanotechnol.* **2007,** *2,* 209–215.

72. Kasianowicz, J. J.; Brandin, E.; Branton, D.; Deamer, D. W. Characterization of Individual Polynucleotide Molecules Using a Membrane Channel. *Proc. Natl. Acad. Sci. USA* **1996,** *93,* 13770–13773.

73. Branton, D.; Deamer, D. W.; Marziali, A.; Bayley, H.; Benner, S. A.; Butler, T., et al. The Potential and Challenges of Nanopore Sequencing. *Nat. Biotechnol.* **2008,** *26,* 1146–1153.

74. Ali, M.; Nasir, S.; Ramirez, P.; Cervera J.; Mafe, S.; Ensinger, W. Carbohydrate-Mediated Biomolecular Recognition and Gating of Synthetic Ion Channels. *J. Phys. Chem. C.* **2013,** *117,* 18234–18242.

75. Tagliaucchi, M.; Szleifer, I. How Does Confinement Change Ligand-Receptor Binding Equilibrium? Protein Binding in Nanopores and Nanochannels. *J. Am. Chem. Soc.* **2015,** *137,* 12539–12551.

76. <http://www.merckmillipore.com/> (accessed Jan 14, 2016).

77. Daiguji, H.; Yang, P. D.; Szeri, A. J.; Majumdar, A. Electrochemomechanical Energy Conversion in Nanofluidic Channels. *Nano Lett.* **2004,** *4,* 2315–2321.

78. van der Heyden, F. H. J.; Bonthuis, D. J.; Stein, D.; Meyer, C.; Dekker, C. Power Generation by Pressure-Driven Transport of Ions in Nanofluidic Channels. *Nano Lett.* **2007,** *7,* 1022–1025.

3

Single Ion Track-Etched Nanochannels for Analytical Applications

G. Pérez-Mitta[1], C. Trautmann[2], M.E. Toimil-Molares[2], O. Azzaroni[1]

[1]UNIVERSIDAD NACIONAL DE LA PLATA–CONICET, LA PLATA, ARGENTINA
[2]GSI HELMHOLTZZENTRUM FÜR SCHWERIONENFORSCHUNG GmbH, DARMSTADT, GERMANY

CHAPTER OUTLINE

3.1 Introduction

Ion track technologies for membrane applications have been extensively explored during the past decade for diverse applications such as molecule size separation or templates for nanoarchitectures.[1,2] However, it is the use of single etched ion-track membranes for sensing what is pushing this technology toward cutting-edge advances in material sciences. The interest of the scientific community on track-etched foils for analytical applications is fairly gained, their unique fluidic properties make them an excellent substrate for the design of complex sensing devices.[3] One interesting characteristic of the fluidic properties of ion-track nanochannels is that they resemble in a highly comparable manner the fluidic properties of biological protein nanopores which are already in the market being part of sophisticated devices for DNA sequencing technologies. Therefore, the use of artificial nanopores-channels could be the next step toward fully abiotic, mechanically resistant sequencing devices. Another particular property of ion-track nanochannels is their versatility for obtaining different channel geometries with a variety of fluidic properties.[4] This is a distinctive characteristic of these systems compared with other abiotic nanochannels.

Up to date, most of the work done in order to use ion-track nanochannels as sensing devices is related to their ion-transport properties when a transmembrane voltage is applied through them.[5] The electrokinetic behavior of confined ions is different from their bulk behavior and several characteristic phenomena stem from these differences. One example of this is the ion current rectification that can be easily observed by applying a triangular zero average transmembrane potential at constant scan rate while simultaneously measuring the current through the membrane.[6] This effect is clearly seen in conical single etched ion-track foils due to its high impedance. In membranes with a high density of pores this effect is less clear due to the emergence of other electrochemical processes. Since it has been proved that this effect is dependent on the charge and shape of the channel, it is an interesting phenomenon for sensing species that can change either the geometry or the surface charge of the nanochannels.

Another well-documented phenomenon that occurs in single nanochannels is the partial or total blockage of the pore by molecules or particles. By recording these blockages and a correct statistical treatment of the data an idea of the size and quantity of molecules in solution can be obtained. This method is called Resistive Pulse and it was previously developed for biosensing with micropores in the Coulter counter device which is still largely used for biochemical essays.[7]

Over the past 20 years several techniques for construction of nanochannels have been developed, for example, other type of nanopores that have been heavily used for these types of applications are inorganic silicon oxide and silicon nitride nanopores. These nanopores can be obtained in a very reproducible manner by electron or ion beam methods. Other fabrication techniques for single nanopores include the use of a Femtosecond-pulsed laser to produce glass single nanopores, soft-lithography techniques, and hybrid biological–artificial nanopores by combining protein channels with solid-state nanopores. All of these types of nanopores have been reviewed elsewhere and will not be treated in this chapter.[8]

Even though extensive study has been done to develop stable and reproducible sensing devises based on ion-track-etched solid-state nanochannels there is still a long way to go to fully understand the physicochemical behavior of these nanochannels. The existing models to explain nanochannels transport properties had been discussed elsewhere and will not be discussed in this chapter. However, for the sake of discussion we will refer occasionally to the continuous model based on the Poisson–Nernst–Planck equations to address certain points.[9]

3.1.1 Etched Single-Ion-Track Membranes

Ion track technology is based on the irradiation of dielectric materials with swift heavy ions that produces irreversible and highly localized damage in the material. This damage left by heavy ions is commonly called latent track and its chemical nature for commonly used polymer foils has been found to be really wide, from formation of alkynes in the region 2–5 nm from the center of the track so-called the track *core*, to conventional degradation processes occurring in the range between this core radius and about 100 nm, a region called

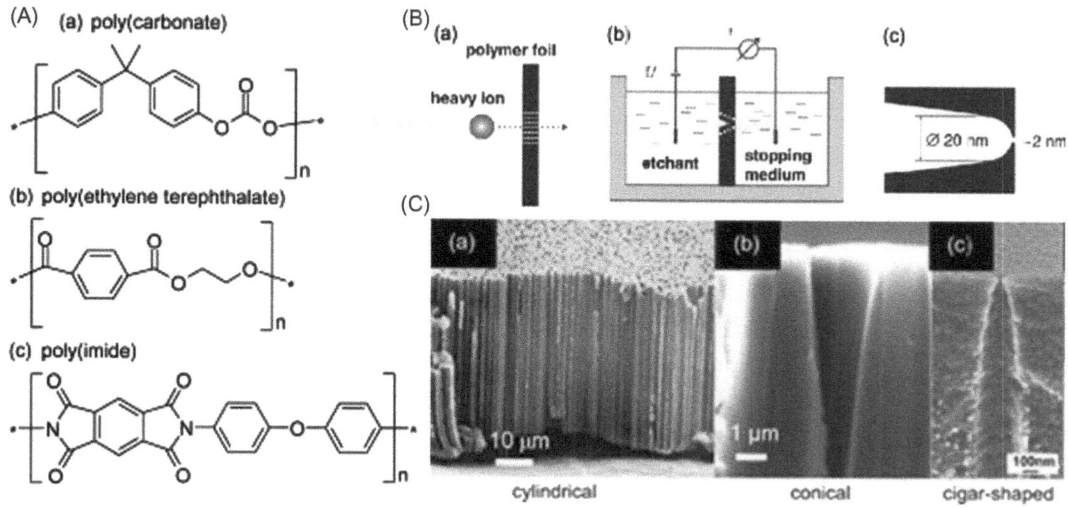

FIGURE 3.1 (A) Different polymers used for ion-track applications. (B) Scheme of the ion-track irradiation and etching process of a polymer foil. (C) Scanning Electron Microscopy (SEM) images of cross sections of nanochannels having different geometries, from left to right (a) cylindrical, (b) conical, and (c) Cigar-shaped.

halo.[10] Normal polymeric materials used for ion-track applications are commercially available polymers like poly(carbonate) (PC), poly(ethyleneterephtalate) (PET), or poly (imide) (Fig. 3.1A).[11]

Fig. 3.1B depicts the full fabrication process of single nanochannels by means of the ion track-etching technique.[12] As can be seen in the figure, after irradiation chemical etching is performed in an electrochemical cell. Due to the higher sensitivity of the material in the latent track toward chemical reactions, decomposition reactions are used to selectively remove the material in the track, thus forming an open nanochannel. For example, in common condensation polymers like PET or PC, an alkaline hydrolysis in highly concentrated NaOH solutions is generally performed. It was proven that the velocity of the etching reaction, i.e., etching rate, occurs about 1000 times faster in the direction of the track that in the bulk material, this allows to reach the opening of the channel without significantly changing the thickness of the supporting foil. The velocity in the direction of the track is normally called, *track etching rate* (v_t) while the velocity of etching over the polymeric membrane is called *bulk etching rate* (v_b), the geometry of the channels is related to the ratio between these two values.[13]

As aforementioned, one important characteristic of the etching procedure of ion-track membranes is that it allows obtaining a large number of different channel geometries by simply adjusting some experimental conditions during the process. By these means, geometries ranging from asymmetric and tapered, like conical, bullet-like, or funnel-like to symmetric geometries like cylindrical or cigar-like can be obtained (Fig. 3.1B).[14] The importance of this control over the geometry is related to the ion transport properties of nanochannels which are highly dependent on the geometry of the channel. Therefore, a fine control over this factor is critical for all the applications of nanochannels.

An important characteristic of the ion-track-etched nanopores in polymer foils is their relatively ease to perform secondary chemical modifications.[6] After the etching of condensation polymers like PET or PC, carboxylic groups are present at the surface. These groups allow the use of diverse chemical reactions to functionalize and incorporate other functional groups over the whole membrane surface, inside and outside the pores. Activation with carbodiimide, especially with 1-ethyl-3-(3-dimethylaminopropyl) carbodiimide (EDC), represents a routine procedure to incorporate functional groups over the whole surface inside and outside the pores modifying the properties of the nanopores.[15] This modification method has been thoroughly used for multiple modifications like the growth of polymer brushes from the surface by anchoring the polymerization initiator to the surface. Several interesting functional devices have been developed by using this technique. Other modification methods do not imply the use of covalent chemistries but the charge of the surface stemming from the surface groups. For example for PET nanochannel the net surface charge after etching is negative at pH above ~3 due to the carboxylate groups and this characteristic was used as a substrate for electrostatic layer-by-layer deposition.[16]

Also we can mention the use of physical techniques like metal sputtering, atomic layer deposition, or electron-beam evaporation for modifying the performance of nanochannels or for further modifications. In a recent work by Perez-Mitta et al. a gold sputtered PC conical nanochannel was used as a substrate for electrochemical polymerization of aniline in order to confer pH and voltage responsiveness to the nanochannel.[17] By combining different surface chemistry approaches with abiotic nanochannels, there is also an interest in mimicking the properties of biological pores. These properties include gating under certain stimuli or a particular selectivity toward certain types of molecules or ions. In this context, it was recently shown the construction of a potassium-dependent nanofluidic diodes by using host–guest supramolecular chemistry.[18] This system showed a potassium-dependent ionic selectivity which is in close resemblance to the behavior of gated ionic channels like the calcium-activated potassium channels. Moreover, chemical functionalization can be also used to eliminate or modify features that are not useful, like changing the surface charge from positive to negative or neutral, or eliminating electrical noise coming from the chemical groups present on the polymeric substrate.

3.1.2 Nanofluidics

The commonly accepted definition for nanofluidics is *the study of motion of fluids or dissolved species through or past structures with at least one of its dimensions between 1 and 100 nm.* Due to the fact that the characteristic dimensions that control the mechanical and electrochemical behavior, as for example the thickness of the electrical double layer (EDL), are also within this range, nanofluidics study the emergence of very unique phenomena arising from the closeness between these dimensions and the size of the fluidic structures.[19]

When the EDLs at the walls of a given nanochannel are close enough to overlap significantly, a unipolar solution forms, and the composition (ion concentration) within the channel is related to the composition inside the double layer which is not neutral and with a nonzero

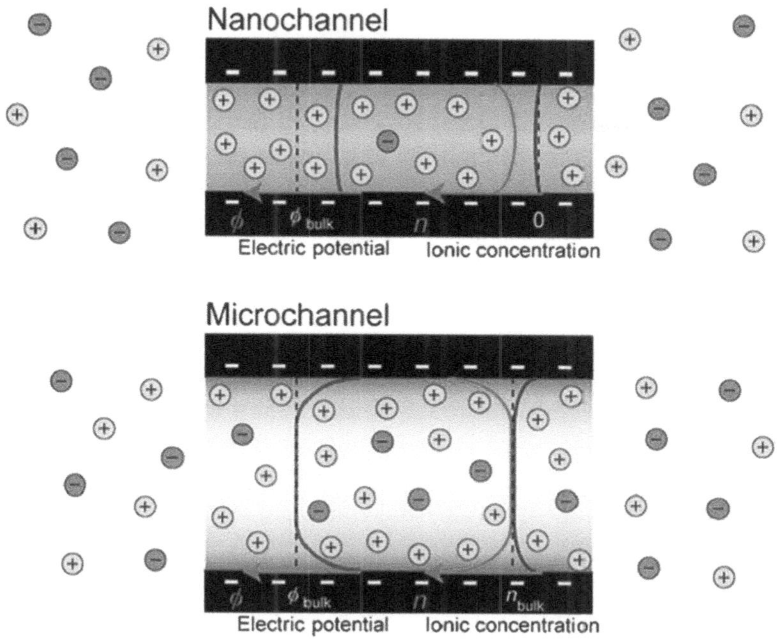

FIGURE 3.2 Scheme depicting the emergence of a unipolar solution in nanochannel, curves shown represent the change in concentration and electric potential as a function of the radius of the channels. Source: *Adapted from reference Daiguji, H. Ion Transport in Nanofluidic Channels.* Chem. Soc. Rev. ***2010***, 39, *901–911.*

potential. Smaller pores will lead to an even higher overlapping of the layers driving an enrichment of counter-ions inside the nanochannel and a more unipolar solution (Fig. 3.2).[20] The double layer depends on the electrolyte concentration in the solution and is given by the Debye screening length defined as:

$$\lambda_d = \sqrt{\varepsilon\varepsilon_0 k_B T / 2c_{bulk} z^2 e^2} \tag{3.1}$$

where e is the charge of an electron, ε the dielectric constant of water and ε_0 the permittivity of vacuum, k_B is the Boltzmann constant, T is the absolute temperature, c_{bulk} is the bulk concentration of ions, and z is the valency of the ions. As it can be seen, the Debye length depends inversely with the square root of the electrolyte concentration which means that at higher concentrations the DL becomes smaller. Therefore a 10 nm nanochannel will not present a significant overlapping of the double layer for electrolyte concentrations of 0.1 M ($\lambda_d \sim 1$ nm) but it would present a large overlapping at a concentration of 0.001 M ($\lambda_d \sim 10$ nm).

The transport of ions within nanochannels is thus different from the transport in microchannels or bulk systems. As it was stated before, the composition inside nanochannels is different from the bulk system because inside the pores the counter-ion concentration is enhanced due to electrostatic interactions with the surface. Therefore, bulk properties of the solution like the conductivity cannot be taken into account in a straightforward way when addressing the behavior of ionic transport in nanochannels. In order to understand these problems Schoch and Renaud developed a model for the conductance of a nanochannel by using nanoslits as a model system.[21] The conductance of the nanochannels was expressed as the superimposition of the bulk conductance and the excess ion conductance related to the excess of counter-ions needed to neutralize the surface charge of the channel,

$$G = 10^3 N_A e(\mu_{K+} + \mu_{Cl-})c\frac{wh}{l} + 2\mu_{K+}\sigma_s^*\frac{w}{l} \tag{3.2}$$

where w is the width and l is the length of the nanoslits, μ_i is the mobility of ion i, c is the concentration of ions, N_A is the Avogadro constant, and e is the electron charge.

This model predicts that at high electrolyte concentrations the conductance of the channels is dependent mostly on the geometry of the channels as it is in bulk systems because of the salt concentration being higher than the excess mobile counter-ion concentration, but if the salt concentration falls below this value the conductance of the nanochannel starts to be regulated by the surface charge density of the channel. This is observed experimentally as a plateau in the conductance value as a function of the electrolyte concentration in a log–log scale, this is why additional care must be taken into account when analyzing experimental data regarding the conductance of a nanochannel, specially at low electrolyte concentrations.

The phenomenon that arises when there is a significant surface charge, low electrolyte concentrations and when a variable transmembrane potential is applied across asymmetrically shaped nanochannels is a rectification of the ionic current through the channels. In order to measure this effect in ion-track-etched nanochannels, previously etched foils are collocated in an electrochemical cell that is further filled with an electrolyte solution, and then current–voltage (I–V) characteristics are obtained (Fig. 3.3).

This effect was proven to occur due to a longitudinal asymmetrical distribution of the surface charge density through the nanochannels.[22] The most common way to obtain this asymmetry has been by using etching protocols that produce asymmetric geometries like conical or bullet-like (Fig. 3.1C); since the surface charge of the polymer is supposed to be homogeneous after the etching procedure, an asymmetry in the geometry of the channel would provide a spatial asymmetry in the charge distribution and the broken symmetry necessity for the current rectification phenomenon would be fulfilled. For the case of nanochannels with a symmetric geometry it was proven that an anisotropic functionalization can also produce the occurrence of current rectification. For example, Karnik et al. built up a rectifying nanochannel by modifying a nanochannel fabricated in a sacrificial polysilicon with biotin and then using a diffusion-limited patterning method to selectively modify part of the channels with Avidin, a protein that renders a net positive charge (Fig. 3.4).[23] In this work it was shown that I-V curves presented current rectification after the modification took place.

The occurrence of the rectification phenomenon is closely related both to the sign and magnitude of the surface charge of the nanochannels. It has been found theoretically and experimentally that there is a strong direct correlation between the magnitude of the surface

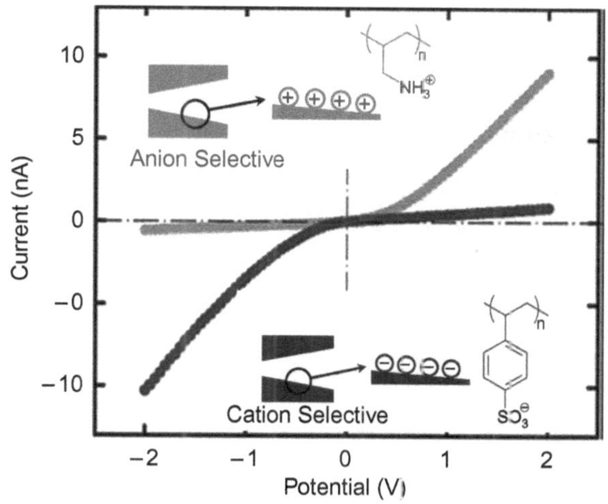

FIGURE 3.3 Current–voltage (*I*–*V*) characteristics for a single nanochannel. The red (gray in the print version) curve corresponds to a negative surface charge while the blue (black in the print version) one corresponds to a positive charge. Source: *Reproduced with permission from reference Ali, M.; Yameen, B.; Cervera, J.; Ramírez, P.; Neumann, R.; Ensinger, W.; et al. Layer-by-Layer Assembly of Polyelectrolytes into Ionic Current Rectifying Solid-State Nanopores: Insights from Theory and Experiment.* J. Am. Chem. Soc. ***2010***, 132, *8338–8348.*

charge and the rectification efficiency and that the sign of the charge modulates the direction of the rectification by changing the selectivity of the channels toward either cations or anions. Positive surface charge would produce current–voltage curves inverse to that of a negative charged surface (Fig. 3.3). Several models have been developed to explain this effect and one of the most successful is the continuum model based on the coupled equations of Poisson–Nernst–Planck. It has been proved that for pores over 10 nm the continuum model can be applied while for pores over 10 nm stochastic dynamic models need to be used. By using this model it was shown that the contribution for the current in the high conductive state, which in the case of a conical negatively charged nanochannel occurs when cations are transported from the small opening of the channel toward the big opening, is given by both anions and cations, this means that the fraction of the total current transported by each type of ion, i.e., transport number, for the cations and anions is ~0.5. On the other hand, for the low conductive state, when cations are transported from the large opening toward the small opening, the main contribution to the total current is from the cations, which yields transport number of more than 0.95. This means that almost the totality of the current is transported by cations, since the only cations available are the ones in close contact with the walls of the channel, the conductance of this state is much lower and this triggers the appearance of the rectification phenomenon. Therefore, the current rectification arises from a voltage modulated ionic selectivity of the nanochannels. This non-ohmic behavior can be then used for sensing applications since as it has been demonstrated that the rectification of the current is highly sensitive to changes either in the internal volume of the nanochannel or the surface.

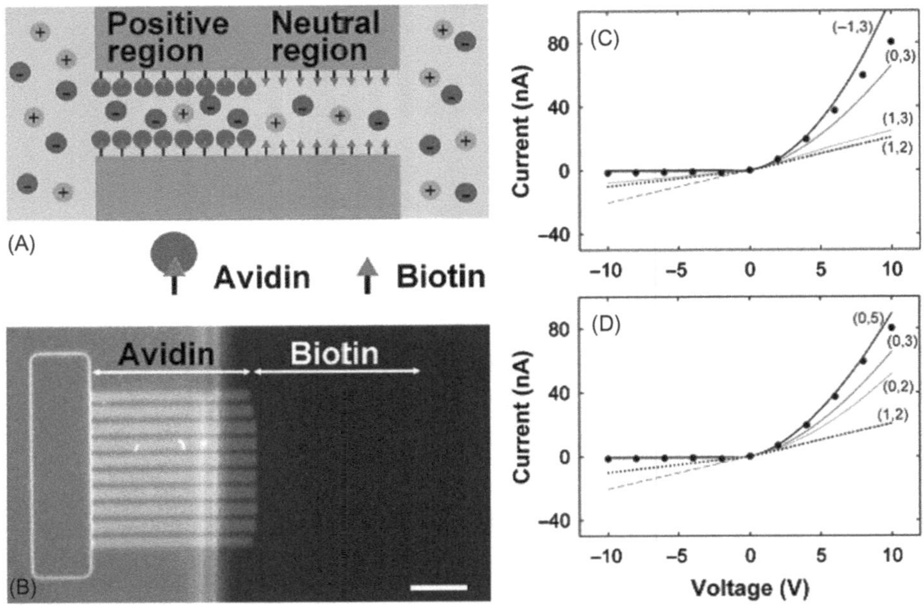

FIGURE 3.4 (A) Scheme showing the nanofluidic diode fabricated by selective modification with Avidin. (B) Nanofluidic diode imaged by epifluorescence, the image shows a fluorescence-labeled Avidin in one half of the nanochannels. Scale bar 20 mm. (C) and (D) Current–voltage characteristics at a KCl concentration of 10 mM. Solid circles represent experimental data, dashed line is a linear fit of the experimental points between −50 mV and 50 mV to calculate the channel conductance, and solid lines correspond to theoretical modeling for a 120 μm × 40 μm × 30 nm nanoslit with different combinations of surface charge densities for the Avidin and Biotin parts, the first number corresponds to the Biotin and the second to the Avidin. Source: *Reproduced with permission from reference Karnik, R.; Duan, C.; Castelino, K.; Daiguji, H.; Majumdar, A. Rectification of Ionic Current in a Nanofluidic Diode.* Nano Lett. ***2007**, 7, 547–551.*

3.2 Resistive Pulse Sensing

Resistive pulse sensing is currently the most common nanofluidic technique for sensing applications. It is based on the concept of the Coulter counter device which is widely used in clinical biochemistry and in particular in hematology (Fig. 3.5). The Coulter counter works with a simple size-based principle in which analyte molecules are forced through a microporous membrane while the transmembrane current is recorded.[24] A blocking of the pore, either partial or complete, would produce a characteristic drop of the current. By knowing the size of the pores, the size of the molecules can be inferred, while the frequency of the blockings provides information about the concentration. This device is commonly used to obtain, for example, the complete blood cell value. Other applications of the Coulter counter can be found also in the industry of paints, food, and ceramics among others.

By combining etched ion-track nanochannels and resistive pulse sensing much smaller molecules than for the traditional Coulter device can potentially be sensed. The promising outlook of this unique combination is the development of a nanochannel-based single molecule sensing method.[25]

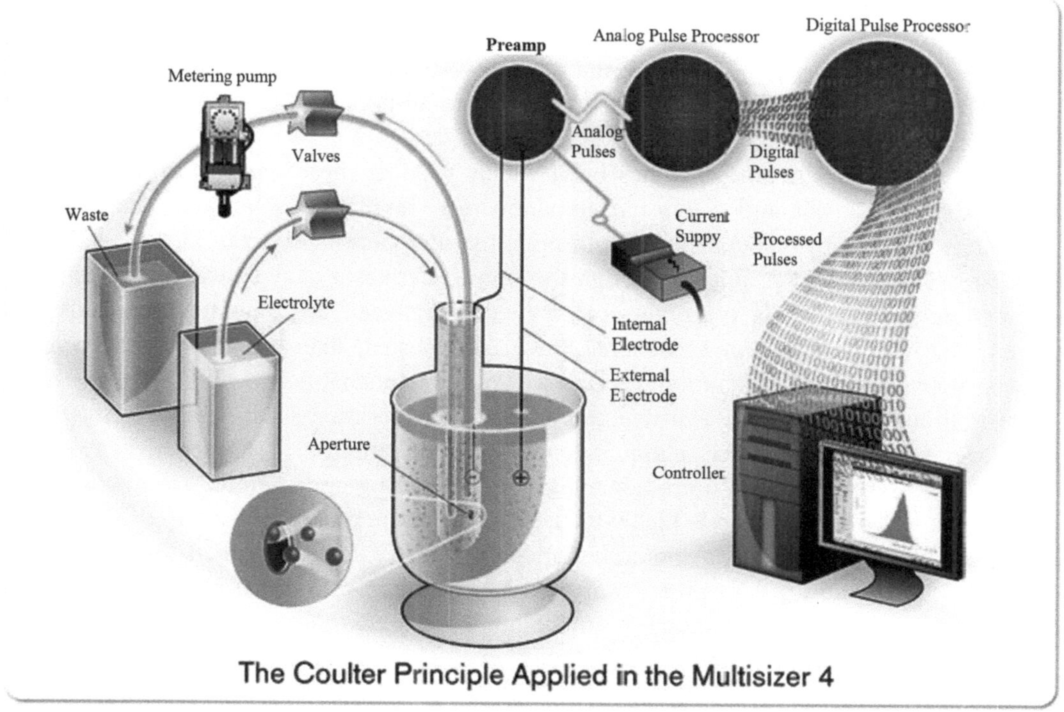

FIGURE 3.5 Scheme showing the functioning of the Coulter counter. Source: *Reproduced from https://www.beckmancoulter.com/wsrportal/wsr/industrial/particle-technologies/coulter-principle/index.htm.*

One of the most studied applications of the resistive pulse method is the sensing of DNA. The ultimate objective is to be able to identify the nucleic acid sequence by passing a single strand of DNA trough a narrow nanochannel and by differentiating the resistive peak of every nucleotide. Since each nucleic acid has a different size, the blockage that each one produces is characteristic so that by driving electrophoretically a DNA strand through a narrow nanochannel the complete sequence can be obtained.[26] This presents an easy, fast, and inexpensive way to obtain genomic information. In 1996, Kasianowicz et al. proved for the first time that nanopores could be used to sequence a DNA strand[27] and in 2004 the National Human Genome Research Institute National Institutes of Health offered grants for a total of 38 million dollars to whom could provide a route toward low-cost DNA sequencing (<1000 dollars). Since then several improvements were accomplished. In 2012 Oxford Nanopore Technologies announced the first DNA sequencing technology based on nanopore systems.

However, until now commercially available nanopore applications are mostly based on biological nanopores, channel transmembrane proteins, for example, alpha-hemolysin which is a toxin from *Staphylococcus aureus* that spontaneously forms an ionic channel after interacting with a lipidic membrane.[25] Even though biological nanopores ensure a reproducible behavior there are problems associated with the use of this family of nanopore systems.

They are difficult to maintain in nonphysiological conditions since proteins are unstable under extreme environmental conditions like acidic or basic pH or high temperatures. Also, their mechanical properties are not as good as in artificial materials like polymers or ceramics, this being a problem that becomes an important challenge for commercial applications.

Another type of nanopore that has been widely used in these types of applications are inorganic silicon dioxide and silicon nitride nanopores. The main problem with these systems is the control of the passage of the DNA through the nanopore in order to identify each part. For example, driving DNA quickly through the pore generates a convolution of the signals of different nucleotides losing the ability to distinguish between the different nucleotides. To prevent this problem several approaches have been tested, like adding a lipidic layer to increase the viscosity of the solution over the wall of the nanochannels.[28]

Even though in principle resistive pulse sensing experiments can be performed with nanopores with any geometry, either symmetric or asymmetric, it has been proved that more tapered geometries are more advantageous for sensing applications. The main reason for this is that the electric field is highly focused toward the tip of the nanopores. That is why this region of the nanopores having an enhanced sensitivity has been called "Sensing Zone." Other reasons include the lower resistance as compared to cylindrical nanopores with the same opening diameter and also lower probability of clogging.[29]

Ion-track-etched nanochannels have been applied for sensing applications with the resistive pulse technique. Even though the current signal in this type of channels is not as stable as it is for inorganic materials, their mechanical properties are of high interest for developing portable sensing devices.

In 2005 Heins et al. reported a 4,4′,4″,4‴-(porphine-5,10,15,20-tetrayl)- tetrakis(benzenesulfonic acid) (TPPS) sensor by using an asymmetrically shaped Kapton nanopore. The TPPS is a tetravalent molecule that was electrophoretically driven through the nanochannel producing a downward peak in the current every time a molecule passed through the pore (Fig. 3.6).[30] This occurred due to a partial blockage of the tip of the pore and the frequency of this effect was demonstrated to be correlated with the concentration of TPPS in the solution. This is the first time that a resistive pulse sensing experiment was reported using an artificial nanopore.

Another interesting set of experiments by the same group were carried out in order to sense DNA, both single and double stranded. A conical PC nanochannel with a tip diameter of 40 nm was used. Due to the fact that the DNA chains are anionic it was possible to drag them electrostatically through conical PC single nanochannels while recording the transmembrane currents. In the absence of analyte a steady-state current of ~5 nA was obtained and it started to show downward peaks as soon as an analyte solution was introduced in the cell (Fig. 3.7A).[31] The analytes that were used for these experiments were a single-stranded phage DNA and a double-stranded plasmid DNA. A linear relation between the concentration of the ss-DNA and the frequency of the current-block events was found (Fig. 3.7B). It is interesting to mention that for this experiment as well as for the TPPS, the

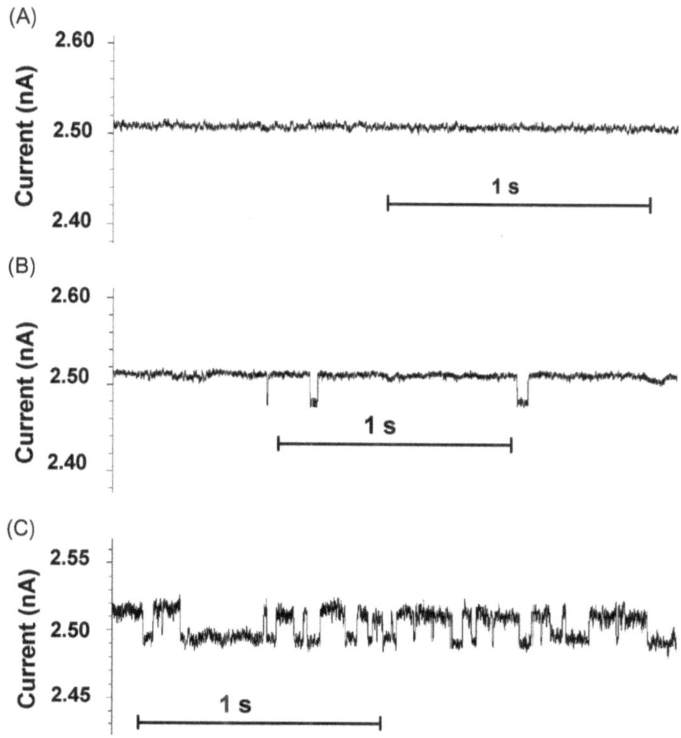

FIGURE 3.6 Response of a nanopore to a concentration of (A) 0 nM, (B) 20 nM, and (C) 60 nM of TTPS. Source: *Reproduced with permission from reference Heins, E. A.; Siwy, Z. S.; Baker, L. A.; Martin, C. R. Detecting Single Porphyrin Molecules in a Conically Shaped Synthetic Nanopore.* Nano Lett. ***2005***, 9, *1824–1829.*

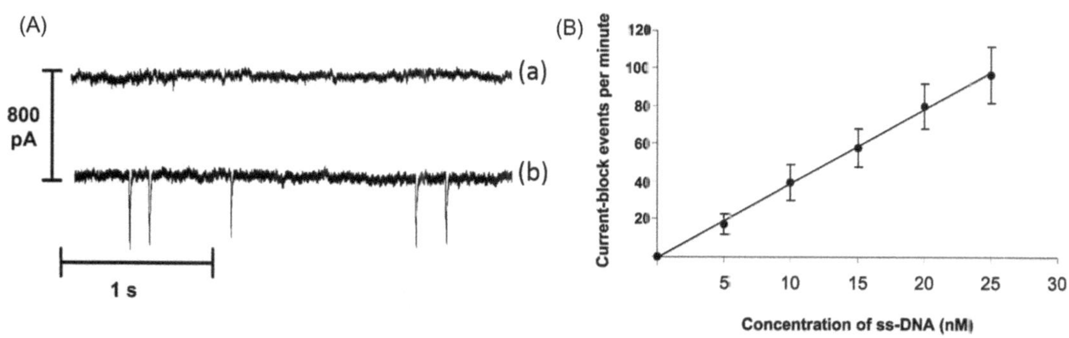

FIGURE 3.7 (A) Current behavior of a nanochannel with (a) 0 nM and (b) 10 nM of DNA in solution. (B) Current-block events per minute versus the concentration of ss-DNA, a linear relation is observed. Source: *Reproduced with permission from reference Harrell, C. C.; Choi, Y.; Horne, L. P.; Baker, L. A.; Siwy, Z. S.; Martin, C. R. Resistive-Pulse DNA Detection with a Conical Nanopore Sensor.* Langmuir ***2006***, 22, *10837–10843.*

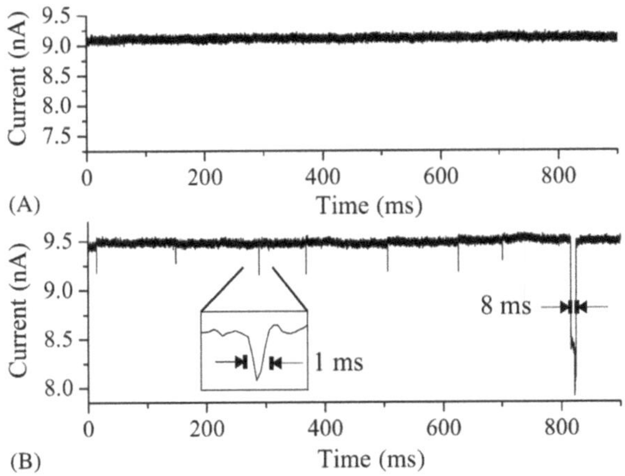

FIGURE 3.8 Resistive pulse sensing experiments performed in a Kapton nanochannel, before (A) and after (B) the addition of 10 mM solution of DNA. Source: *Reproduced with permission from reference Schiedt, B.; Healy, K.; Morrison, A. P.; Neumann, R.; Siwy, Z. Transport of Ions and Biomolecules Through Single Asymmetric Nanopores in Polymer Films. Nucl. Instrum.* Methods Phys. Res. B ***2005***, 236, *109–116.*

analyte was introduced only on the tip side of the membrane in order to translocate the molecules from the tip to the base of the nanochannels.

Regarding the resistive pulse sensing of DNA with ion-track-etched nanochannels, one pioneering experiment was reported by Schiedt et al. using Kapton single conical nanochannels.[32] The use of a Kapton nanochannel was preferred due to the stability of the current signal in this material. In these experiments double-stranded PhiX174 plasmid DNA was used. A clear difference in the signal was observed upon the addition to the DNA in the solution (Fig. 3.8). One important outcome of this work was the finding that conical nanochannels are capable of distinguishing between DNA chains of different length and conformation in a comparable manner to previous reports using protein nanopores.

3.3 Sensing via Current–Voltage Characteristics (or *I*–*V* Characteristics for Sensing)

Another approach for sensing with single nanochannels is based on the analysis of the current–voltage characteristics. In this case the geometry of the channel has a critical role in defining the features of the current–voltage response. When the nanochannels are cylindrical they behave in a pure ohmic way, showing a linear relation between the current and the voltage. In this configuration the conductance of the channel is fully regulated by the size of the channel as well as for the inner surface charge of the channel. Therefore, the conductance of the channel is sensitive to changes in the inner volume, like adsorption of a certain molecule, or in the surface charge. Any blockage of the channel will be translated into a decrease of the channel conductance. This effect can be used for sensing applications.

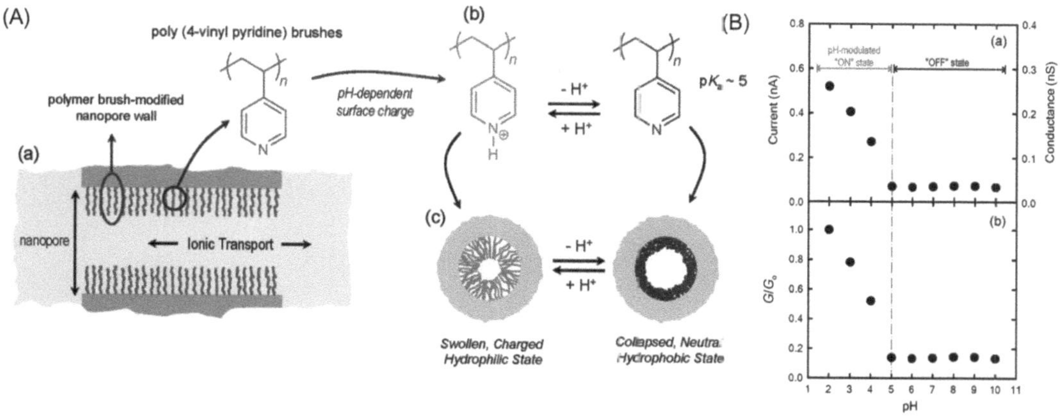

FIGURE 3.9 (A) Scheme showing a cylindrical nanochannel after the modification with pH-responsive PVP polymer brushes, a front cross section of the channel is also depicted showing the change in the swelling state of the polymer under acidic and basic conditions. (B) Change in the transmembrane current and conductivity of the channels as a function of pH. Source: *Reproduced with permission from reference Yameen, B.; Ali, M.; Neumann, R.; Ensinger, W.; Knoll, W.; Azzaroni, O. Synthetic Proton-Gated Ion Channels via Single Solid-State Nanochannels Modified with Responsive Polymer Brushes.* Nano Lett. ***2009***, 9, *2788–2793.*

Yameen et al. demonstrated the construction of a pH-sensitive device by incorporating poly (4-vinyl pyridine) (PVP) brushes inside cylindrical PET nanochannels.[33] After the polymerization was accomplished, the system showed a gating behavior toward acidic pH conditions (Fig. 3.9).

As it occurs in resistive pulse sensing techniques, there is a great interest in working with more tapered geometries because they show lower resistivities and display very unusual and interesting non-ohmic current–voltage characteristics.

The emergence of current rectification was demonstrated to be of great relevance in order to design nanofluidic devices with sensing applications. Since their rectification properties are more sensitive to changes in the volume and charge of the nanochannels, the use of asymmetric geometries is more appropriate for sensing applications. Up to date, several examples of analytical devices based on asymmetric solid-state nanopores using the current–voltage characteristics have been developed.

One of the first sensing experiments using ion-track-etched conical nanochannels was reported by Siwy and coworkers using gold-coated PET conical nanochannels.[34] The as-obtained Au nanotubes exhibited with tip diameters in the range 5–9 nm thus allowing the authors to identify different analytes. After the Au metallization the pore surface was modified with different molecular recognition agents to achieve sensitivity toward different specific analytes. The examples included biotin/streptavidin (Fig. 3.10A), protein-G/immunoglobulin G (IgG) (Fig. 3.10B), and also anti-ricin/ricin (Fig. 3.10C). For all the analytes the first step was to attach a thiolated Biotin to the gold and for the particular cases of IgG and Ricin, first Streptavidin was added to subsequently add a biotinylated protein G or anti/ricin. For these experiments "on/off" curves were obtained upon the addition of the analyte

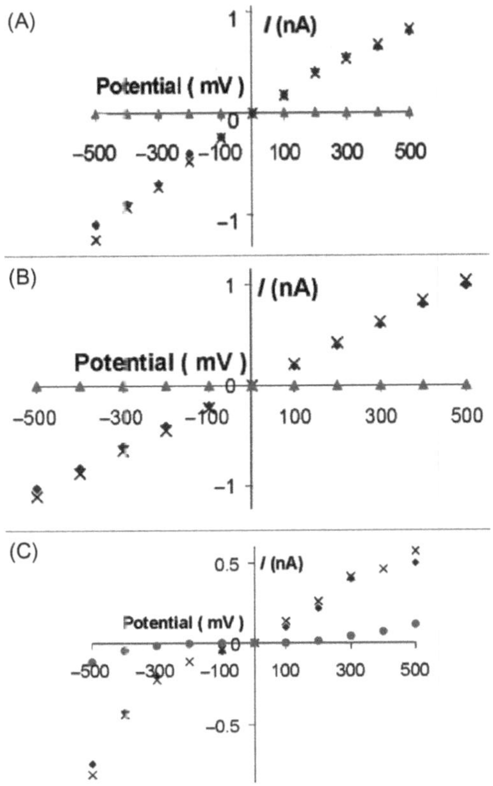

FIGURE 3.10 Current–voltage characteristics for a nanopore functionalized with biotin for sensing of (A) streptavidin, (B) IgG, and (C) anti-ricin. The curves show a decrease of the current after the analyte solution was used. Source: *Reproduced with permission from reference Siwy, Z.; Trofin, L.; Kohli, P.; Baker, L. A.; Trautmann, C.; Martin, C. R. Protein Biosensors Based on Biofunctionalized Conical Gold Nanotubes.* J. Am. Chem. Soc. ***2005***, 127, *5000–5001.*

finding high selectivities toward the respective analytes and low limits of detection, e.g., for IgG a 10 nM concentration was enough to fully decrease the current.

In another, more simplistic approach, Wang and coworkers designed a nanofluidic sensor by using an unmodified conical PET nanochannel.[35] These authors showed that the unmodified nanochannels display larger sensitivities toward smaller and more hydrophobic molecules, like Hoechst 33342, propidium iodide, and bupivacaine (Fig. 3.11). These results were analyzed using a Langmuir model for the adsorption of the analytes on the nanochannels finding a good fitting between the results and the model. The larger constant was found for Hoechst 33342 with a value of 2.53×10^6 L/mol which is the smallest and more hydrophobic molecule. The authors related these results to the capacity of the molecule to have pi-stacking interactions with the molecules present in the nanochannel surface.

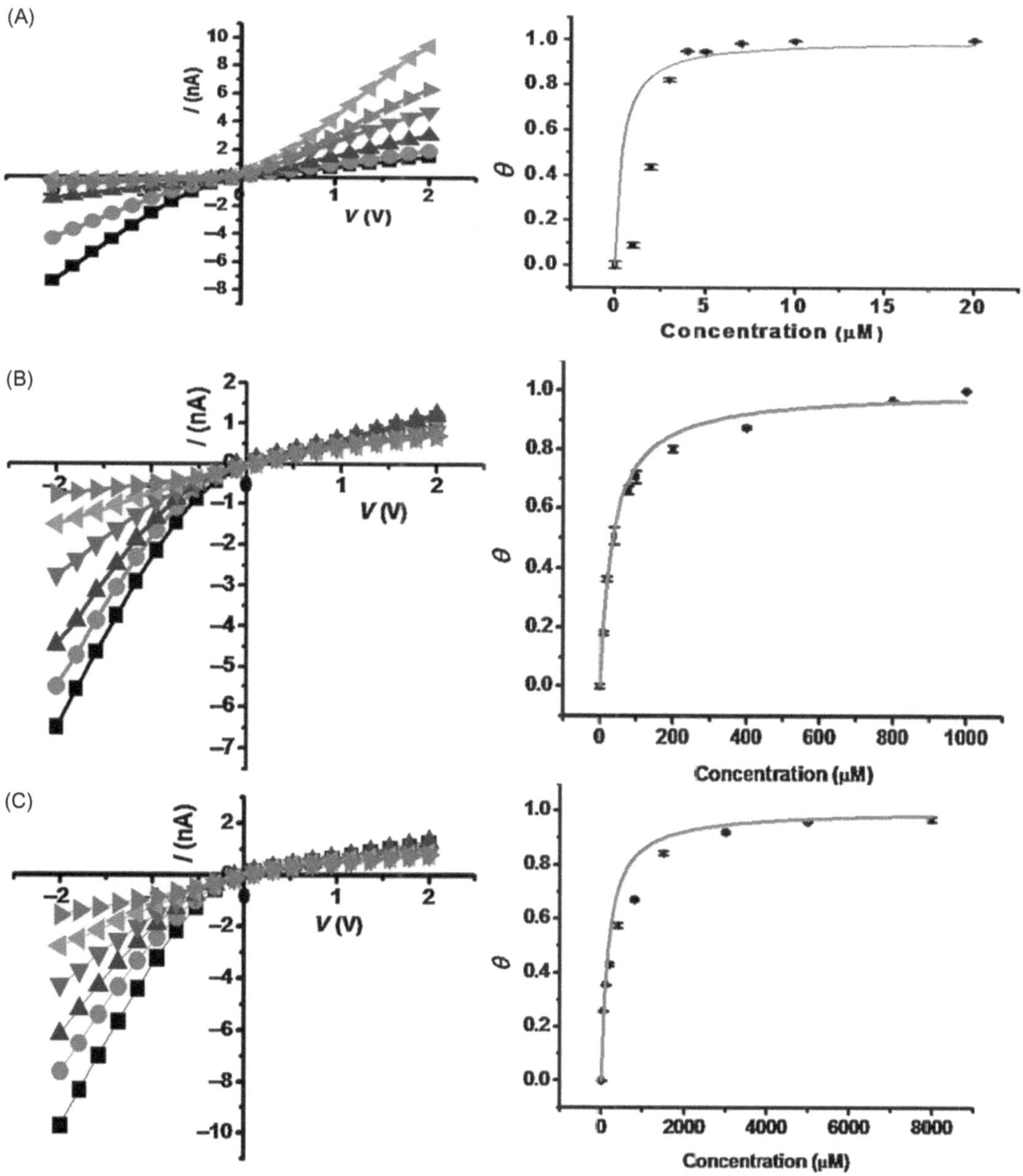

FIGURE 3.11 Current–voltage curves and surface coverage versus the concentration of analyte for the system of (A) Hoechst 33342, (B) propidium iodide, and (C) bupivacaine. Source: *Reproduced with permission from reference Ali, M.; Tahir, M. N.; Siwy, Z.; Neumann, R.; Tremel, W.; Ensinger, W. Hydrogen Peroxide Sensing with Horseradish Peroxidase-Modified Polymer Single Conical Nanochannels.* Anal. Chem. ***2011***, *83, 1673–1680.*

One important advantage of the current–voltage characteristics approach toward the sensing of molecules is the possibility of sensing smaller molecules than the resistive pulse sensing. In asymmetric nanopores the sensing mechanism is not exclusively limited to a size reduction of the nanochannel as in the case of resistive pulse sensing but also to changes in the surface characteristics of pore walls. Ali et al. described the construction of a hydrogen peroxide sensor by using a PET nanochannel decorated with horseradish peroxidase (HRP).[36] In order to regenerate the enzyme a redox system containing 2,2′-azino-bis(3-ethylbenzothiazoline-6-sulfonate) (ABTS) as a substrate was used. The ABTS becomes protonated in presence of hydrogen peroxide and accumulates inside the nanochannel, therefore reducing the negative surface charge of the channel and consequently reducing the rectification efficiency of the device. It was hypothesized that the presence of ABTS could also reduce the current by steric obstruction of the channel. A good reproducibility was observed for this system as well as a sensitivity close to 10 nM which is higher than typical values reported for electrochemical sensors (Fig. 3.12A).

Another related approach described by Sun et al. implied the use of amino phenyl boronic acid (APBA) to render saccharide selectivity to nanochannels.[37] Ion-track PET conical nanochannels were modified with APBA molecules via carbodiimide coupling. Due to the fact that the acid–base equilibrium of the boronic acid depends on the binding to a diol molecule it is possible to observe a change in the rectification efficiency upon the binding between the boronic acid and a saccharide (at pH ~ 8). These authors were able to show that the selectivity toward glucose was greater than to other similar saccharides (Fig. 3.12B).

In other interesting biosensing experiments reported by Siwy et al. both an anisotropic etching procedure and an anisotropic chemical functionalization were performed to enhance the sensing capacities of the nanochannel. In this way a sensor for γ-linked D-glutamic acid (γDPGA) was developed by modifying a nanochannel selectively on the tip with the monoclonal antibody for γDPGA (Fig. 3.13).[38] Then, the authors extended their approach to construct a streptavidin sensor. These authors also demonstrated by simulations that bipolar diodes exhibit superior capabilities for sensing applications due to the increased sensitivity of the "sensing zone" close to the tip. For these type of channels the quantity of analyte molecules needed to change the surface charge is less due to the reduced interaction zone that is more concentrated toward the tip (Fig. 3.13C).

Along these lines, one interesting approach described by Sun et al. involves the use of click chemistry to incorporate biorecognition elements into nanofluidic devices.[39] In the mentioned work, a thiol-yne coupling technique was used to obtain cysteine-responsive nanochannel by first modifying the surface of the channels with propargylamine so that it could react with cysteine in the presence of ultraviolet light (Fig. 3.14A). The outcome of this work was a sensor with a high sensitivity toward cysteine. This method is useful only with analytes that present a thiol group within the molecule but since there is a large amount of commercially available molecules with such characteristics it represents an interesting approach for integrating sensing motifs into nanopores.

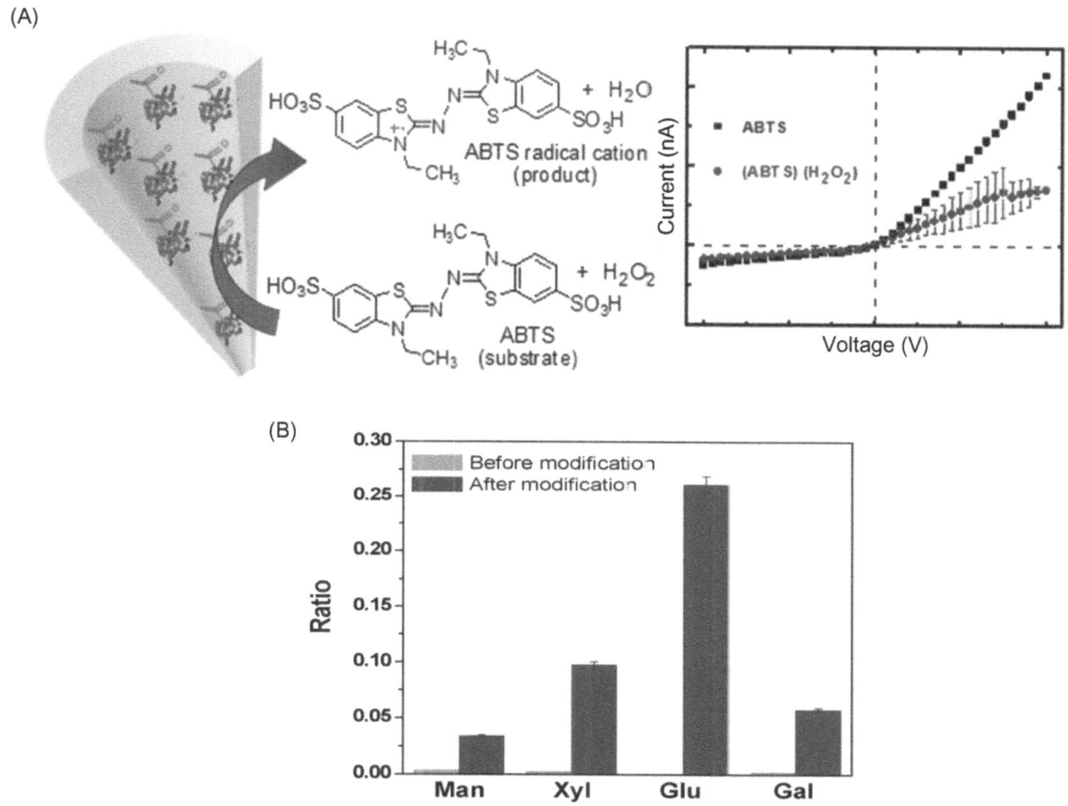

FIGURE 3.12 (A) Scheme showing a PET nanochannel modified with HRP and the formation of ABTS radical cation that ultimately produce the change in the ionic current shown in the plot. (B) Rectification efficiency obtained for different saccharides before (orange (grey in print version)) and after (purple (black in print version)) the modification of the nanochannel with APBA. A higher sensitivity toward glucose is observed. Source: *Reproduced with permission from references Ali, M.; Tahir, M. N.; Siwy, Z.; Neumann, R.; Tremel, W.; Ensinger, W. Hydrogen Peroxide Sensing with Horseradish Peroxidase-Modified Polymer Single Conical Nanochannels.* Anal. Chem. ***2011****, 83, 1673–1680; Sun, Z.; Han, C.; Wen, L.; Tian, D.; Li, H.; Jiang, L. pH Gated Glucose Responsive Biomimetic Single Nanochannels.* Chem. Commun. ***2012****, 48, 3282–3284.*

The use of covalent chemistries is not the only suitable method for chemical functionalization of nanochannels. Electrostatic layer-by-layer method can be used to modify ion-track-etched nanochannel due to the fact that after the etching procedure the nanochannels bear carboxylate groups.

The use of layer-by-layer assembly for sensing applications in nanopores was extended to the use of polyethyleneimine to invert the surface charge of the channels from negative to positive in order to assemble sulfonate calixarene (SCX4) in the nanochannel.[40] The SCX4 presents an affinity for acetylcholine, generating in this way a sensor for this molecule (Fig. 3.15).

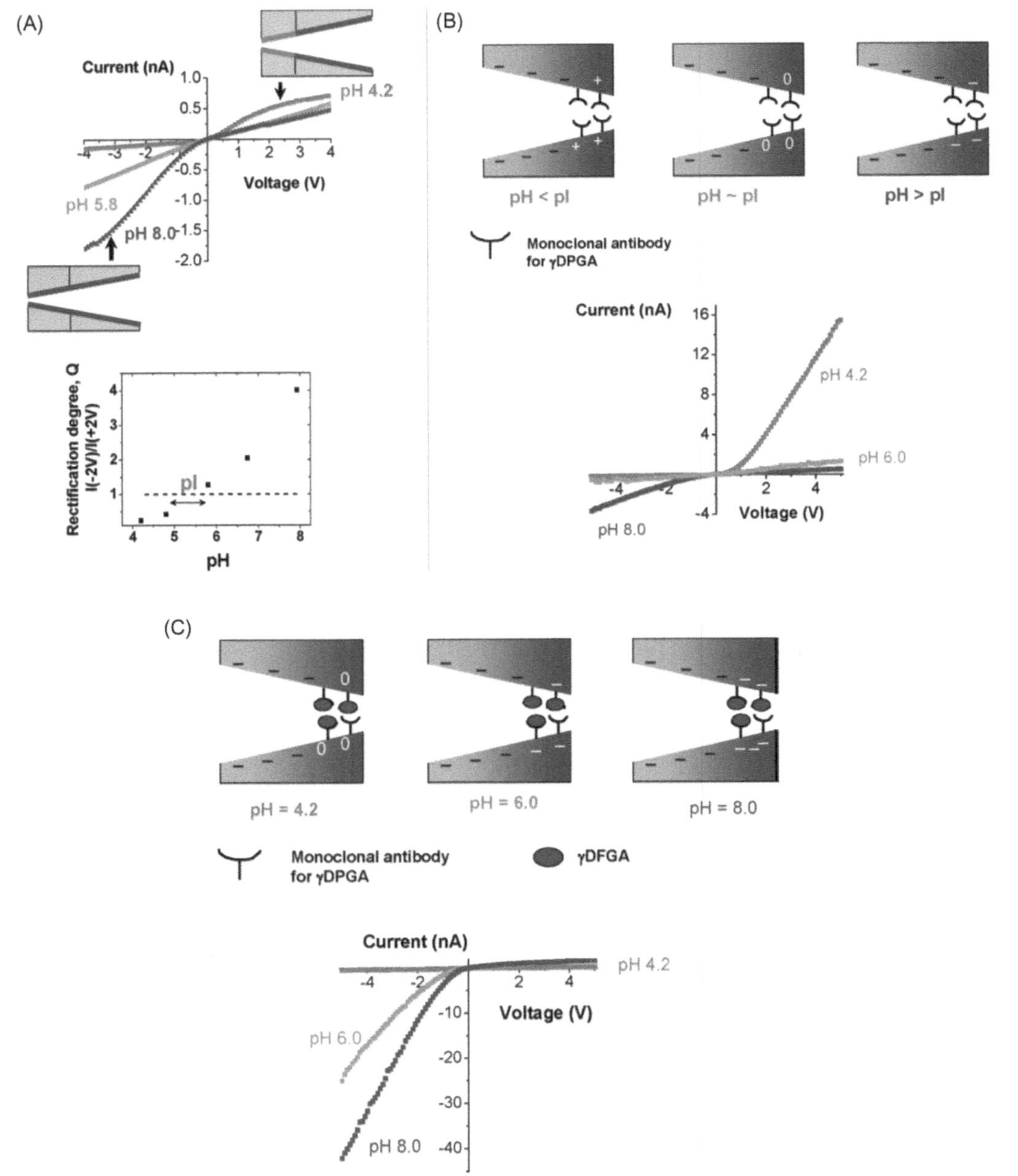

FIGURE 3.13 (A) Biosensor for streptavidin, the current–voltage curves are shown with a respective diagram of the charge changes produced by the chemical modification of the channel, also the rectification efficiency versus pH is shown. Current–voltage curves for (B) a nanochannel modified with the antibody for γDPGA and (C) after the incorporation of γDPGA. Source: *Reproduced with permission from reference Vlassiouk, I.; Kozel, T. R.; Siwy, Z. S. Biosensing with Nanofluidic Diodes.* J. Am. Chem. Soc. ***2009***, 131, *8211–8220.*

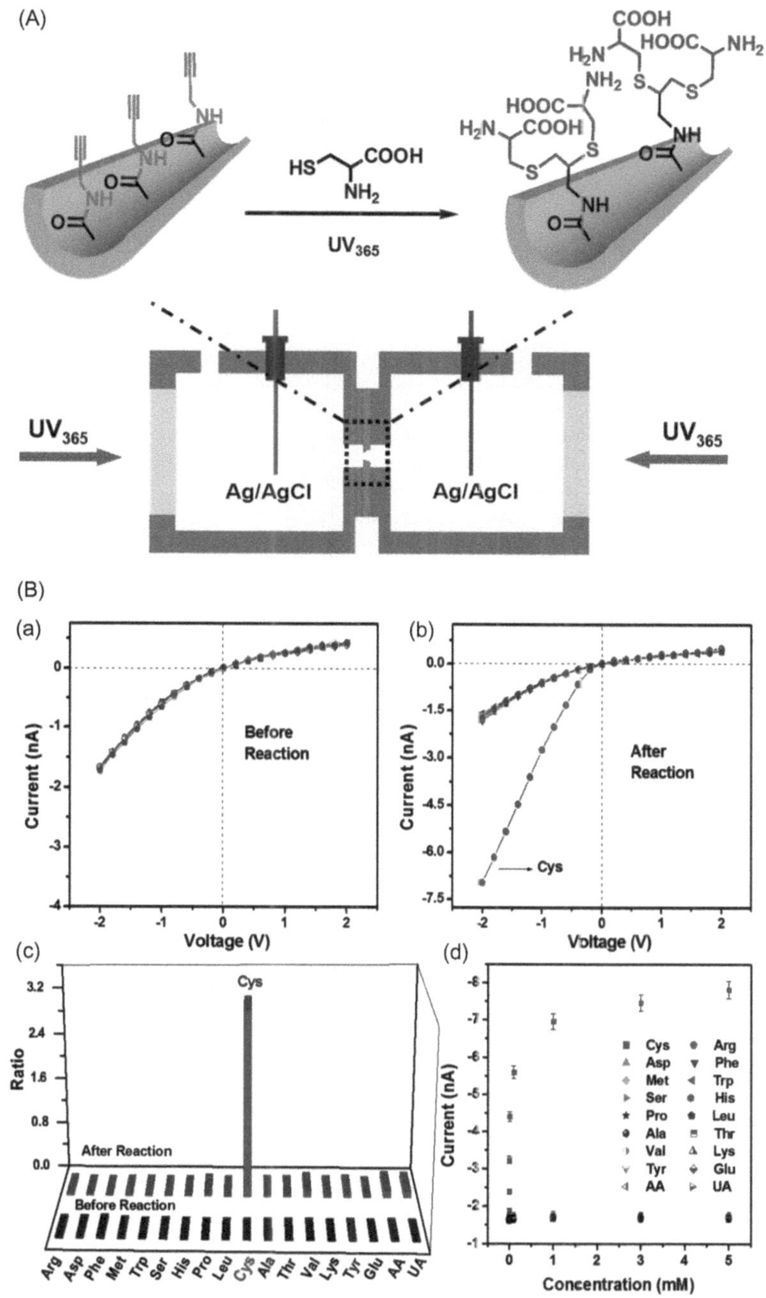

FIGURE 3.14 (A) Scheme showing the experimental setup for the sensing of cysteine. (B) Experimental results showing the current–voltage curves for the system before and after the reaction. Plot (a) shows the rectification ratio obtained for each different amino acid and plot (d) shows the current at a potential of −2 V. Source: *Reproduced with permission from reference Sun, Z.; Han, C.; Song, M.; Wen, L.; Tian, D.; Li, H.; et al. Fabrication of Cysteine-Responsive Biomimetic Single Nanochannels by a Thiol-yne Reaction Strategy and Their Application for Sensing in Urine Samples.* Adv. Mater. ***2014***, 26, *455–560.*

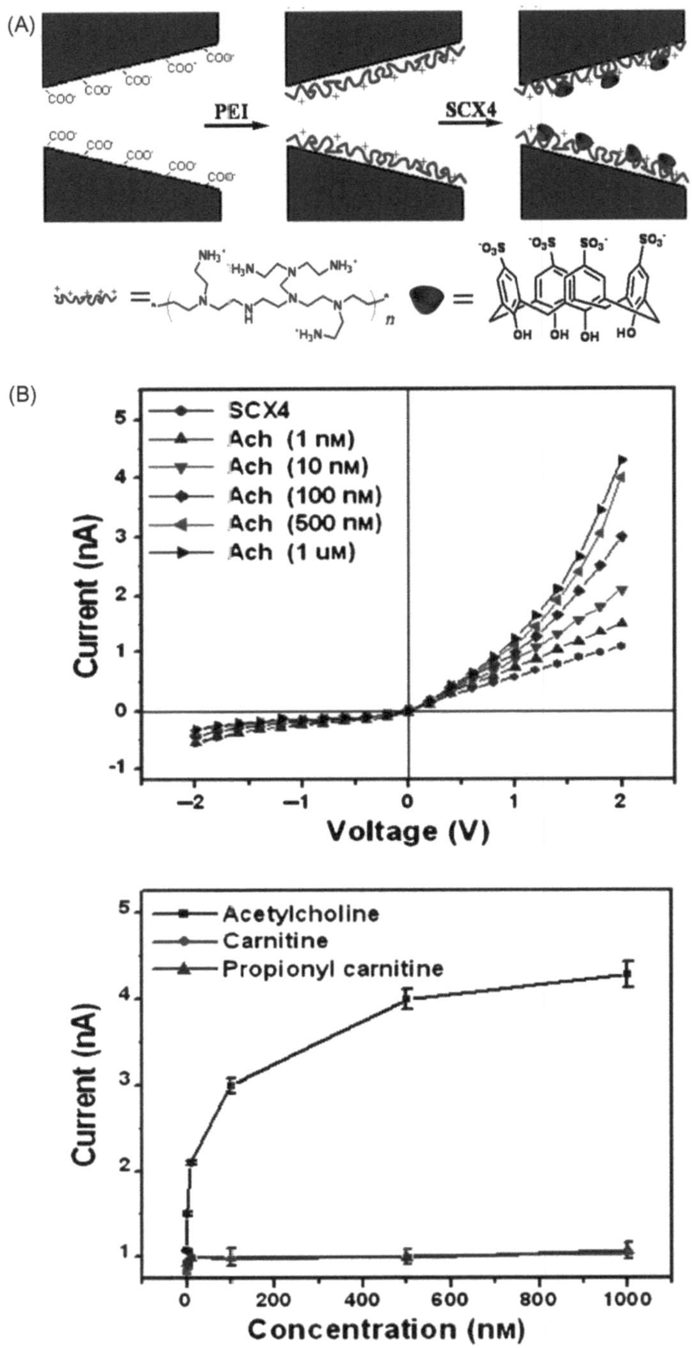

FIGURE 3.15 (A) Scheme of the modification procedure. (B) Experimental result for the sensing of acetycholine, the top figure shows the current–voltage curves and the bottom figure shows the current measured at a voltage of +2 V for different analytes. Source: *Reproduced with permission from reference Wen, L.; Sun, Z.; Han, C.; Imene, B.; Tian, D.; Li, H.; et al. Fabrication of Layer-by-Layer Assembled Biomimetic Nanochannels for Highly Sensitive Acetylcholine Sensing.* Chem. Eur. J. ***2013***, 19, *7686–7690.*

3.4 Conclusions and Outlook

In this chapter, we described the analytical capabilities of solid-state nanochannels fabricated using the single ion-track-etching method. We have introduced general concepts of ion-track nanotechnology for the fabrication of polymeric single nanochannels and nanofluidic devices. We discussed the most common sensing approaches: the resistive pulse sensing, in which a constant potential is applied while recording the current, and the analysis of current–voltage characteristics. The capabilities of resistive pulse sensing have been demonstrated for large molecules and nanoparticles. The analysis of the current–voltage characteristics has been widely used to sense small molecules like hydrogen peroxide, saccharides, or amino acids. For both applications asymmetric nanopore geometries are preferred as they are more sensitive to analyte molecules. In addition, the current rectification properties exhibited by asymmetric nanochannels are highly sensitive to the surface charge of the nanochannels.

References

1. Apel, P. Track Etching Technique in Membrane Technology. *Radiat. Meas.* **2001,** *34,* 559–566.
2. Spohr, R. Status of Ion Track Technology—Prospects of Single Tracks. *Radiat. Meas.* **2005,** *40,* 191–202.
3. Howorka, S.; Siwy, Z. Nanopore Analytics: Sensing of Single Molecules. *Chem. Soc. Rev.* **2009,** *38,* 2360–2384.
4. Hou, X.; Zhang, H.; Jiang, L. Building Bio-Inspired Artificial Functional Nanochannels: From Symmetric to Asymmetric Modification. *Angew. Chem. Int. Ed.* **2012,** *51,* 2–14.
5. Kocer, A.; Tauk, L.; Dejardin, P. Nanopore Sensors: From Hybrid to Abiotic Systems. *Biosens. Bioelectron.* **2012,** *38,* 1–10.
6. Hou, X.; Guo, W.; Jiang, L. Biomimetic Smart Nanopores and Nanochannels. *Chem. Soc. Rev.* **2011,** *40,* 2385–2401.
7. Dekker, C. Solid-State Nanopores. *Nat. Nanotechnol.* **2007,** *2,* 209–215.
8. Rhee, M.; Burns, M. A. Nanopore Sequencing Technology: Nanopore Preparations. *Trends Biotechnol.* **2007,** *25,* 174–181.
9. Cervera, J.; Schiedt, B.; Neumann, R.; Mafé, S.; Ramírez, P. Ionic Conduction, Rectification, and Selectivity in Single Conical Nanopores. *J. Chem. Phys.* **2006,** *124,* 104706.
10. Steckenreiter, T.; Balanzas, E.; Fuess, H.; Trautmann, C. Chemical Modifications of PET Induced by Swift Heavy Ions. *Nucl. Instrum. Methods Phys. Res. B* **1997,** *131,* 159–166.
11. Choi, Y.; Baker, L. A.; Hillebrenner, H.; Martin, C. R. Biosensing with Conically Shaped Nanopores and Nanotubes. *Phys. Chem. Chem. Phys.* **2006,** *8,* 4976–4988.
12. Siwy, Z.; Gu, Y.; Spohr, H. A.; Baur, D.; Wolf-Reber, A.; Spohr, R., et al Rectification and Voltage Gating of Ion Currents in a Nanofabricated Pore. *Europhys. Lett.* **2002,** *60,* 349.
13. Fleischer, R. L., Price, P. B., Walker, R. M. Nuclear Tracks in Solids: Principles and Applications, University of California Press, 1975.
14. Toimil-Molares, M. E. Characterization and Properties of Micro- and Nanowires of Controlled Size, Composition, and Geometry Fabricated by Electrodeposition and Ion-Track Technology. *Beilstein J. Nanotechnol.* **2012,** *3,* 860–883.

15. Xia, F.; Guo, W.; Mao, Y.; Hou, X.; Xue, J.; Xia, H., et al. Gating of Single Synthetic Nanopores by Proton-Driven DNA Molecular Motors. *J. Am. Chem. Soc.* **2008,** *130*, 8345–8350.

16. Ali, M.; Yameen, B.; Cervera, J.; Ramírez, P.; Neumann, R.; Ensinger, W., et al. Layer-by-Layer Assembly of Polyelectrolytes into Ionic Current Rectifying Solid-State Nanopores: Insights from Theory and Experiment. *J. Am. Chem. Soc.* **2010,** *132*, 8338–8348.

17. Pérez-Mitta, G.; Marmisollé, W. A.; Trautmann, C.; Toimil-Molares, M. E.; Azzaroni, O. Nanofluidic Diodes with Dynamic Rectification Properties Stemming from Reversible Electrochemical Conversions in Conducting Polymers. *J. Am. Chem. Soc.* **2015,** *137*, 15382–15385.

18. Pérez-Mitta, G.; Albesa, A. G.; Knoll, W.; Trautmann, C.; Toimil-Molares, M. E.; Azzaroni, O. Host–Guest Supramolecular Chemistry in Solid-State Nanopores: Potassium-Driven Modulation of Ionic Transport in Nanofluidic Diodes. *Nanoscale* **2015,** *7*, 15594–15598.

19. Eijkel, J. C. T.; van den Berg, A. Nanofluidics: What Is It and What Can We Expect from It? *Microfluid. Nanofluid.* **2005,** *1*, 249–267.

20. Daiguji, H. Ion Transport in Nanofluidic Channels. *Chem. Soc. Rev.* **2010,** *39*, 901–911.

21. Schoch, R. B.; Renaud, P. Ion Transport Through Nanoslits Dominated by the Effective Surface Charge. *Appl. Phys. Lett.* **2005,** *86*, 253111.

22. Siwy, Z.; Heins, E.; Harrell, C. C.; Kohli, P.; Martin, C. R. Conical-Nanotube Ion-Current Rectifiers: The Role of Surface Charge. *J. Am. Chem. Soc.* **2004,** *126*, 10850–10851.

23. Karnik, R.; Duan, C.; Castelino, K.; Daiguji, H.; Majumdar, A. Rectification of Ionic Current in a Nanofluidic Diode. *Nano Lett.* **2007,** *7*, 547–551.

24. Coulter Wallace H. US Patent 2,656,508; 20 Oct 1953.

25. Bayley, H.; Martin, C. R. Resistive-Pulse Sensing from Microbes to Molecules. *Chem. Rev.* **2000,** *100*, 2575–2594.

26. Wanunu, M. Nanopores: A Journey Towards DNA Sequencing. *Phys. Life Rev.* **2012,** *9*, 125–158.

27. Kasianowicz, J. J.; Brandin, E.; Branton, D.; Deamer, D. W. Characterization of Individual Polynucleotide Molecules Using a Membrane Channel. *Proc. Natl. Acad. Sci. USA* **1996,** *93*, 13770–13773.

28. Albrecht, T. Nanobiotechnology: A New Look for Nanopore Sensing. *Nat. Nanotechnol.* **2011,** *6*, 195–196.

29. Mara, A.; Siwy, Z.; Trautmann, C.; Wan, J.; Kamme, F. An Asymmetric Polymer Nanopore for Single Molecule Detection. *Nano Lett.* **2004,** *4*, 497–501.

30. Heins, E. A.; Siwy, Z. S.; Baker, L. A.; Martin, C. R. Detecting Single Porphyrin Molecules in a Conically Shaped Synthetic Nanopore. *Nano Lett.* **2005,** *9*, 1824–1829.

31. Harrell, C. C.; Choi, Y.; Horne, L. P.; Baker, L. A.; Siwy, Z. S.; Martin, C. R. Resistive-Pulse DNA Detection with a Conical Nanopore Sensor. *Langmuir* **2006,** *22*, 10837–10843.

32. Schiedt, B.; Healy, K.; Morrison, A. P.; Neumann, R.; Siwy, Z. Transport of Ions and Biomolecules Through Single Asymmetric Nanopores in Polymer Films. *Nucl. Instrum. Methods Phys. Res. B* **2005,** *236*, 109–116.

33. Yameen, B.; Ali, M.; Neumann, R.; Ensinger, W.; Knoll, W.; Azzaroni, O. Synthetic Proton-Gated Ion Channels via Single Solid-State Nanochannels Modified with Responsive Polymer Brushes. *Nano Lett.* **2009,** *9*, 2788–2793.

34. Siwy, Z.; Trofin, L.; Kohli, P.; Baker, L. A.; Trautmann, C.; Martin, C. R. Protein Biosensors Based on Biofunctionalized Conical Gold Nanotubes. *J. Am. Chem. Soc.* **2005,** *127*, 5000–5001.

35. Guo, Z.; Wang, J.; Wang, E. Selective Discrimination of Small Hydrophobic Biomolecules Based on Ion-Current Rectification in Conically Shaped Nanochannel. *Talanta* **2012,** *89*, 253–257.

36. Ali, M.; Tahir, M. N.; Siwy, Z.; Neumann, R.; Tremel, W.; Ensinger, W. Hydrogen Peroxide Sensing with Horseradish Peroxidase-Modified Polymer Single Conical Nanochannels. *Anal. Chem.* **2011,** *83*, 1673–1680.

37. Sun, Z.; Han, C.; Wen, L.; Tian, D.; Li, H.; Jiang, L. pH Gated Glucose Responsive Biomimetic Single Nanochannels. *Chem. Commun.* **2012,** *48,* 3282–3284.
38. Vlassiouk, I.; Kozel, T. R.; Siwy, Z. S. Biosensing with Nanofluidic Diodes. *J. Am. Chem. Soc.* **2009,** *131,* 8211–8220.
39. Sun, Z.; Han, C.; Song, M.; Wen, L.; Tian, D.; Li, H., et al. Fabrication of Cysteine-Responsive Biomimetic Single Nanochannels by a Thiol-yne Reaction Strategy and Their Application for Sensing in Urine Samples. *Adv. Mater.* **2014,** *26,* 455–560.
40. Wen, L.; Sun, Z.; Han, C.; Imene, B.; Tian, D.; Li, H., et al. Fabrication of Layer-by-Layer Assembled Biomimetic Nanochannels for Highly Sensitive Acetylcholine Sensing. *Chem. Eur. J.* **2013,** *19,* 7686–7690.

4

Biomimetic Smart Nanopores and Nanochannels

Q. Kang, W. Guo

CHINESE ACADEMY OF SCIENCES, BEIJING, P.R. CHINA

CHAPTER OUTLINE

4.1 Biological Nanopores

Biological pores made of transmembrane proteins have diverse and essential functions such as providing pathways for the ions or other polar molecules across the lipid barrier, participating in cell signaling, and so on (Fig. 4.1). The most important applications of such bio-nanopores are for DNA sequencing and single-molecular detection. Early in the 1990s, Kasianowicz et al. reported that single-stranded DNA (ssDNA) passes through a reconstituted biological nanopore in lipid membranes driven by electric field.[1] It was the first time that the length of single DNA molecules could be identified with the biological nanopores.[2] In this section, we will introduce several classic biological nanopores in order to allow comparison with biomimetic ones later.

The α-hemolysin (α-HL) protein pore has a simple structure composed of seven identical polypeptides. It is currently the most widely used biological nanopore.[4,5] The α-HL protein pore can be spontaneously inserted into lipid bilayers forming mushroom-shaped channels. The length and diameter of the stem is 5.2 and 2.6 nm, respectively (Fig. 4.2A).[1] The passage of each molecule is detected as a transient decrease of ionic current. The duration of the

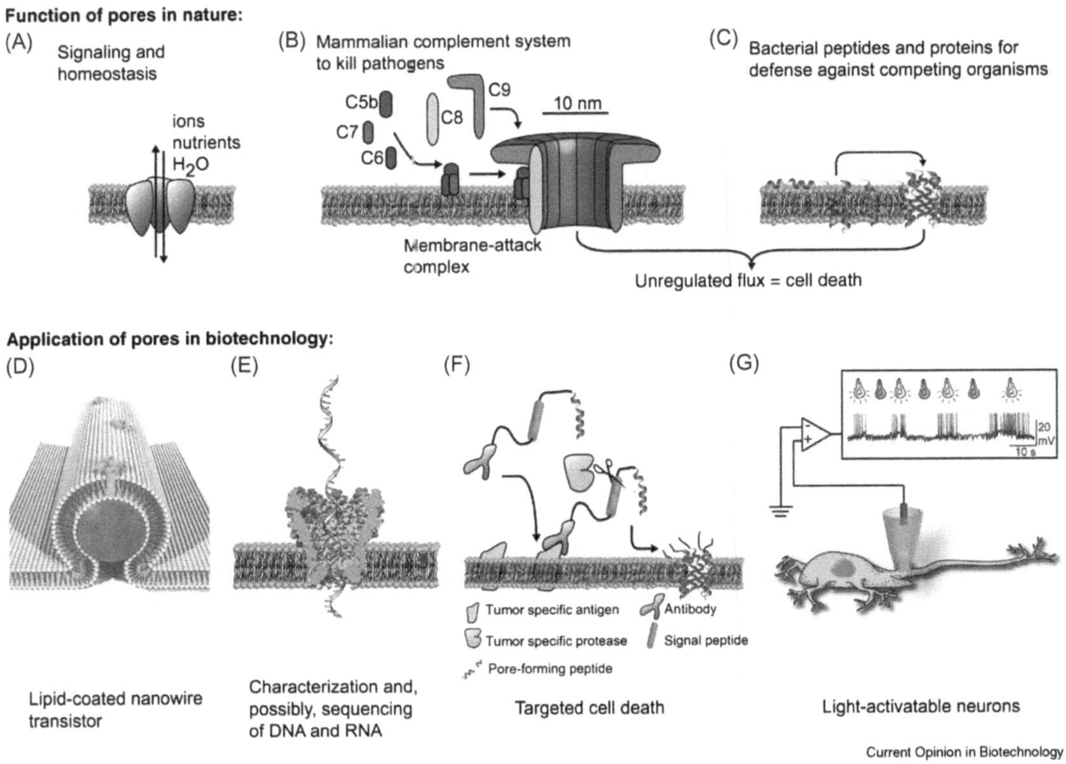

FIGURE 4.1 Functions of biological nanopores.[3] (A) Ion channel proteins transport ionic species across the plasma membrane to maintain homeostasis. (B) Membrane-attack complexes form a lytic pore in the plasma membrane of a pathogen. (C) Antibiotic peptides insert into the membrane of target microbes and form lytic pores. (D) Bionanoelectronic device consisting of a silicon nanowire coated with lipid bilayer with peptide pores. (E) Translocation of an ssDNA through a bacterial mycobacterium smegmatis porin A (MspA) pore. (F) Activation of multimeric pores by a tumor-specific targets to kill malignant cells. (G) Activation of neurons by light-activated ion channels. Source: *Reproduced with permission from reference Majd, S.; Yusko, E. C.; Billeh, Y. N.; Macrae, M. X.; Yang, J.; Mayer, M. Applications of Biological Pores in Nanomedicine, Sensing, and Nanoelectronics.* Curr. Opin. Biotechnol. ***2010***, 21, *439–4763.*

blockade events is proportional to the polymer length. Combining the nanopore detector with a support vector machine, the stem length, base pair mismatches, and loop length are able to be discriminated. Engineering artificial binding sites into the protein pores makes selective biosensors to detect divalent metal ions, e.g., Refs. 6,7.

Besides α-HL nanopores, other types of biological nanopores are also emerging. A structural drawback of the α-HL is that the cylindrical β-barrel can accommodate up to ~10 nucleotides at a time, which significantly reduces the resolution toward single-nucleotide discrimination in sequencing applications. The MspA protein pores do not suffer from this problem as they have a funnel-shaped channel of merely ~0.5 nm long and a constriction diameter of ~1.2 nm (Fig. 4.2B).[8] In this regard, the MspA's short and narrow constrictions can be used to discriminate individual nucleotides.[9,10]

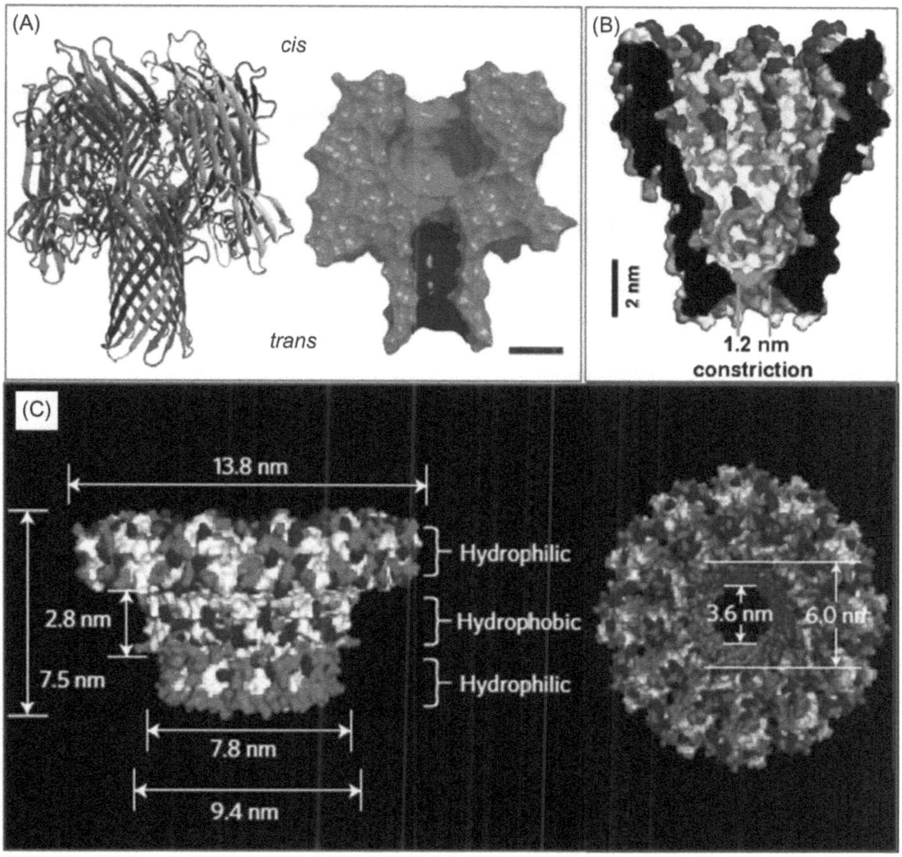

FIGURE 4.2 The structures of three kinds of protein nanopores. (A) The α-HL pore. The cross-sectional view (right) displays the inner cavity (up), inner constriction (middle), and β-barrel (down). The scale bar is 2 nm. (B) Crystal structure of the MspA. The cross-sectional view displays the classes of amino acids, including positively charged, negatively charged, polar, hydrophobic-aliphatic, and hydrophobic-aromatic parts. (C) Structure of phi29 connector. Top view shows the diameter of the narrow and wide parts along the channel axis.

α-HL and MspA pores allow the translocation of ssDNA and RNA. Alternatively, the bacteriophage phi29 DNA-packaging motor allows double-stranded DNA to enter or exit during the maturation or infection processes (Fig. 4.2C).[11] It contains a connector protein with a channel between 3.6 and 6 nm wide.[12] In 1986, Guo et al. found that the bacteriophage phi29 uses six RNAs strung together in the shape of a hexagon to create a motor that transports DNA in and out of the virus. Recently, they reconstituted the central channel component of phi29 connector protein into the planar lipid bilayer and the pore allowed the translocation of double-stranded DNA.[11]

Biological nanopores have fixed geometry that assures the reproducibility of the measurements. Biological nanopores offer great advantages for single-molecule detection and have promising applications in biosensing, DNA sequencing, and drug delivery.[13]

However, they also have inherent disadvantages, such as mechanical instability and difficulty in integrating with other bio- or nanodevices. Therefore, in recent years, more robust and durable solid-state nanopores are emerging and becoming alternatives to biological nanopores.

4.2 Biomimetic Smart Nanopores

Compared with artificial nanopores embedded in lipid membrane, solid-state nanopores offer great advantages: chemical and physical stability, the flexibility to tune the size and shape, ease of integration into high-density arrays, and other electronic or optical devices.[2,13,14] Many kinds of materials can be used to construct artificial and smart nanofluidic systems including biological, inorganic (SiN), organic (polyimide, polycarbonate, and polyethylene terephthalate), and biological/solid-state composite materials (Fig. 4.3, see also chapter: Introduction to Chemically Modified Nanochannels and Nanopores).[14] Various fabrication technologies are used to process these different materials, such as molecular self-assembly, electrochemical etching, anodic oxidation, electron beam lithography, laser beam lithography, and ion-track-etching technology.[15–19]

The mass and charge-transport properties in these smart nanofluidic systems can be manipulated in a controlled fashion by external stimuli such as stress, temperature, moisture, pH, electric or magnetic field, etc. There are several classical approaches to fabricate biomimetic smart nanopores. One is to control the geometry of the nanopores; and the other is to control the chemical composition of the inner surfaces of the nanopores. Generally, once the nanopores have been prepared, functional molecules are immobilized onto the nanopores to realize the responsibility to external stimuli.[14] In this section, we introduce several stimuli-responsive smart nanopores prepared by chemical modification.

4.2.1 pH-Responsive Nanopores

pH-responsive nanosystems are widely studied, since pH is a very important parameter in biochemical reactions and is easy to adjust.[20] Compared to its biological counterpart, the chemistry on the inner nanopore surface can be easily modified. Here, we take the polyethylene terephthalate (PET) nanopores for example. During chemical etching, negatively charged carboxyl groups are created on the inner pore wall. The activated carboxyl groups on the nanopore surface can be further functionalized through amidation or esterification reactions (see chapter: Introduction to Chemically Modified Nanochannels and Nanopores). Chemical modification broadens the functionality of the solid-state nanopores. When the pH of the electrolyte is higher than isoelectric point of the etched polymer surface, the surface is negatively charged. The ionic current rectification (ICR) phenomenon is observed in geometrically asymmetric nanopores, showing asymmetric current–voltage (I–V) properties.[21] This diode-like $I-V$ behavior indicates a preferential direction for ion transportation. Similar effects have been observed in other polymer nanopores (polyimide and polycarbonate) and

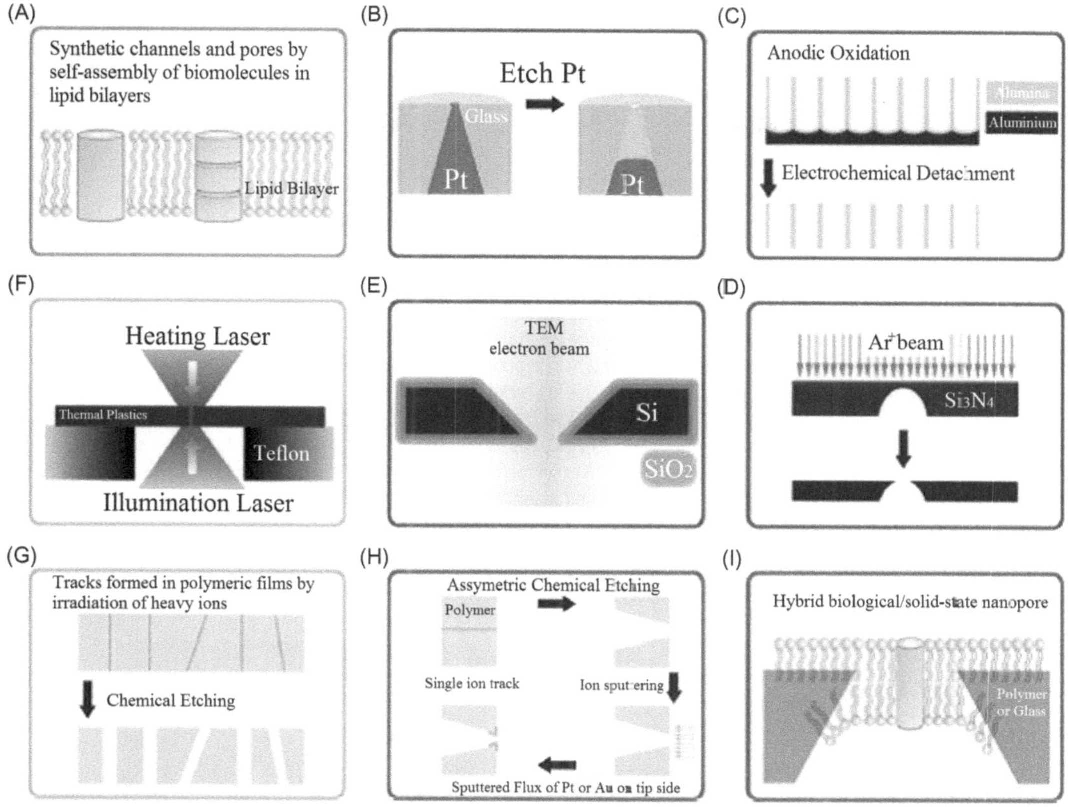

FIGURE 4.3 Fabrication of solid-state nanopores. (A) Artificial nanopores in lipid bilayer. (B) Nanopore electrodes prepared by electrochemical etching. (C) Nanopore array in anodic alumina membrane. (D) Nanopores in Si_3N_4 membrane by ion beam sculpting. (E) Nanopores prepared by electron beam irradiation and anisotropic etching. (F) Laser-induced shrinking of micron scale pores in thermoplastic materials. (G) Chemical etching of single or multitracked polymer membrances. (H) Metal–polymer composite nanochannels fabricated by ion-track-etching and ion sputtering. (I) Hybrid biological/solid-state nanopores assembled in lipid bilayers.

glass nanopipette (Fig. 4.4). The rectification ratio depends on solution concentration and environmental pH.[22,23]

In 2007, we developed a pH-gated, asymmetric single nanopore by grafting pH-sensitive DNA molecular motor onto the inner surface of the nanopore (Fig. 4.5).[24] In low-pH solution (pH ~ 4.5), the motor DNA folds into a densely packed quadruplex i-motif structure that reduces the effective diameter of the nanopore. In high-pH solution (pH ~ 8.5), the motor DNA relaxes to a loosely packed single-stranded structure that enhances total ion conductivity through the nanopore. The concept of DNA motor-driven nanofluidic switch could be used to construct more complex nanopore machinery with other functionalities.

Azzaroni et al. reported several pH-tunable, asymmetric single nanopores by integration of polymer brushes or amphoteric molecules with solid-state nanopores (Fig. 4.6A–C).[25–27] The ionic rectification properties can be finely tuned by immobilizing zwitterionic polymer brushes

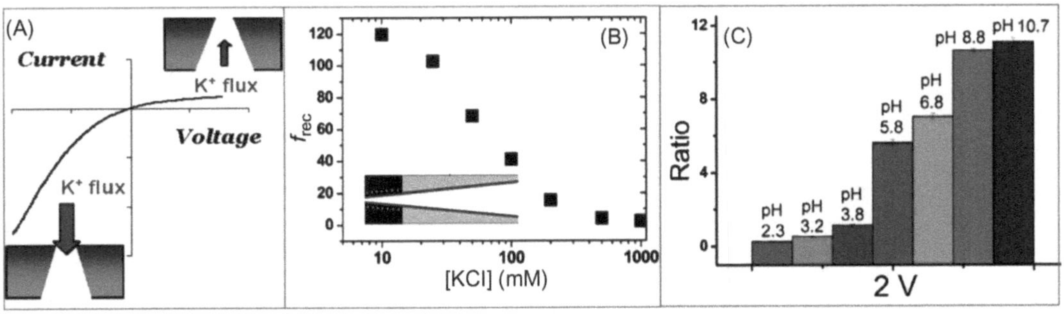

FIGURE 4.4 The effect of ICR in pH-responsive asymmetric nanopores. (A) Ionic rectification in conical polymeric nanopores. (B) Dependence of rectification ratio (f_{rec}) on electrolyte concentration at pH = 5.5. The inset shows pattern of surface charge distribution. (C) Dependence of rectification ratio on pH. Source: *Reproduced with permission from references (A) Siwy, Z.S. Ion-Current Rectification in Nanopores and Nanotubes with Broken Symmetry. Adv. Funct. Mater.* ***2006****,16, 735–746; (B) Hou, X.; Dong, H.; Zhu, D.; Jiang, L. Fabrication of Stable Single Nanochannels with Controllable Ionic Rectification.* Small ***2010****, 6, 361–365; (C) Alcaraz, A.; Ramírez, P.; García-Giménez, E.; López, M.L.; Andrio, A.; Aguilella, V.M. A pH-Tunable Nanofluidic Diode: Electrochemical Rectification in a Reconstituted Single Ion Channel. J. Phys. Chem. B* ***2006****, 110, 21205–21209.*

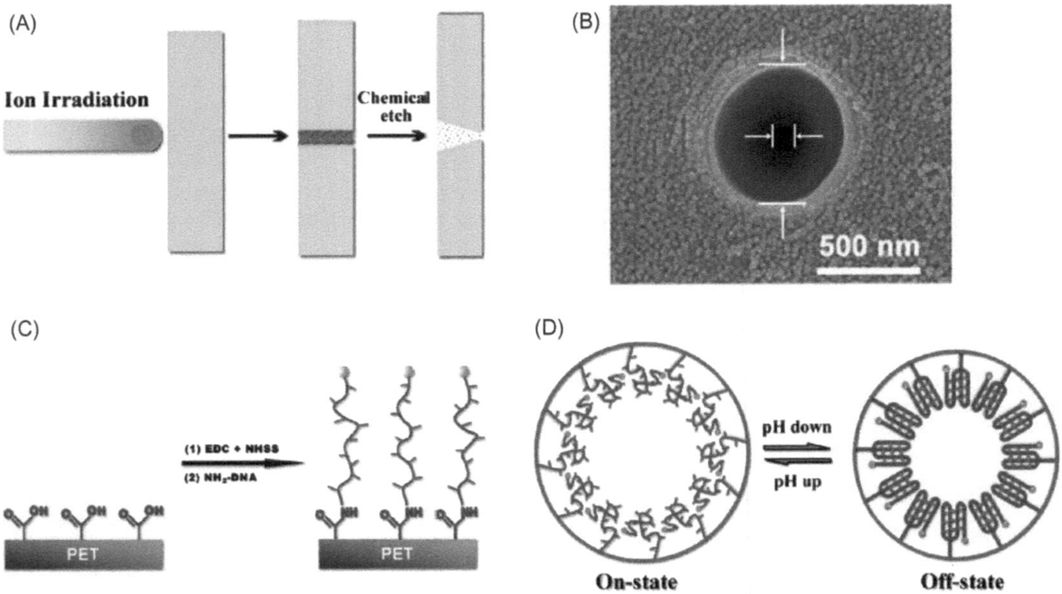

FIGURE 4.5 DNA molecular motor-driven, pH-sensitive single nanopore in PET film.[24] (A) Nanopore fabrication by ion-track-etching technique. (B) SEM image of the conical nanopore from the base side. (C) Immobilization of DNA motor onto the nanopore through a two-step chemical reaction. (D) The effective cross section of chemically modified nanopore can be regulated by conformational change of the DNA motor in responsive to pH. Source: *Reproduced with permission from reference Xia, F.; Guo, W.; Mao, Y. D.; Hou, X.; Xue, J. M.; Xia, H. W.; et al. Gating of Single Synthetic Nanopores by Proton-Driven DNA Molecular Motors.* J. Am. Chem. Soc. ***2008****,* 130, *8345–8350.*

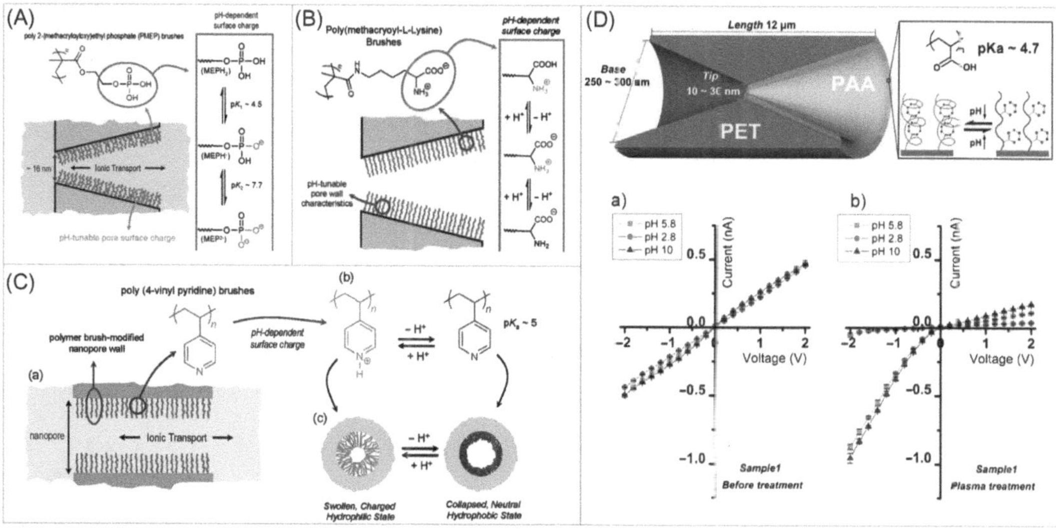

FIGURE 4.6 pH-responsive smart nanopores. (A) PMEP brush modified conical nanopores bearing two pK_a (B) Zwitterionic polymer brushes modified conical nanopores that rectify ionic current in a wide range of pH. (C) Poly (4-vinyl pyridine) brushes modified cylindrical nanopores with pH-responsive ionic gating function. (D) Asymmetrically PAA-modified hour-glass shaped nanopore. The PAA brushes were modified on one end of the nanopore, resulting in high ionic rectification ratio. Source: *Reproduced with permission from references (A) Yameen, B.; Ali, M.: Neumann, R.; Ensinger, W.; Knoll, W.; Azzaroni, O. Proton-Regulated Rectified Ionic Transport through Solid-State Conical Nanopores Modified with Phosphate-Bearing Polymer Brushes. Chem. Commun.* ***2010****, 46, 1908–1910; (B) Yameen, B.; Ali, M.; Neumann, R.; Ensinger, W.; Knoll, W.; Azzaroni, O. Single Conical Nanopores Displaying pH-Tunable Rectifying Characteristics. Manipulating Ionic Transport with Zwitterionic Polymer Brushes. J. Am. Chem. Soc.* ***2009****, 131, 2070–2071; (C) Yameen, B.; Ali, M.; Neumann, R.; Ensinger, W.; Knoll, W., Azzaroni, O. Synthetic Proton-Gated Ion Channels via Single Solid-State Nanochannels Modified with Responsive Polymer Brushes. Nano Lett.* ***2009****, 9, 2788–2793; (D) Hou, X.; Liu, Y.; Dong, H.; Yang, F.; Li, L.; Jiang, L., et al. A pH-Gating Ionic Transport Nanodevice: Asymmetric Chemical Modification of Single Nanochannels,* Adv. Mater. ***2010****, 22, 2440–2443.*

onto the inner surface of the nanopores. In addition to amphoteric molecules, other pH-responsive polymer brushes bearing phosphate groups or 4-vinyl pyridine units were grafted onto the nanopores to achieve ionic gating characteristics. The transmembrane ionic currents could be accurately controlled by manipulating the proton concentration in the surrounding environment. We further developed the asymmetric pH responsibility by asymmetrically modifying the hour-glass-shaped single nanochannel.[28] This chemical modification strategy results in asymmetric ion transport property in a geometrically symmetric nanopore system by bringing asymmetry in both chemical composition and surface charge. Aside from organic nanopores in polymer films, there are also some reports on pH-responsive inorganic nanopores.[29,30]

4.2.2 Temperature-Responsive Nanopores

Some biological ion channels are activated by sensing temperature and translate this signal into conformational changes of the protein complex.[31] Inspired by such thermo-sensitive ion channels, scientists have prepared temperature-responsive smart nanopores (Fig. 4.7). Poly

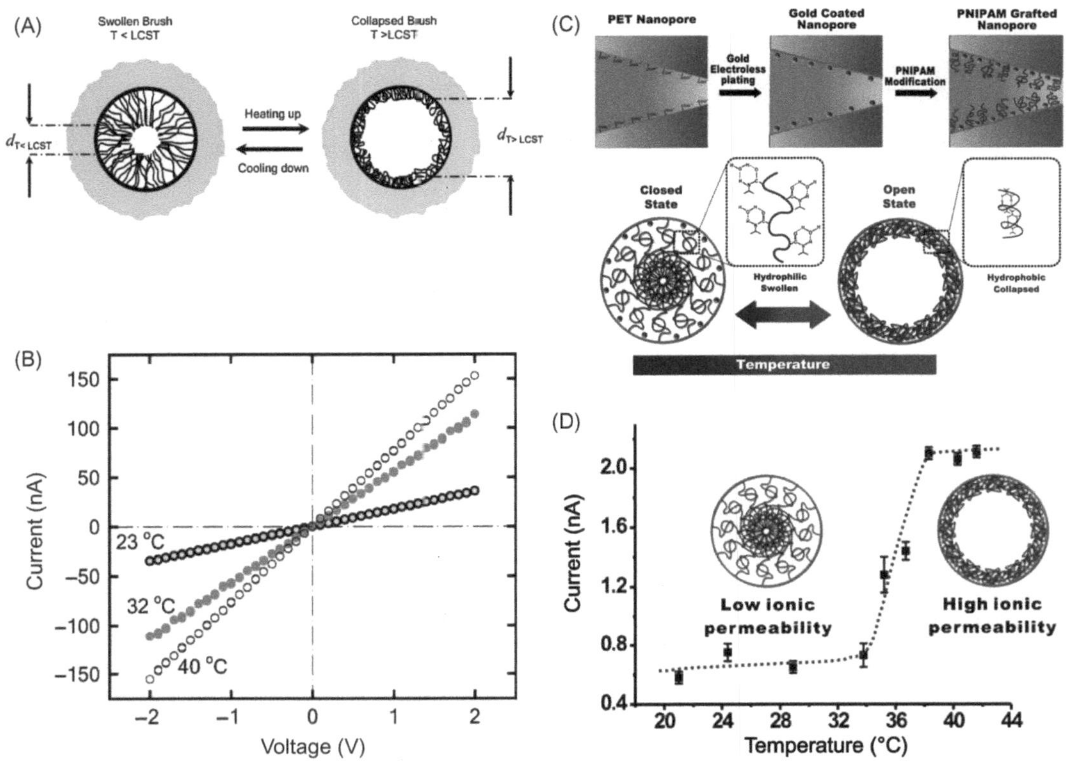

FIGURE 4.7 Temperature-responsive ionic gates and rectifiers. (A and B) Thermally-driven nanoactuation of the PNIPAAm brushes in the nanopore. (C and D) PNIPAAm-modified gold-coated conical nanopore exhibits both temperature-controlled ionic gating and ionic rectifier behaviors. The switching properties are triggered by phase transition of the attached PNIPAAm brushes between the swollen hydrophilic state (the closed state, below the LCST) and collapsed hydrophobic state (the open state, above the LCST). Source: *Reproduced with permission from reference Yameen, B.; Ali, M.; Neumann, R.; Ensinger, W.; Knoll, W.; Azzaroni, O. Ionic Transport through Single Solid-State Nanopores Controlled with Thermally Nanoactuated Macromolecular Gates.* Small ***2009***, 5, *1287–1291.*

(*N*-isopropylacrylamide) (PNIPAAm) is a representative temperature-sensitive polymer which exhibits temperature-driven conformational transitions from a swollen state to collapsed state. Azzaroni's group prepared single polyimide nanopores modified with PNIPAAm brushes.[32] At room temperature (23°C, below the lower critical solubility temperature (LCST) of PNIPAAm), the polymer brushes remain swollen, decreasing the effective cross section of the nanopore. Raising the temperature above the LCST leads to a change in the conformation of PNIPAAm brushes. The polymer brushes transit into a collapsed state, leading to an increase in the effective cross section of the nanopore. The thermo-responsive polymer brushes act as a thermally driven macromolecular gate controlling the ionic flow through the nanopore. We further studied the ICR properties in a conically shaped metal-coated nanopore modified with PNIPAAm.[33] The current rectification properties in such nanofluidic systems could be switched between a rectifying state and a nonrectifying state controlled by the phase transition of the

PNIPAAm brush as well as the surface charge properties. Chu et al. prepared AAO-g-PNIPAM membranes bearing thermo-responsive gating properties by grafting PNIPAAm to the pores of anodic aluminum oxide (AAO). They adjusted the length and density of the grafted PNIPAAm chains to control gating characteristics and mass transport through the nanopores.[34]

4.2.3 Light-Responsive Nanopores

Irradiation with light is an effective mean for uptaking or releasing cargo materials. Thus, light-responsive molecular devices have attracted much attention. The light-responsive molecular switches have opened up a challenging field for gating artificial nanopores (Fig. 4.8). By using structure-based design, Trauner's group have successfully developed a novel chemical gate that confers light sensitivity to an ion channel.[35] Brinker et al. modified the nanopores with photoresponsive molecules with azobenzene ligands.[36] The optically switchable conformation (*trans* or *cis*) of azobenzene ligands controls the effective pore size and transport behavior on the nanoscale. White's group reported a photon-gated glass nanopore by chemical modification of the interior of nanopore with spiropyran. It provides a highly efficient mean to impart photochemical control of molecular transport through the pore orifice.[37] Smirnov et al. modified the surface of nanoporous alumina using a

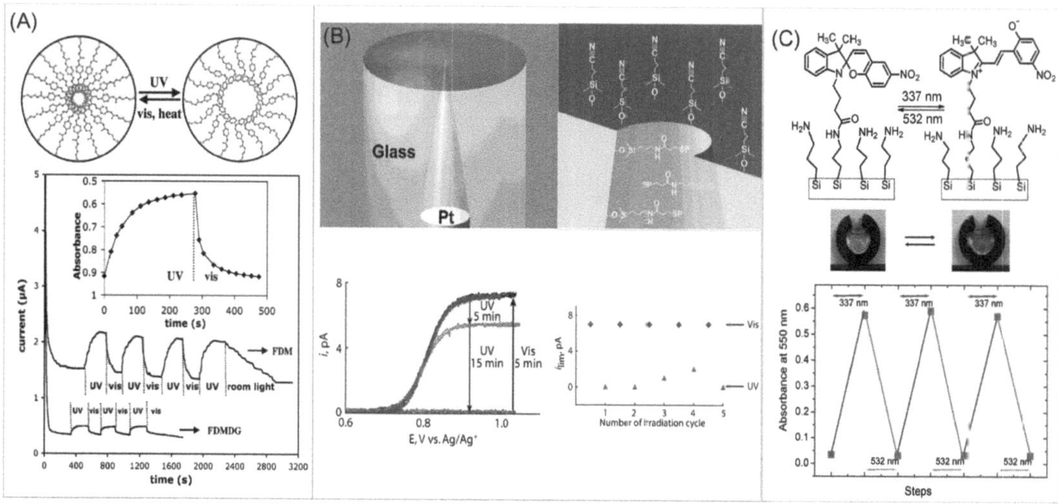

FIGURE 4.8 Photo-responsive smart nanopores. (A) The reversible change in size of azobenzene-modified nanopores in response to UV (360 nm) and visible light (435 nm). (B) Photon-gated transport through glass nanopores. (C) Spiropyran-modified nanoporous filter membrane shows a reversible on–off behavior under UV and vis irradiation. Source: *Reproduced with permission from references (A) Liu, N.G.; Dunphy, D.R.; Atanassov, P.; Bunge, S.D.; Chen, Z.; López, G.P., et al. Photoregulation of Mass Transport through a Photoresponsive Azobenzene-Modified Nanoporous Membrane,* Nano Lett. ***2004****, 4, 551–554; (B) Wang, G.; Bohaty, A. K.; Zharov, I.; White, H. S. Photon Gated Transport at the Glass Nanopore Electrode.* J. Am. Chem. Soc. ***2006****, 128, 13553–13558; (C) Vlassiouk, I.; Park, C.-D.; Vail, S.A.; Gust D.; Smirnov, S. Control of Nanopore Wetting by a Photochromic Spiropyran: A Light-Controlled Valve and Electrical Switch,* Nano Lett. ***2006****, 6, 1013–1017.*

photochromic spiropyan and hydrophobic molecules.[38] The modified nanopores were able to control the penetration of water using light irradiation.

4.2.4 Electric Potential Responsive Nanopores

Membrane potential is the difference in electrical potential between the interior and the exterior of a biological cell membrane. It is noninvasive, tunable, and acts over a negligible time scale. Inspired by these features, the scientific community builds biomimetic electric potential responsive nanopores (Fig. 4.9). Martin et al. created artificial ion channels that rectify the ionic current via an electromechanical mechanism.[39] These artificial channels are based on conical gold nanochannels. The electromechanical responsiveness is provided by ssDNA molecules attached to the pore orifice.[40] Before DNA modification, the gold nanochannels do not rectify. After DNA attachment, the nanochannels rectify the ionic current, which shows an on-state at negative potentials and an off-state at positive potentials. It was postulated that the ionic rectification in these nanochannels results from the electrophoretic insertion of the DNA chains into (off-state) and out of (on-state) the nanochannel. Siwy's group has further developed an electric potential controlled nanofluidic diode with tunable ionic rectification properties by a metal gate placed on the narrow entrance of the nanopore.[41]

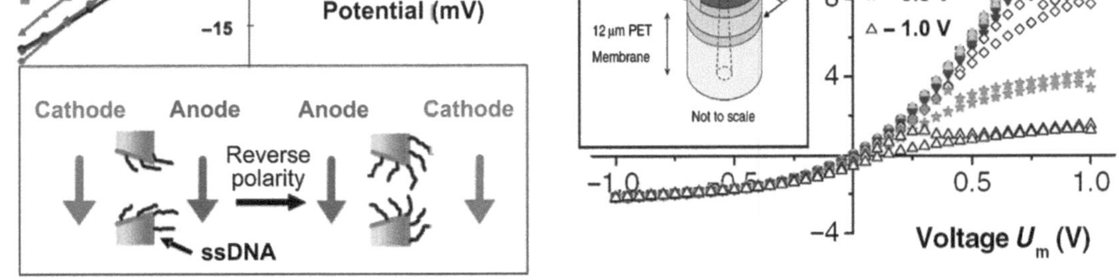

FIGURE 4.9 Electric potential responsive nanochannels. (A) ssDNA modified metal-coated nanochannel shows an electromechanical gating behavior in response to the direction of electric field. (B) *I–V* curves of a single conical nanochannel with a gate placed at the narrow opening show an electric potential responsiveness with respect to the gate voltage. The inset shows structure of the composite nanochannel with metal and silicon dioxide layers. Source: *Reproduced with permission from references (A) Harrell, C. C.; Kohli, P.; Siwy, Z.; Martin, C. R. DNA-Nanotube Artificial Ion Channels. J. Am. Chem. Soc.* ***2004****, 126, 15646–15647; (B) Kalman, E.; Sudre, O.; Vlassiouk, I.; Siwy, Z. Control of Ionic Transport through Gated Single Conical Nanopores. Anal. Bioanal. Chem.* ***2009****, 394, 413–419.*

4.2.5 Ion-Responsive Nanopores

Metallic ions are crucial in modulating the activity of muscles and nerves. In these cells, ion channels for signal transduction are turned on and off by particular kind of metallic ions such as sodium, potassium, and calcium. Body functions rely on the regulation of ion concentrations across the cell membrane through the controlled opening and closing of the ion channels. Therefore, ion-responsive artificial nanopores have become important and provide a platform to study and simulate these physiological processes in living organisms (Fig. 4.10).

Siwy et al. discovered calcium-induced voltage gating in an original asymmetric PET single nanopore. The artificial ion channel produced voltage-dependent ionic current fluctuations in the presence of sub-millimolar concentrations of calcium ions.[42] Millimolar concentrations of calcium reverse the rectification and a negative incremental resistance is

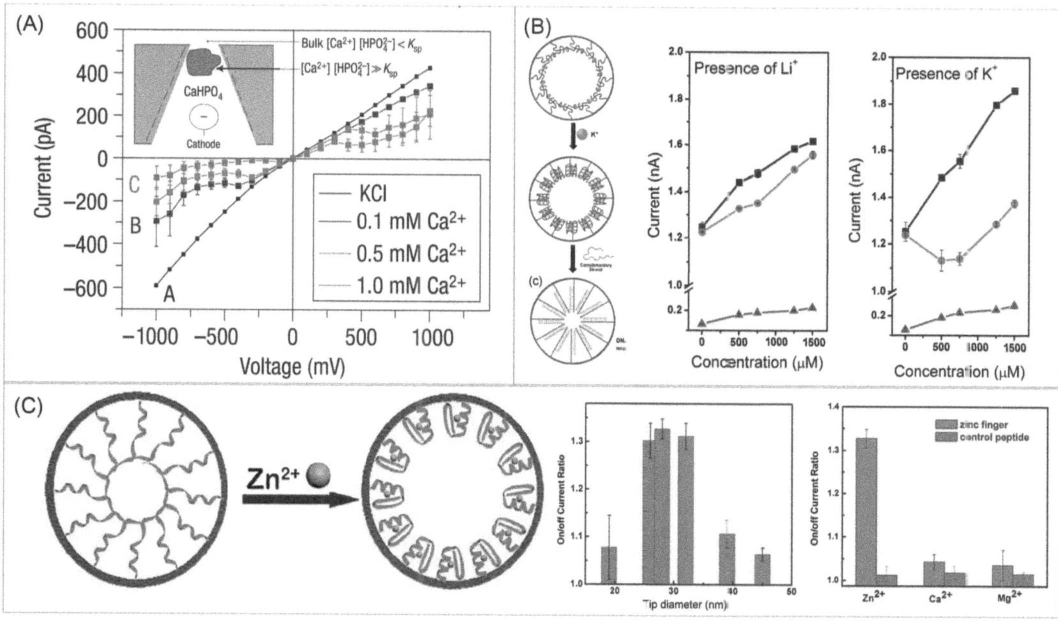

FIGURE 4.10 Ion-responsive nanopores. (A) Calcium-induced voltage gating in a single conical nanopore. I–V curves were recorded in 0.1 M KCl with 2 mM phosphate buffer saline (PBS) and with varying calcium concentrations. (B) G4 DNA was immobilized onto the inner surface of a single nanopore yielding a biomimetic potassium-responsive nanopore. Current–concentration properties before and after DNA attachment show nonlinear response to potassium ion at K^+ concentration ranging from 0 to 1500 mM. (C) Zinc finger protein immobilized single nanopore shows zinc ion responsive gating behaviors with different tip diameters. Source: *Reproduced with permission from references (A) Powell, M. R.; Sullivan, M.; Vlassiouk, I.; Constantin, D.; Sudre, O.; Martens, C. C., et al. Nanoprecipitation-Assisted Ion Current Oscillations. Nat. Nanotechnol.* ***2008****, 3, 51–57; (B) Hou, X.; Guo, W.; Xia, F.; Nie, F.-Q.; Dong, H.; Tian, Y., et al. A Biomimetic Potassium Responsive Nanochannel: G-Quadruplex DNA Conformational Switching in a Synthetic Nanopore. J. Am. Chem. Soc.* ***2009****, 131, 7800–7805; (C) Tian, Y.; Hou, X.; Wen, L.P.; Guo, W.; Song, Y.L.; Sun, H.Z., et al. A Biomimetic Zinc Activated Ion Channel,* Chem. Commun. ***2010****, 46, 1682–1684.*

observed. The addition of small amount of other divalent cations also results in an oscillating ionic current through the asymmetric nanopores.[43] This phenomenon is caused by the transient formation and re-dissolution of calcium nanoprecipitates, which temporarily block the pathway of the ionic flow.[44] The frequency and duration of ionic current instabilities are regulated by the electric potential across the nanopore and the chemistry of the precipitate.

Jiang et al. have developed a biomimetic potassium-responsive single nanochannel that shows nonlinear response to potassium ion at a concentration ranging from 0 to 1500 mM.[45] G-quadruplex (G4) DNA was immobilized onto the inner surface of a single nanochannel. The G4 DNA experiences potassium-responsive conformational change, and therefore induces the change in effective diameter at the narrowest end of the nanochannel. The responsiveness can be regulated by the stability of G4 structure through adjusting potassium ions. Based on similar working mechanism, they further developed a zinc ion responsive nanopore by immobilizing zinc finger peptides into a single asymmetric nanopore.[46] Similar to the biological zinc activated ion channel, the modified nanopore is opened or closed by zinc ions owing to the conformational changes of the zinc fingers.

4.2.6 Molecular Responsive Nanopores

Molecular responsive nanopore materials gain increasing interest for their applications in molecular recognition, biosensors, and drug separation. Martin et al. reported a protein analyte binding to a molecular recognition agent immobilized at the small opening of a single gold nanochannel (Fig. 4.11A).[47] Since the protein and the mouth of the nanochannel are comparable in size, the protein molecules bind onto the nanochannel and block the pathway for ion conduction. $I-V$ curves for the biotinylated nanochannel before and after exposure to 100 nM lysozyme are identical, indicating that the chemically modified nanochannel does not respond to nonspecific protein targets. In contrast, the ionic current is completely shut off after exposure to a solution containing 180 pM streptavidin. Total ionic current blockage occurs because diameter of the streptavidin molecule, about 5 nm, is comparable to the mouth of the biotinylated nanochannel.

Xie et al. has developed an enantioselective sensing device based on a single artificial β-cyclodextrin modified nanopore.[48,49] This nanodevice shows highly selective recognition of histidine enantiomers through monitoring transmembrane ionic current signatures (Fig. 4.11B). Ali et al. reported a biosensing platform for detection of lactoferrin (LFN) via modifying metal–organic frameworks in nanopores.[50] In such chemically modified nanofluidic systems, the metal–organic frameworks structure provides accessible free coordination sites for metal binding. Thus, the presence of target LFN is recognized as the change in both pore conductance and rectification properties (Fig. 4.11C).

4.2.7 Multiple Responsive Smart Nanopores

Single responsive nanopores have been well developed; however, imparting these nanopores with more intellect is still challenging. Several gating membranes using multiple stimuli-responsive materials have been reported. The chemical properties and geometry of the

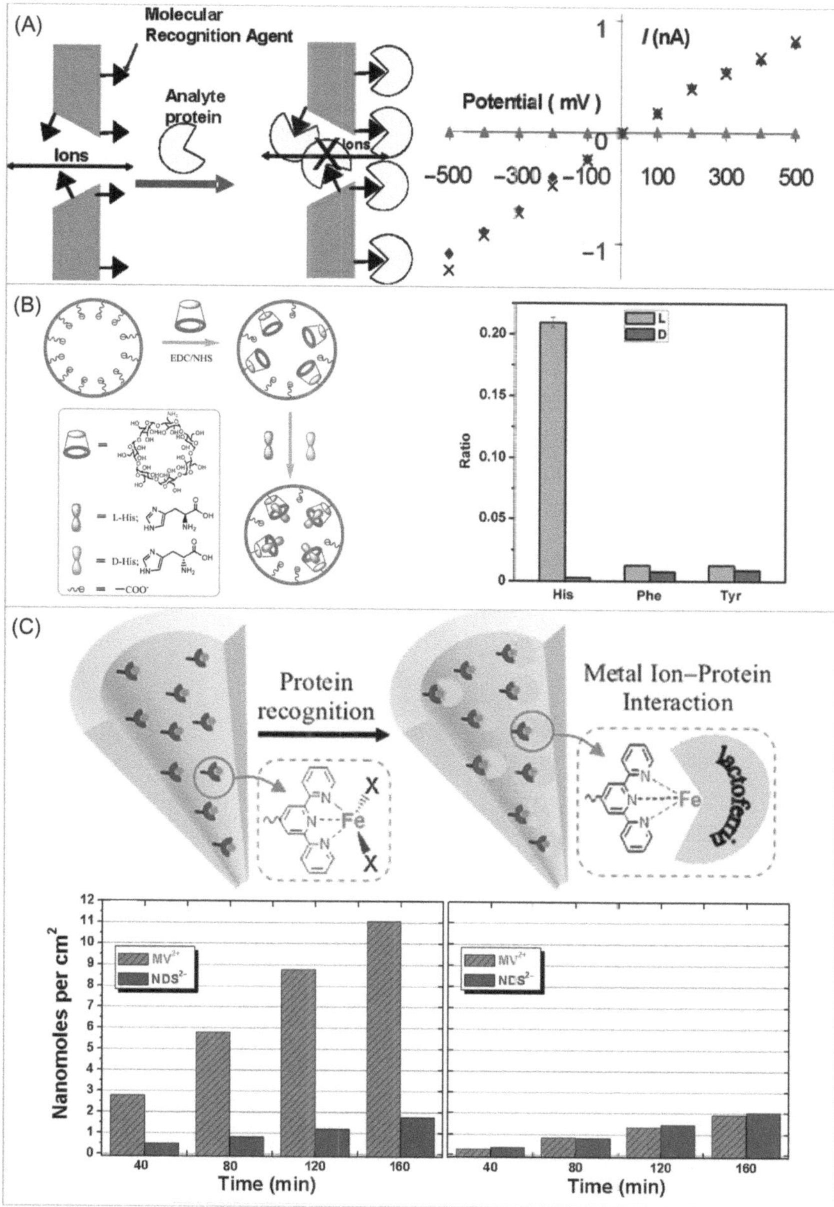

FIGURE 4.11 Molecular responsive nanopores. (A) Protein-responsive conical gold nanochannels. (B) Biomimetic chiral-selective single naonochannel system. (C) Metal ion affinity-based biomolecular recognition within metal–organic frameworks. Source: *(A) Siwy, Z.; Trofin, L.; Kohli, P.; Baker, L. A.; Trautmann, C.; Martin, C. R. Protein Biosensors Based on Biofunctionalized Conical Gold Nanotubes. J. Am. Chem. Soc.* ***2005****, 127, 5000–5001; (B) Xie, G.; Tian, W.; Wen, L.; Xiao, K.; Zhang, Z.; Liu, Q.; Hou, G.; Li, P.; Tian, Y.; Jiang, L. Chiral Recognition of L-Tryptophan with Beta-Cyclodextrin-Modified Biomimetic Single Nanochannel. Chem. Commun.* ***2015****, 51, 3135–3138; (C) Ali, M.; Nasir, S.; Nguyen, Q. H.; Sahoo, J. K.; Tahir, M. N.; Tremel, W.; Ensinger, W. Metal Ion Affinity-Based Biomolecular Recognition and Conjugation inside Synthetic Polymer Nanopores Modified with Iron-Terpyridine Complexes.* J. Am. Chem. Soc. ***2011***, 133, *17307–17314.*

nanopores are the two key factors to control ionic transport properties. The first strategy focuses on design and synthesis of functional molecules with multiple responsive properties on the inner surface of the nanopores. The second one is to prepare symmetric and asymmetric nanopores for different chemical modification.[14]

Based on the first strategy, Ulbricht et al. developed a nanochannel array that showed responsiveness to pH and temperature via modifying dual responsive poly(*N*-isopropyl acrylamide-co-acrylic acid) copolymer brushes.[51,52] We presented an integrated ionic gate and rectifier within an asymmetric single nanochannel via modifying pH and temperature dual-responsive copolymer bushes (Fig. 4.12A and B).[53] The thermal gating ratios remain approximately the same in the pH range from 3.6 to 9.4. But the ICR ratio is remarkably higher in basic environment. Li et al. modified a single glass conical nanopore with homopolymer poly(2-dimethylamino ethyl methacrylate) which undergoes both pH- and temperature-induced conformational transitions. This nanopore system achieves a smart

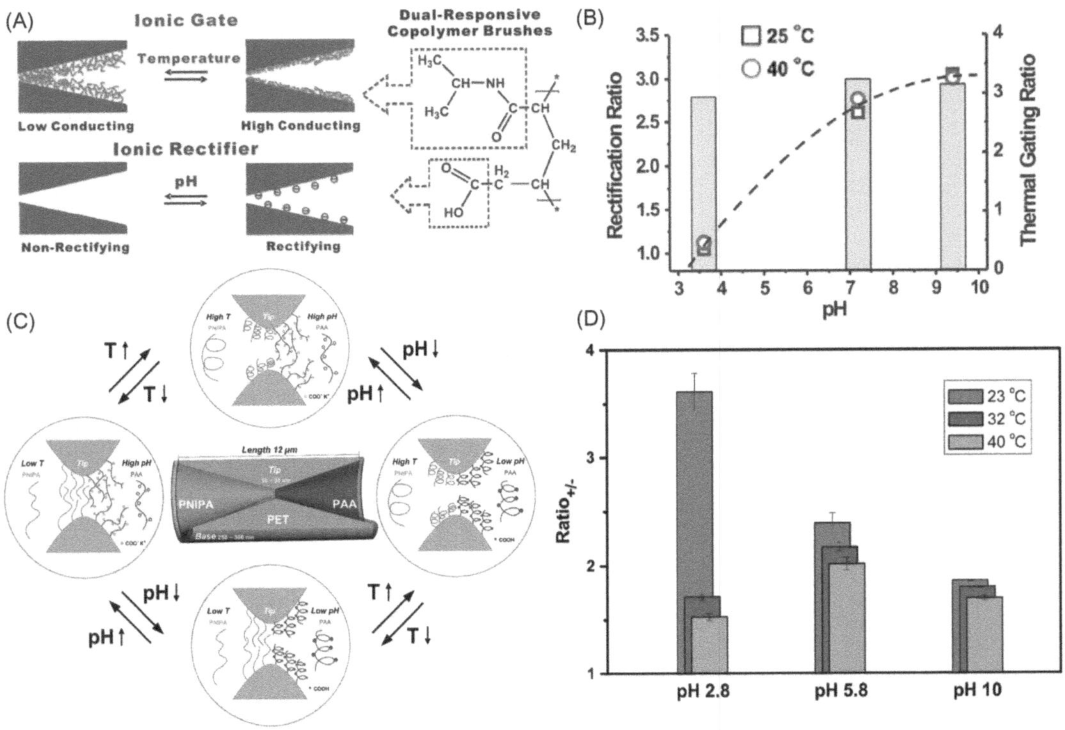

FIGURE 4.12 pH/temperature dual responsive nanopores.[53] (A and B) Homogenous modification of the cone-shaped polyimide nanopore with poly(NIPAAm-co-AAC) copolymer brushes, yielding pH and temperature dual responsive nanopores. (C and D) Asymmetrically modified PNIPAAm and PAAC homopolymer brushes on different parts of the nanopore result in an asymmetrically responsive nanopore system. Source: *Guo, W.; Xia, H.; Cao, L.; Xia, F.; Wang, S.; Zhang, G.; Song, Y.; Wang, Y.; Jiang, L.; Zhu, D. Integrating Ionic Gate and Rectifier within One Solid-State Nanopore Via Modification with Dual-Responsive Copolymer Brushes.* Advanced Functional Materials ***2010***, 20, *3561–3567.*

nanodevice that can be reversibly switched between high and low conducting states with high gating efficiency due to conformational change of the polymer brushes.[48]

Based on the second strategy, a biomimetic asymmetric responsive single nanopore system is fabricated (Fig. 4.12C and D).[54] By using asymmetric chemical modification on different parts of the nanochannel, a cooperative control over pH and temperature responsibility is realized by monitoring the asymmetric ionic transport property through the nanopores.

4.3 Conclusions and Outlook

In summary, biomimetic smart nanopores provide a new route to realize functionalization (Fig. 4.13). The idea is inspired by the gating functions of biological ion channels on cell membrane. Initially, scientists built reconstructed protein pores in lipid membrane. However, the lipid-based artificial nanopores are very fragile and are difficult to be integrated with other nanodevices, which prevent them from large-scale utilization. Therefore, scientists began to fabricate artificial nanoporous systems in solid-state substrates including organic, inorganic, or composite materials. After chemical modification, these solid-state nanofluidic

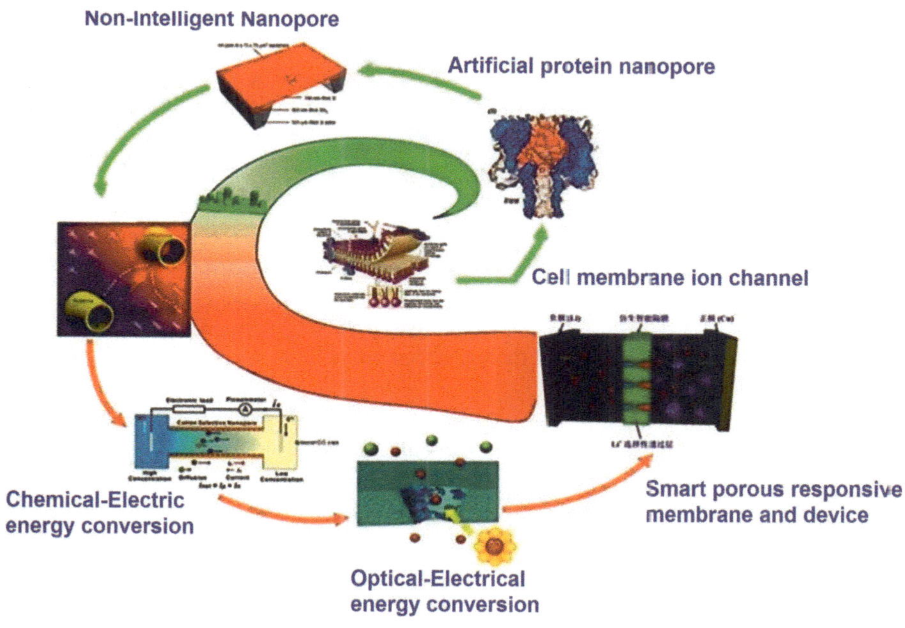

FIGURE 4.13 The roadmap for the development of biomimetic smart nanopores and their future applications in energy-related fields. The concept of biomimetic nanopores originated from the bionic study on membrane ion channels. Initially, artificial protein channels were reconstituted in lipid membrane, and afterward, the solid-state nanopores (nonintelligent) have been invented as artificial alternatives. Recently, through chemical modification, the synthetic nanopores become smart and show response to the environmental stimuli. From a perspective view, biomimetic smart nanopores are expected to find promising applications in many aspects of functional devices and materials for energy conversion and smart membranes.

systems become "smart" and show response to the environmental stimuli. In the process of nanopore functionalization, scientists accumulate a wealth of experience on material selection and nanopore fabrication and gained an in-depth understanding of the nature of mass and charge transport on the nanoscale. Based on those points, smart solid-state nanofluidic systems provide a reliable platform for the study of biological ion channels and pave a way for future functional nanofluidic devices for biochemical sensing, energy conversion, and smart membranes.

References

1. Kasianowicz, J. J.; Brandin, E.; Branton, D.; Deamer, D. W. Characterization of Individual Polynucleotide Molecules Using a Membrane Channel. *Proc. Natl. Acad. Sci. USA* **1996,** *93*, 13770–13773.
2. Venkatesan, B. M.; Bashir, R. Nanopore Sensors for Nucleic Acid Analysis. *Nat. Nanotechnol.* **2011,** *6*, 615–624.
3. Majd, S.; Yusko, E. C.; Billeh, Y. N.; Macrae, M. X.; Yang, J.; Mayer, M. Applications of Biological Pores in Nanomedicine, Sensing, and Nanoelectronics. *Curr. Opin. Biotechnol.* **2010,** *21*, 439–476.
4. Nelson, D. L.; Lehninger, A. L.; Cox, M. M. *Lehninger Principles of Biochemistry;* Macmillan, 2008.
5. Cox, M. M. *Lehninger Principles of Biochemistry;* Worth Publishers: New York, 2000.
6. Braha, O.; Gu, L.-Q.; Zhou, L.; Lu, X.; Cheley, S.; Bayley, H. Simultaneous Stochastic Sensing of Divalent Metal Ions. *Nat. Biotechnol.* **2000,** *18*, 1005–1007.
7. Schibel, A. E.; An, N.; Jin, Q.; Fleming, A. M.; Burrows, C. J.; White, H. S. Nanopore Detection of 8-Oxo-7,8-dihydro-2′-deoxyguanosine in Immobilized Single-Stranded DNA via Adduct Formation to the DNA Damage Site. *J. Am. Chem. Soc.* **2010,** *132*, 17992–17995.
8. Derrington, I. M.; Butler, T. Z.; Collins, M. D.; Manrao, E.; Pavlenok, M.; Niederweis, M., et al. Sequencing with MspA. *Proc. Natl. Acad. Sci. USA* **2010,** *107*, 16060–16065.
9. Butler, T. Z.; Pavlenok, M.; Derrington, I. M.; Niederweis, M.; Gundlach, J. H.; Single-Molecule, D. N. A. Detection with an Engineered MspA Protein Nanopore. *Proc. Natl. Acad. Sci. USA* **2008,** *105*, 20647–20652.
10. Stoddart, D.; Heron, A. J.; Mikhailova, E.; Maglia, G.; Bayley, H. Single-Nucleotide Discrimination in Immobilized DNA Oligonucleotides with a Biological Nanopore. *Proc. Natl. Acad. Sci. USA* **2009,** *106*, 7702–7707.
11. Wendell, D.; Jing, P.; Geng, J.; Subramaniam, V.; Lee, T. J.; Montemagno, C., et al. Translocation of Double-Stranded DNA through Membrane-Adapted Phi29 Motor Protein Nanopores. *Nat. Nanotechnol.* **2009,** *4*, 765–772.
12. Jing, P.; Haque, F.; Shu, D.; Montemagno, C.; Guo, P. One-Way Traffic of a Viral Motor Channel for Double-Stranded DNA Translocation. *Nano Lett.* **2010,** *10*, 3620–3627.
13. Kowalczyk, S. W.; Blosser, T. R.; Dekker, C. Biomimetic Nanopores: Learning from and About Nature. *Trends Biotechnol.* **2011,** *29*, 607–614.
14. Hou, X.; Guo, W.; Jiang, L. Biomimetic Smart Nanopores and Nanochannels. *Chem. Soc. Rev.* **2011,** *40*, 2385–2401.
15. Zhang, B.; Zhang, Y.; White, H. S. The Nanopore Electrode. *Anal. Chem.* **2004,** *76*, 6229–6238.
16. Yuan, J. H.; He, F. Y.; Sun, D. C.; Xia, X. H. A Simple Method for Preparation of Through-Hole Porous Anodic Alumina Membrane. *Chem. Mater.* **2004,** *16*, 1841–1844.
17. Storm, A. J.; Chen, J. H.; Ling, X. S.; Zandbergen, H. W.; Dekker, C. Fabrication of Solid-State Nanopores with Single-Nanometre Precision. *Nat. Mater.* **2003,** *2*, 537–540.

18. Wu, S.; Park, S. R.; Ling, X. S. Lithography-Free Formation of Nanopores in Plastic Membranes Using Laser Heating. *Nano Lett.* **2006,** *6,* 2571–2576.
19. Apel, P. Track Etching Technique in Membrane Technology. *Radiat. Meas.* **2001,** *34,* 559–566.
20. Alcaraz, A.; Ramírez, P.; García-Giménez, E.; López, M. L.; Andrio, A.; Aguilella, V. M. A pH-Tunable Nanofluidic Diode: Electrochemical Rectification in a Reconstituted Single Ion Channel. *J. Phys. Chem. B* **2006,** *110,* 21205–21209.
21. Siwy, Z. S. Ion-Current Rectification in Nanopores and Nanotubes with Broken Symmetry. *Adv. Funct. Mater.* **2006,** *16,* 735–746.
22. Hou, X.; Dong, H.; Zhu, D.; Jiang, L. Fabrication of Stable Single Nanochannels with Controllable Ionic Rectification. *Small* **2010,** *6,* 361–365.
23. Vlassiouk, I.; Siwy, Z. S. Nanofluidic Diode. *Nano Lett.* **2007,** *7,* 552–556.
24. Xia, F.; Guo, W.; Mao, Y. D.; Hou, X.; Xue, J. M.; Xia, H. W., et al. Gating of Single Synthetic Nanopores by Proton-Driven DNA Molecular Motors. *J. Am. Chem. Soc.* **2008,** *130,* 8345–8350.
25. Yameen, B.; Ali, M.; Neumann, R.; Ensinger, W.; Knoll, W.; Azzaroni, O. Single Conical Nanopores Displaying pH-Tunable Rectifying Characteristics. Manipulating Ionic Transport with Zwitterionic Polymer Brushes. *J. Am. Chem. Soc.* **2009,** *131,* 2070–2071.
26. Yameen, B.; Ali, M.; Neumann, R.; Ensinger, W.; Knoll, W.; Azzaroni, O. Proton-Regulated Rectified Ionic Transport through Solid-State Conical Nanopores Modified with Phosphate-Bearing Polymer Brushes. *Chem. Commun.* **2010,** *46,* 1908–1910.
27. Yameen, B.; Ali, M.; Neumann, R.; Ensinger, W.; Knoll, W.; Azzaroni, O. Synthetic Proton-Gated Ion Channels via Single Solid-State Nanochannels Modified with Responsive Polymer Brushes. *Nano Lett.* **2009,** *9,* 2788–2793.
28. Hou, X.; Liu, Y.; Dong, H.; Yang, F.; Li, L.; Jiang, L., et al. A pH-Gating Ionic Transport Nanodevice: Asymmetric Chemical Modification of Single Nanochannels. *Adv. Mater.* **2010,** *22,* 2440–2443.
29. Siwy, Z.; Apel, P.; Dobrev, D.; Neumann, R.; Spohr, R.; Trautmann, C., et al. Ion Transport through Asymmetric Nanopores Prepared by Ion Track Etching. *Nucl. Instrum. Methods Phys. Res. Sect. B Beam Interact. Mater. Atoms* **2003,** *208,* 143–148.
30. Wei, C.; Bard, A. J.; Feldberg, S. W. Current Rectification at Quartz Nanopipet Electrodes. *Anal. Chem.* **1997,** *69,* 4627–4633.
31. Huang, J.; Zhang, X.; McNaughton, P. A. Modulation of Temperature-Sensitive TRP Channels. *Semin. Cell. Dev. Biol.* **2006,** *17,* 638–645.
32. Yameen, B.; Ali, M.; Neumann, R.; Ensinger, W.; Knoll, W.; Azzaroni, O. Ionic Transport through Single Solid-State Nanopores Controlled with Thermally Nanoactuated Macromolecular Gates. *Small* **2009,** *5,* 1287–1291.
33. Guo, W.; Xia, H.; Xia, F.; Hou, X.; Cao, L.; Wang, L., et al. Current Rectification in Temperature-Responsive Single Nanopores. *ChemPhysChem* **2010,** *11,* 859–864.
34. Li, P.-F.; Xie, R.; Jiang, J.-C.; Meng, T.; Yang, M.; Ju, X.-J., et al. Thermo-Responsive Gating Membranes with Controllable Length and Density of Poly(*N*-Isopropylacrylamide) Chains Grafted by ATRP Method. *J. Membr. Sci.* **2009,** *337,* 310–317.
35. Banghart, M.; Borges, K.; Isacoff, E.; Trauner, D.; Kramer, R. H. Light-Activated Ion Channels for Remote Control of Neuronal Firing. *Nat. Neurosci.* **2004,** *7,* 1381–1386.
36. Liu, N. G.; Dunphy, D. R.; Atanassov, P.; Bunge, S. D.; Chen, Z.; López, G. P., et al. Photoregulation of Mass Transport through a Photoresponsive Azobenzene-Modified Nanoporous Membrane. *Nano Lett.* **2004,** *4,* 551–554.
37. Wang, G.; Bohaty, A. K.; Zharov, I.; White, H. S. Photon Gated Transport at the Glass Nanopore Electrode. *J. Am. Chem. Soc.* **2006,** *128,* 13553–13558.

38. Vlassiouk, I.; Park, C.-D.; Vail, S. A.; Gust, D.; Smirnov, S. Control of Nanopore Wetting by a Photochromic Spiropyran: A Light-Controlled Valve and Electrical Switch. *Nano Lett.* **2006,** *6,* 1013–1017.
39. Siwy, Z. S.; Howorka, S. Engineered Voltage-Responsive Nanopores. *Chem. Soc. Rev.* **2010,** *39,* 1115–1132.
40. Harrell, C. C.; Kohli, P.; Siwy, Z.; Martin, C. R. DNA – Nanotube Artificial Ion Channels. *J. Am. Chem. Soc.* **2004,** *126,* 15646–15647.
41. Kalman, E.; Sudre, O.; Vlassiouk, I.; Siwy, Z. Control of Ionic Transport through Gated Single Conical Nanopores. *Anal. Bioanal. Chem.* **2009,** *394,* 413–419.
42. Siwy, Z. S.; Powell, M. R.; Petrov, A.; Kalman, E.; Trautmann, C.; Eisenberg, R. S. Calcium-Induced Voltage Gating in Single Conical Nanopores. *Nano Lett.* **2006,** *6,* 1729–1734.
43. Siwy, Z. S.; Powell, M. R.; Kalman, E.; Astumian, R. D.; Eisenberg, R. S. Negative Incremental Resistance Induced by Calcium in Asymmetric Nanopores. *Nano Lett.* **2006,** *6,* 473–477.
44. Powell, M. R.; Sullivan, M.; Vlassiouk, I.; Constantin, D.; Sudre, O.; Martens, C. C., et al. Nanoprecipitation-Assisted Ion Current Oscillations. *Nat. Nanotechnol.* **2008,** *3,* 51–57.
45. Hou, X.; Guo, W.; Xia, F.; Nie, F.-Q.; Dong, H.; Tian, Y., et al. A Biomimetic Potassium Responsive Nanochannel: G-Quadruplex DNA Conformational Switching in a Synthetic Nanopore. *J. Am. Chem. Soc.* **2009,** *131,* 7800–7805.
46. Tian, Y.; Hou, X.; Wen, L. P.; Guo, W.; Song, Y. L.; Sun, H. Z., et al. A Biomimetic Zinc Activated Ion Channel. *Chem. Commun.* **2010,** *46,* 1682–1684.
47. Siwy, Z.; Trofin, L.; Kohli, P.; Baker, L. A.; Trautmann, C.; Martin, C. R. Protein Biosensors Based on Biofunctionalized Conical Gold Nanotubes. *J. Am. Chem. Soc.* **2005,** *127,* 5000–5001.
48. Zhang, L.-X.; Cai, S.-L.; Zheng, Y.-B.; Cao, X.-H.; Li, Y.-Q. Smart Homopolymer Modification to Single Glass Conical Nanopore Channels: Dual-Stimuli-Actuated Highly Efficient Ion Gating. *Adv. Funct. Mater.* **2011,** *21,* 2103–2107.
49. Xie, G.; Tian, W.; Wen, L.; Xiao, K.; Zhang, Z.; Liu, Q., et al. Recognition of L-Tryptophan with Beta-Cyclodextrin-Modified Biomimetic Single Nanochannel. *Chem. Commun.* **2015,** *51,* 3135–3138.
50. Ali, M.; Nasir, S.; Nguyen, Q. H.; Sahoo, J. K.; Tahir, M. N.; Tremel, W., et al. Metal Ion Affinity-Based Biomolecular Recognition and Conjugation Inside Synthetic Polymer Nanopores Modified with Iron–Terpyridine Complexes. *J. Am. Chem. Soc.* **2011,** *133,* 17307–17314.
51. Friebe, A.; Ulbricht, M. Cylindrical Pores Responding to Two Different Stimuli via Surface-Initiated Atom Transfer Radical Polymerization for Synthesis of Grafted Diblock Copolymers. *Macromolecules* **2009,** *42,* 1838–1848.
52. Geismann, C.; Tomicki, F.; Ulbricht, M. Block Copolymer Photo-Grafted Poly(Ethylene Terephthalate) Capillary Pore Membranes Distinctly Switchable by Two Different Stimuli. *Sep. Sci. Technol.* **2009,** *44,* 3312–3329.
53. Guo, W.; Xia, H.; Cao, L.; Xia, F.; Wang, S.; Zhang, G., et al. Integrating Ionic Gate and Rectifier within One Solid-State Nanopore via Modification with Dual-Responsive Copolymer Brushes. *Adv. Funct. Mater.* **2010,** *20,* 3561–3567.
54. Hou, X.; Yang, F.; Li, L.; Song, Y.; Jiang, L.; Zhu, D., et al. Responsive Single Nanochannel. *J. Am. Chem. Soc.* **2010,** *132,* 11736–11742.

5

Chemically Modified Nanopores in Colloidal Crystals

E. Fullwood, E. Green, J. Selden, I. Zharov
UNIVERSITY OF UTAH, SALT LAKE CITY, UT, UNITED STATES

CHAPTER OUTLINE

5.1 Colloidal Crystals as Nanoporous Materials

Chemically modified nanopores have been the subject of the studies with the purpose of elucidating the effects of size, surface interactions, and confinement on the transport through the nanopores.[1,2] The ability to control the transport by chemical modification of the nanopores surfaces[3] leads to applications of nanoporous materials in separations of biomacromolecules[4] and pharmaceuticals,[5] in sensing,[6] and in drug release.[7–9] The nanopore surface modifications included monolayers capable of non-covalent interactions with the diffusing species and polymer brushes that provide sterically control of the transport.[10,11]

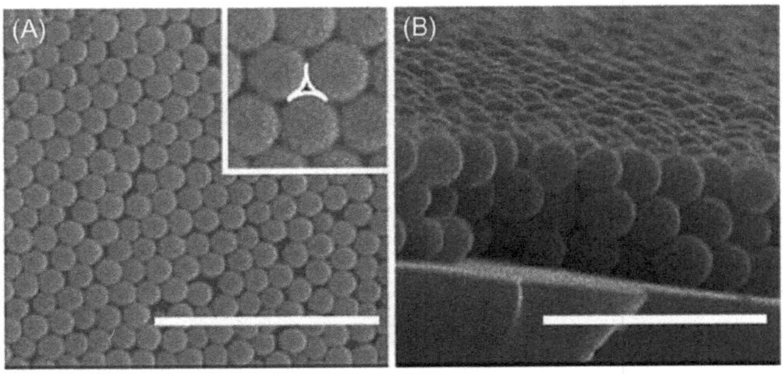

FIGURE 5.1 SEM images of silica colloidal crystals prepared on glass from 440 nm diameter silica spheres deposited (A) top view (size bar = 4 μm) and (B) side view (size bar = 2 μm). The geometric projection of a pore observed from the (111) plane is outlined in the inset in (A). Source: *Reproduced with permission from reference Newton, M.R.; Bohaty, A.K.; White, H.S.; Zharov, I. Chemically Modified Opals as Thin Permselective Nanoporous Membranes. J. Am. Chem. Soc. 2005, 127, 7268–7269. Copyright 2005 American Chemical Society.*

While polymeric nanoporous materials are widely used,[12] the chemical modification of their nanopores is limited. On the other hand, inorganic nanoporous materials, such as zeolites,[13] silicon nitride,[14] silicon,[15] silica,[16] and alumina,[17] provide rich surface chemistry. Many of these materials possess low nanopore density, limited available nanopore size range, and difficult preparation. In contrast, silica nanoporous colloidal films and membranes, developed in our group in the past decade, are prepared by self-assembly of easy-to-make nanoparticles, they possess high molecular flux, easily controllable nanopore size in the range of 5–100 nm, and most importantly, facile surface chemistry. The latter allowed us to prepare and study a variety of chemically modified nanopores, leading to materials with controlled ionic and molecular transport, as described in this chapter.

Silica colloidal crystals are widely used as templates for the preparation of photonic[18] and magnetic materials,[19] macroporous polymer membranes,[20] and sensors.[21] Silica colloidal crystals comprise a close-packed face-centered cubic (fcc) lattice of silica spheres of a submicrometer diameter (Fig. 5.1) with ordered arrays of interconnected three-dimensional nanoscale voids.[22] The distance from the center of the nanopore projection to the nearest silica sphere surface is ca. 15% of the sphere radius. Because of the three-dimensional geometry of the nanopores, we use this projection (Fig. 5.1A) as a simplified description of the nanopore geometry, and the distance described above as the nanopore "radius."

The highly ordered structure of colloidal crystals allows using accurate mathematical descriptions of the molecular transport.[23–25] An estimate of the molecular flux J_{fcc} in the fcc lattice can be obtained using Eq. (5.1):

$$J_{fcc} = (\Delta C/L) \times (\varepsilon/\tau) D_{sol} \tag{5.1}$$

where D_{sol} is the diffusivity of molecules in free solution, the void fraction ε (0.26) and the tortuosity τ (~3.0) are intrinsic geometrical parameters independent of the size of the silica spheres used to prepare the colloidal crystal, ΔC is the concentration gradient, and L is the thickness of the lattice. Importantly, the diffusive flux of small molecules normal to the

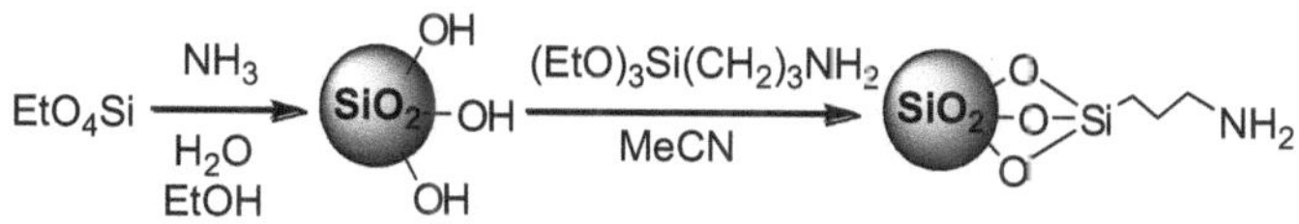

SCHEME 5.1 Preparation and surface amination of silica spheres.

(111) plane of a colloidal crystal is only ca. 10 times smaller relative to the free solution value, independent of the size of the spheres used to assemble the crystal. Thus, the flux remains significant, even when the pore size is reduced to the nanoscale to impart molecular transport selectivity. For a typical D_{sol} of 10^{-5} cm^2/s, L of 10 μm, and ΔC of 10 mM, J_{fcc} is $\sim 10^{-8}$ $mol/cm^2/s$.

The preparation of silica spheres is straightforward,[26] and the self-assembly of the spheres into the colloidal crystals is well developed.[27] The surface silanol groups in colloidal crystals can be modified by nucleophilic silylation to introduce a variety of functional groups.[28] Alternatively, silica surface can be modified with 3-aminopropyltriethoxysilane (APTES) (Scheme 5.1), followed by treatment with organic molecules carrying electrophilic moieties, such as acyl chloride, isocyanate, isothiocyanate, carboxylic acid, sulfonyl chloride,[29] succinimidyl ester,[30] and 2-bromoisobutyrylbromide. The latter can serve as atom transfer radical polymerization (ATRP) initiator,[31] providing the possibility of growing polymer brushes on the nanopore surface.

5.2 Preparation of Colloidal Nanoporous Materials

The preparation of chemically modified nanopores in colloidal crystals, described in this chapter, followed a multistep procedure, where the first step is the self-assembly of unmodified silica nanospheres. While it might be possible to assemble surface-modified silica spheres, this process is not developed and might be impeded by the surface functionality. Furthermore, in the case of high-temperature sintering of the colloidal crystals, any organic surface functionality would be removed, thus rendering surface modification before the assembly impractical.

Silica colloidal crystals can be assembled on a hydrophilic solid support (glass, oxidized silicon, metal, etc.) using vertical deposition.[27] Using this technique, silica colloidal crystals (Fig. 5.1) are deposited onto the support, which is placed vertically into colloidal solution of silica spheres in ethanol.[32] The solutions are allowed to evaporate for 2–3 days in a vibration-free environment. The thickness of the colloidal film is controlled by varying the weight percentage of silica spheres in the colloidal solution.

In addition to supported silica colloidal crystals, we prepared suspended silica colloidal membranes using silicon[33] and glass[34] supports. For example, silica colloidal membranes suspended in silicon have been prepared using a 0.3-mm-thick Si(100) wafers with 40 × 40, 100 × 100, 250 × 250, or 500 × 500 μm openings shaped as a truncated square pyramid. A wafer was suspended vertically in a colloidal solution of silica spheres in absolute ethanol.

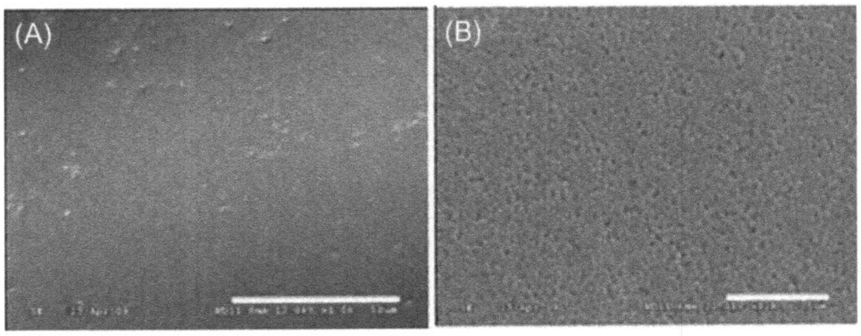

FIGURE 5.2 SEM images of sintered colloidal crystals comprised of 180 nm silica spheres: (A) SEM image showing no major cracks over a large area (size bar = 50 μm) and (B) magnified image displaying the close-packed fcc lattice (size bar = 2.5 μm). Source: *Reproduced with permission from reference Bohaty, A.K.; Smith, J.J.; Zharov, I. Free-Standing Silica Colloidal Nanoporous Membranes. Langmuir 2009, 25, 3096–3101. Copyright 2009 American Chemical Society.*

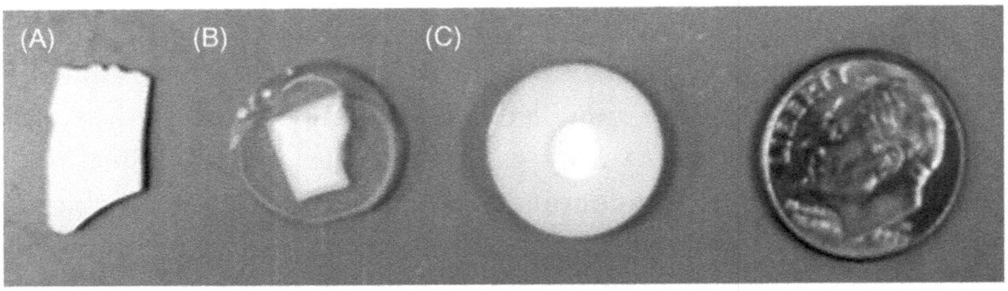

FIGURE 5.3 Photographs of sintered silica colloidal membranes: (A) as-sintered, (B) without PTFE washers showing the sintered colloidal membrane in the epoxy, and (C) with PTFE washers. Source: *Reproduced with permission from reference Bohaty, A.K.; Smith, J.J.; Zharov, I. Free-Standing Silica Colloidal Nanoporous Membranes.* Langmuir ***2009***, 25, *3096–3101. Copyright 2009 American Chemical Society.*

The solvent evaporated at room temperature, resulting in the formation of suspended silica colloidal membranes. Silica colloidal membranes suspended in glass have been prepared in a similar way using a 150-μm-thick glass slide.

We discovered that free-standing silica colloidal membranes can be prepared as a material that is mechanically durable and possesses large area (Figs. 5.2–5.3). To prepare such membranes, the silica spheres assembled into a colloidal crystal were physically bonded together by sintering at 1050°C,[35,36] at which temperature silica flows at the surface. The sintering process can be followed by rehydroxylation with tetrabutylammonium hydroxide in water to restore the surface hydroxyl groups.[37]

We also prepared robust nanoporous membranes with controlled area and uniform thickness by pressing silica colloidal spheres into disks followed by sintering.[38] Three different diameters of silica particles, 390, 220, and 70 nm, were used to prepare the membranes with different pore size. In order to evaluate their size selectivity, we measured the diffusion of polystyrene particles through these membranes. Although pressed silica colloidal membranes did not possess visible order or uniform pore size, they showed size-selective transport.

Finally, we found that gold-coated silica spheres can be self-assembled into colloidal crystals using the same techniques.[39] Moreover, these colloidal crystals can be sintered to obtain free-standing membranes (Fig. 5.4) with gold-coated nanopores, suitable for further surface modifications using thiol chemistry.

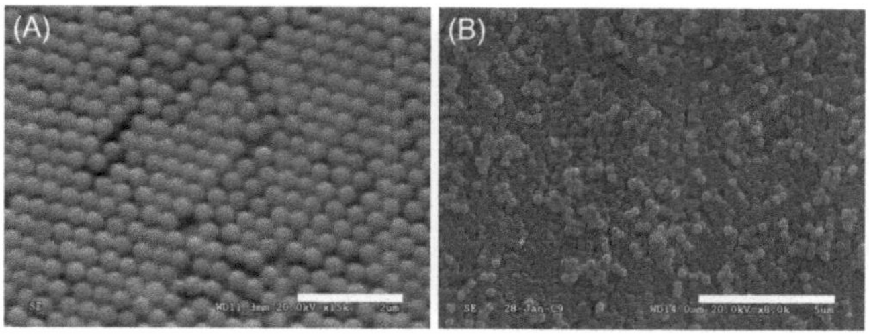

FIGURE 5.4 SEM images of (A) self-assembled and (B) sintered SiO_2@Au colloidal membranes. Scale bar is 2 μm in (A) and 5 μm in (B). Source: *Reproduced with permission from reference Ignacio-de Leon, P.A.; Zharov, I. SiO2@Au Core-Shell Nanospheres Self-Assemble to Form Colloidal Crystals That Can Be Sintered and Surface Modified to Produce pH-Controlled Membranes. Langmuir 2013, 29, 3749–3756. Copyright 2013 American Chemical Society.*

Transport through the chemically modified colloidal nanopores in above materials has been studied using two methods. Supported colloidal films were prepared on the surface of platinum disk microelectrodes shrouded in glass (which we called opal electrodes) and cyclic voltammetry experiments were performed. This allowed measuring the flux of redox-active species across the nanoporous films in quasi-steady-state voltammetric experiments, which resulted in a sigmoidal-shaped voltammograms characteristic of radial transport to a microelectrode (see chapter: Theoretical Basis for Structure and Transport in Nanopores and Nanochannels). For free-standing colloidal membranes, the flux of dye molecules was measured using the standard diffusion cell technique.

5.3 Aminated Colloidal Nanopores

To demonstrate that silica colloidal nanopores can be chemically functionalized to acquire pH- and ionic strength-dependent permselectivity based on electrostatic interactions between the surface groups and the charged permeants, we modified the nanopore surfaces with amino groups.[32,40] Fig. 5.5 shows the voltammetric responses of an electrode coated with silica colloidal films. For amine-modified silica colloidal films, at pH 4 the voltammetric limiting current, i_{lim}, corresponding to the one-electron reduction of $Ru(NH_3)_6^{3+}$ was greatly diminished (Fig. 5.5A), while i_{lim} for $Fe(CN)_6^{4-}$ and $Fc(CH_2OH)_2$ remained approximately the same as before the surface modification (Fig. 5.5B). This result is consistent with the electrostatic repulsion of the cationic species, $Ru(NH_3)_6^{3+}$, by the protonated surface amines. We also investigated the effect of pH on the ionic transport through aminated silica colloidal films and found an S-shaped dependence (Fig. 5.5D).

These observations required an additional analysis in terms of the correlation between the nanopores size, electrostatic potential, and flux. Indeed, for the colloidal crystal comprising 440 nm silica spheres, used in these experiments, the nanopore "radius" is ca. 17 nm. Within this structure, a $\sim$50% blocking of diffusion was observed at an electrolyte concentration of 0.05 M, where the Debye screening length (κ^{-1}) is only ca. 1.5 nm. Assuming that the electric field extends $\sim 5\kappa^{-1}$ from the surface ($\sim$7.5 nm), the electric field extends over only

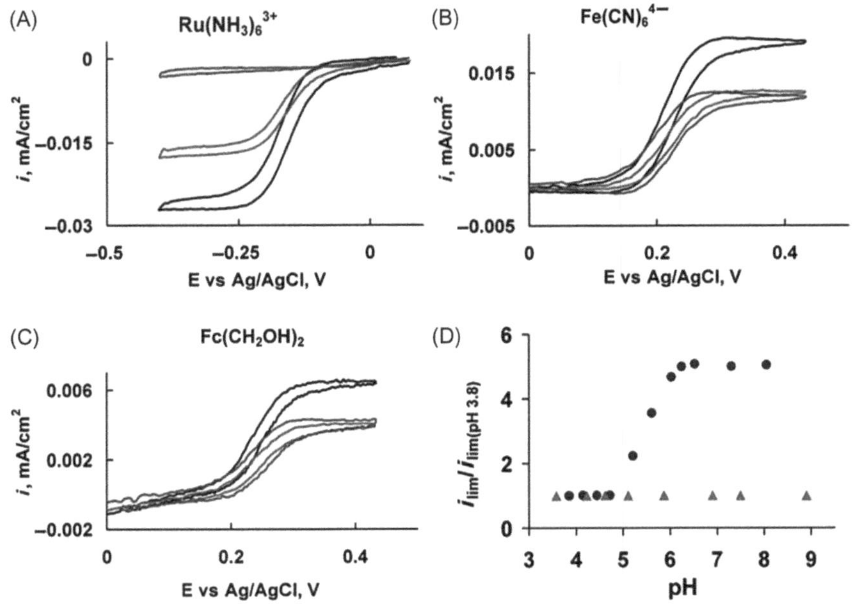

FIGURE 5.5 Voltammetric responses of a Pt electrode: (A) bare (bottom), after opal assembly (middle), and after chemical modification of the thin colloidal membrane with 3-aminopropyltriethoxysilane (top) at pH 4. (B and C) Bare (top), after opal assembly and after surface modification of the thin colloidal film with 3-aminopropyltriethoxysilane (middle) at pH 4. (D) Voltammetric responses of as a function of pH for unmodified (triangles) and modified (circles) opal electrode. Source: *Reproduced with permission from reference Newton, M.R.; Bohaty, A.K.; White, H.S.; Zharov, I. Chemically Modified Opals as Thin Permselective Nanoporous Membranes. J. Am. Chem. Soc. 2005, 127, 7268–7269. Copyright 2005 American Chemical Society.*

a part of the effective pore width. Thus, we speculated that the tortuous path that molecules take to diffuse through the colloidal film to reach the electrode and the high surface area of the colloidal crystal contribute to enhancing the observed electrostatic permselectivity.

We also aminated the nanopores in free-standing silica colloidal membranes.[35] Unlike the thin colloidal films, for the sintered membranes to become modified with amines, their surface needs to be rehydroxylated using tetrabutylammonium hydroxide in water because most of the surface hydroxyl groups are lost during sintering. For the silica colloidal films described above, the surface density of amines was 1.3 groups/nm^2. This number was only 0.45 groups/nm^2 for the rehydroxylated sintered membranes treated once with APTES for 24 h. The amine coverage increased to 0.8 groups/nm^2 when the amine-modified sintered membranes were treated with APTES for 24 h for the second time.

Diffusion rates were used to determine if any transport selectivity can be achieved for positively charged species ($Fe(bpy)_3^{2+}$) through the aminated nanopores in sintered membranes. For the nanopores that underwent a single 24 h amination the experiment showed no change in diffusion rate of $Fe(bpy)_3^{2+}$ when acid was present in solution. This suggested that that there were not enough surface amines to create a positive surface charge sufficient to repel the cations from the colloidal nanopores. For the nanopores aminated twice, the diffusion rate of $Fe(bpy)_3^{2+}$ decreased by 22.4% in the presence of acid. The relatively small decrease in diffusion rate observed for the amine-modified nanopores in sintered membranes, as compared to those in thin colloidal films, could be due to the lower number of amines on the surface.

5.4 Sulfonated Colloidal Nanopores

To introduce permanent negative charges into the colloidal nanopores, we modified the colloidal nanopore surfaces with sulfonic groups.[41] Fig. 5.6 shows the voltammetric responses of an opal electrode with sulfonic groups. For the sulfonated films, i_{lim} for $IrCl_6^{3-}$ was diminished (Fig. 5.6A), i_{lim} for $Ru(NH_3)_6^{3+}$ was increased (Fig. 5.6B), while i_{lim} for $Fc(CH_2OH)_2$ remained the same as before the surface modification (Fig. 5.6C). This is consistent with electrostatic attraction of the cationic species, $Ru(NH_3)_6^{3+}$, and electrostatic repulsion of the anionic species, $IrCl_6^{3-}$, by the negatively charged sulfonated surface. We also investigated the effect of pH on the ionic transport through sulfonated (Fig. 5.7) silica colloidal films.

The diffusion of $Ru(NH_3)_6^{3+}$ and $IrCl_6^{3-}$ through the sulfonated silica colloidal films was also studied as a function of solution ionic strength, using KCl as the electrolyte at pH ~7 in water. Increasing the solution ionic strength reduced the diffusion rates for the positively charged species, $Ru(NH_3)_6^{3+}$ (Fig. 5.8A), while the transport rate of the negatively charged species, $IrCl_6^{3-}$ (Fig. 5.8B), increased with the increasing ionic strength.

We studied the proton transport in surface-sulfonated colloidal nanopores.[42] We prepared two types of materials. First, both unmodified and sulfonated 200 nm silica spheres were pressed into disk-shaped pellets using a hydraulic press. Second, we prepared colloidal assemblies by vertical deposition of both unmodified and sulfonated silica spheres from 4 wt % ethanol colloidal solutions onto the glass substrate. This process led to mechanically robust ~1-mm-thin free-standing colloidal materials. Both materials were free of mechanical

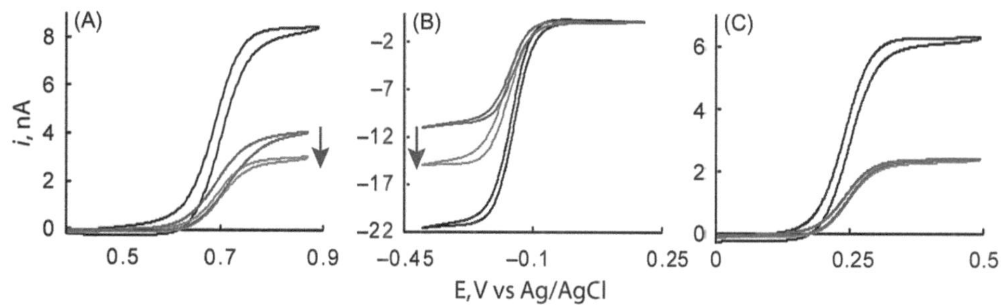

FIGURE 5.6 Representative voltammetric responses of bare electrodes: top in (A) and (C), bottom in (B) and colloidal film electrodes before and after silica surface sulfonation for (A) $IrCl_6^{3-}$ (middle—unmodified film, bottom—sulfonated film); (B) $Ru(NH_3)_6^{3+}$ (top—unmodified film, middle—sulfonated film), and (C) $Fc(CH_2OH)_2$ (middle—unmodified and sulfonated films) at pH 7 with 0.1 M aqueous KCl. The decrease (A) and increase (B) in the limiting current are shown with arrows. Source: *Reproduced with permission from reference Smith, J.J.; Zharov, I. Ion Transport in Sulfonated Nanoporous Colloidal Films. Langmuir 2008, 24, 2650–2654. Copyright 2008 American Chemical Society.*

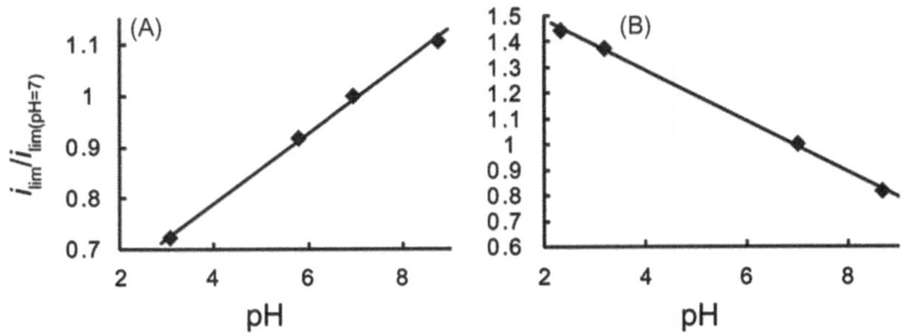

FIGURE 5.7 Relative limiting current of the sulfonated colloidal membrane electrodes as a function of pH for $Ru(NH_3)_6^{3+}$ (A) and for $IrCl_6^{3-}$ (B). Source: *Reproduced with permission from reference Smith, J.J.; Zharov, I. Ion Transport in Sulfonated Nanoporous Colloidal Films.* Langmuir ***2008***, 24, *2650–2654. Copyright 2008 American Chemical Society.*

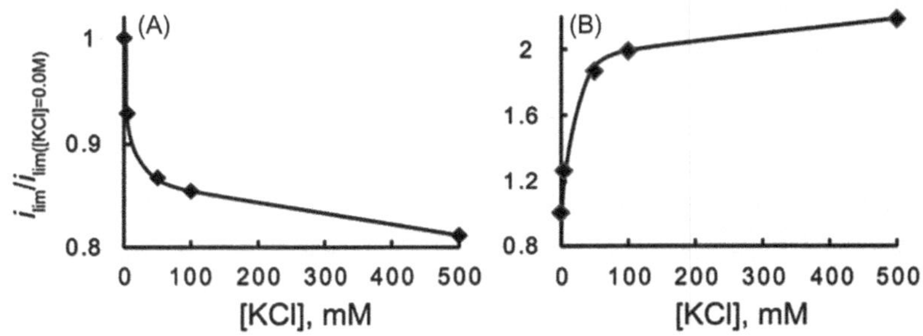

FIGURE 5.8 Relative limiting current of the sulfonated colloidal membrane electrodes as a function of KCl concentration for $Ru(NH_3)_6^{3+}$ (A) and for $IrCl_6^{3-}$ (B). Source: *Reproduced with permission from reference Smith, J.J.; Zharov, I. Ion Transport in Sulfonated Nanoporous Colloidal Films.* Langmuir ***2008***, 24, *2650–2654. Copyright 2008 American Chemical Society.*

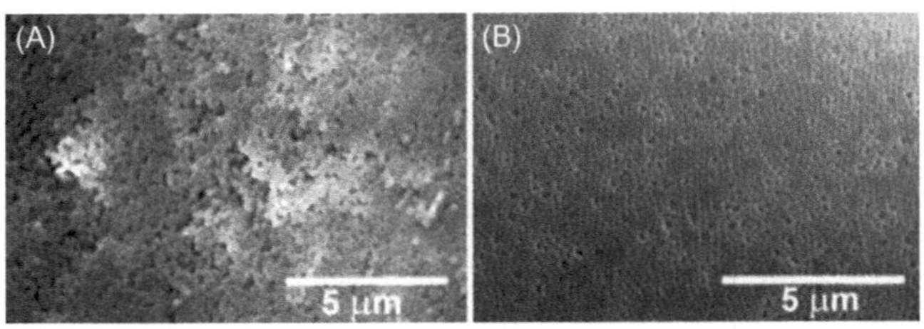

FIGURE 5.9 SEM images of the disordered (A) and self-assembled close-packed (B) materials prepared from ca. 200-nm-diameter sulfonated silica spheres as described in the text. Source: *Reproduced with permission from reference Smith, J.J.; Abbaraju, R.R.; Zharov, I. Proton Transport in Assemblies of Silica Colloidal Spheres. J. Mater. Chem. 2008,* 18, *5335–5338. Copyright 2008 American Chemical Society.*

defects such as cracks, and both possessed similar porosity. However, their structure at the nanoscale was quite different. Based on SEM images (Fig. 5.9) it appears that the pellet had multiple disordered regions compared to the self-assembled material, which seems to possess more structural order.

Table 5.1 Proton Conductivity of Disordered and Close-Packed Colloidal Assemblies Made of Sulfonated Silica Spheres[42]

Material	Proton Conductivity (S/cm) at 100°C			
	Dry Air		98% Relative Humidity	
Disordered	$<10^{-7}$	0.5×10^{-6}	2.2×10^{-6}	7.0×10^{-5}
Close-packed	$<10^{-7}$	0.8×10^{-6}	5.0×10^{-6}	1.4×10^{-2}

We used electrochemical impedance spectroscopy to measure the proton conductivity of the above materials (Table 5.1). The proton conductivity of unmodified pellets at 100°C in dry air was below 10^{-7} S/cm (the lower limit of the instrument sensitivity). Next, we measured the proton conductivity of the self-assembled colloidal materials. We found that at 100°C in dry air the proton conductivity of unmodified self-assembled colloidal materials was also very low. In contrast, under these conditions the proton conductivity of the sulfonated self-assembled colloidal materials was 0.82×10^{-5} S/cm, nearly two times higher compared to the disordered sulfonated pellets. This may result from a higher number of interconnected nanopores and/or a shorter distance the protons have to travel, leading to more efficient proton transport in the self-assembled colloidal material. The lower proton conductivity of the pellets comprising the sulfonated silica spheres might result from the disordered structure of the pellet with a small number of continuous nanoporous channels and a poor contact between the silica spheres. Indeed, it is well known that increasing the network organization of a proton conducting material enhances the proton transport through that material.[43–45]

The proton conductivity of the sulfonated pellets increased ca. 14-fold (to 7.0×10^{-5} S/cm) with increasing relative humidity (RH) from 0 to 98%. Increasing the RH had a much more pronounced effect on the proton conductivity of self-assembled sulfonated colloidal material. At 98% RH and 100°C its proton conductivity was ca. 10^{-2} S/cm, more than two orders of magnitude higher than that of the sulfonated disordered pellets. Since in dry air the difference between the ordered and disordered colloidal materials was much smaller, we speculate that the high proton conductivity of the ordered material at high humidity results from its better ability to retain water.

5.5 Spiropyran-Modified Colloidal Nanopores

We prepared light-responsive spiropyran-modified colloidal nanopores.[46] A spiropyran moiety undergoes a reversible transformation from the neutral form to the charged merocyanine forms and back upon irradiation with UV and visible light, respectively (Scheme 5.2),[47] and therefore could be used to control the ionic transport through colloidal nanopores by light. The succinimidyl ester derivative[48] of the spiropyran was prepared and used to treat the amine-modified colloidal nanopores in thin films assembled on a Pt electrode (Scheme 5.3).

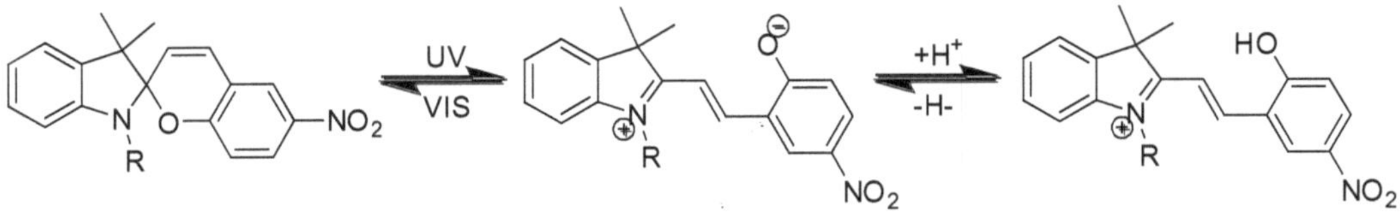

SCHEME 5.2 Interconversions of neutral, zwitterionic, and charged spiropyran forms.

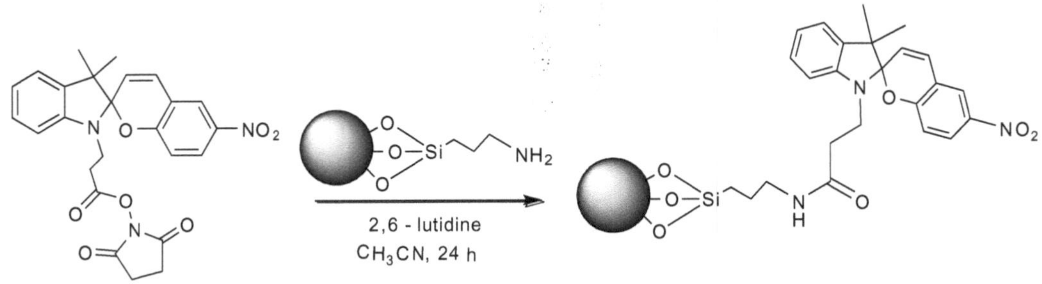

SCHEME 5.3 Attachment of spiropyran moiety to the silica sphere surface.

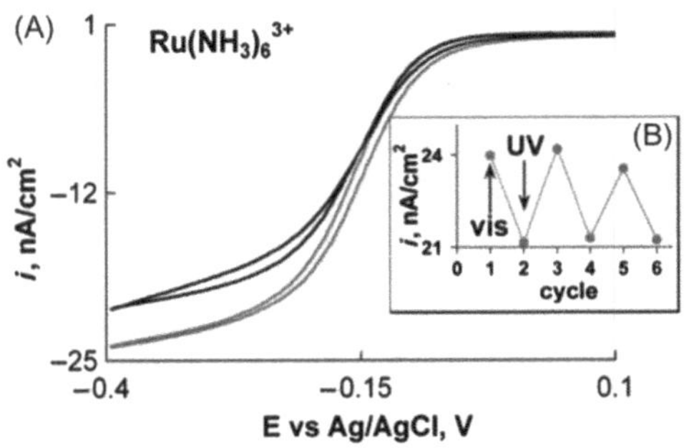

FIGURE 5.10 (A) Voltammetric response of a spiropyran-modified opal Pt electrode after irradiation with UV (top curve) and visible light (bottom curve). (B) Time profile of the response. Source: *Reproduced with permission from reference Bohaty, A.K.; Newton, M.R.; Zharov, I. Light-Controlled Ion Transport Through Spiropyran-Modified Nanoporous Silica Colloidal Films. J. Porous Mater. 2010, 17, 465–473. Copyright 2011 Springer.*

The successful silica surface modification with spiropyran moieties was confirmed using UV spectroscopy.

The flux of charged molecules through the spiropyran-modified nanopores was measured using cyclic voltammetry, in the same way as for the amine-modified nanopores. It was found that upon irradiation with UV light the transport of positively charged species decreased, and that it can be restored to its original value upon irradiating the membrane with visible light (Fig. 5.10). The change in flux of cations was only about 15% for the spiropyran-modified nanopores, compared to nearly 100% in the case of amine-modified nanopores. This may result from having fewer spiropyran molecules on the nanopores surface, or from incomplete conversion of the spiropyran molecules to the merocyanine form.

5.6 Colloidal Nanopores Modified With Chiral Moieties

To prepare silica colloidal nanopores with chiral permselectivity,[49] we modified their surface with a chiral selector moiety (CSM) **1** (Chart 5.1). We then studied the transport of both R and S enantiomers of three chiral ferrocene derivatives (**2**–**4**, Chart 5.1) through the chiral colloidal films of ca. 35 layers of silica spheres, assembled on glass-shrouded Pt microelectrodes. Chiral permselectivity was readily apparent in the voltammetric response of the colloidal film electrode after the attachment of **1R** (Fig. 5.11A, Table 5.2). The observed selectivity was comparable to that reported for the transport of enantiomers in antibody-modified nanotube membranes,[50] and higher than the selectivity reported for optically active polyelectrolyte membranes.[51] Modifying colloidal films with the S enantiomer of **1** led to a reversed selectivity (Fig. 5.11B, Table 5.2). We investigated the transport of two more chiral molecules (**3** and **4**, Chart 5.1) through the chiral colloidal films and in both cases found chiral permselectivity that was reversed with the reversal of the surface chirality (Table 5.2). These results are consistent with surface transport mechanism,[52] where both R and S enantiomers interact with the chiral selector moieties on the nanopore surface, but to a different

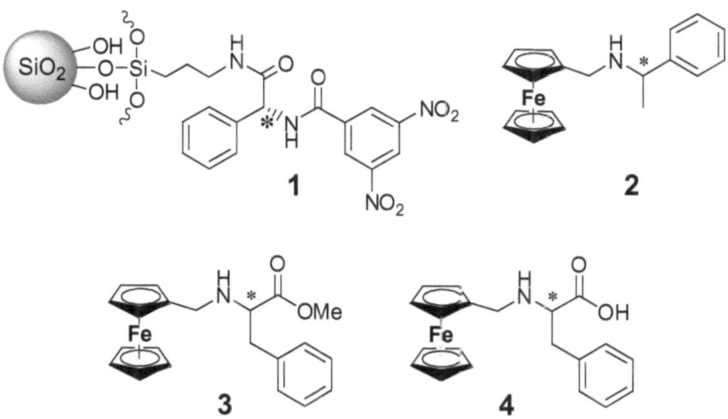

CHART 5.1 Chiral selector moiety used to prepare chiral colloidal nanopores, and chiral ferrocene derivatives used to study their chiral permselectivity.

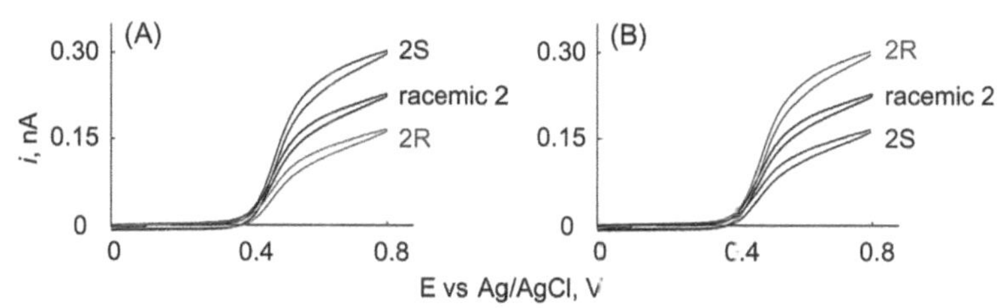

FIGURE 5.11 (A) Overlay of voltammetric responses of a colloidal film electrode modified with **1R** for **2S** and **2R**. (B) Overlay of voltammetric responses of a colloidal film electrode modified with **1S** for **2S** and **2R**.
Source: *Adopted with permission from reference Cichelli, J.; Zharov, I. Chiral Selectivity in Surface-Modified Porous Colloidal Films. J. Am. Chem. Soc. 2006,* 128, *8130–8131. Copyright 2006 American Chemical Society.*

Table 5.2 Selectivities of the Chiral Colloidal Film Electrode for **2–4**

	$i_{lim}(S)/i_{lim}(R)$	
Probe	1R **Colloidal Film**	1S **Colloidal Film**
2	2.16 ± 0.34	0.58 ± 0.01
3	1.33 ± 0.20	0.82 ± 0.11
4	2.01 ± 0.69	0.84 ± 0.04

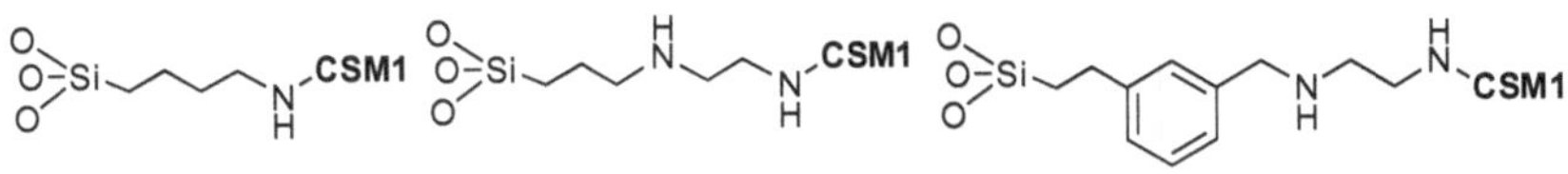

2.72 ± 0.64 1.46 ± 0.03 3.01 ± 0.51

Selectivities obtained using the chiral probe **2**

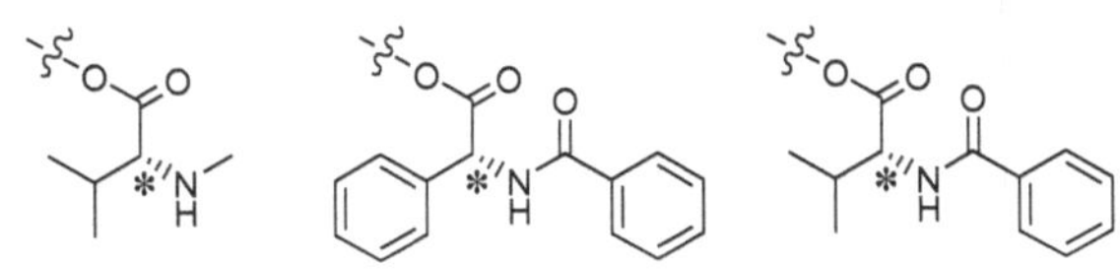

5, 1.46 ± 0.33 **6**, 1.37 ± 0.01 **7**, 4.53 ± 0.57

Selectivities obtained using three carbon atoms linker and chiral probe **2**

CHART 5.2 Linkers and chiral selector moieties used to study the structure–pemselectivity relationships.

extent, leading to selective transport of one enantiomer through the chiral colloidal film compared to the other enantiomer.

We varied the length and structure of the linker attaching the CSM **1** to the silica surface and explored the influence of the chiral selector structure (Chart 5.2) on the permselectivity while maintaining the length of the linker constant (Chart 5.2).[53] The chiral selectivity of 4.5 found for the chiral selector **7** is one of the highest reported for chiral membranes.

5.7 Calixarene-Modified Colloidal Nanopores

We modified the surface of nanopores in colloidal films assembled from 200 nm silica spheres with thiacalix[4]arene moieties (Scheme 5.4).[54] We studied the transport of two redox-active species, ferrocene dimethanol and iron tris(bipyridyl) hexafluorophosphate, through the nanopores using cyclic voltammetry. We observed evidence for two different molecular transport mechanisms. Ferrocene dimethanol was transported via a simple diffusive mechanism while iron tris(bipyridyl) was transported via a surface-hopping mechanism. The difference in transport mechanisms is believed to result from different interaction strength between the diffusing species and the surface-bound organic moieties and sheds light on transport mechanisms in nanoporous systems.

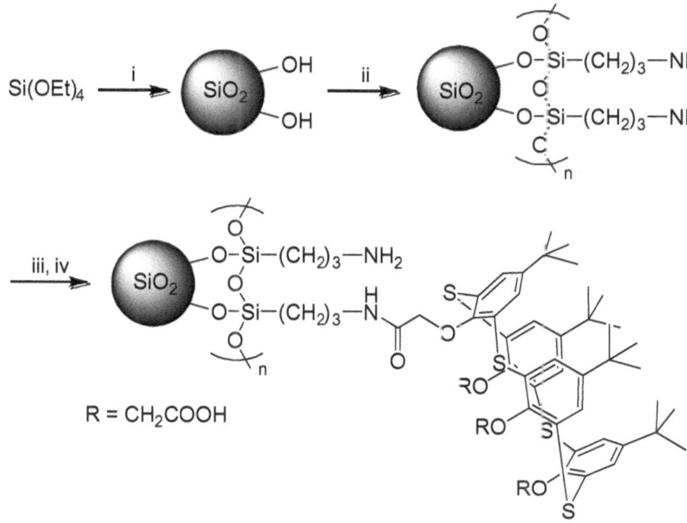

SCHEME 5.4 Modification of colloidal nanopores with thiacalixarenes. Reagents and conditions: (i) NH_4OH, H_2O, EtOH, RT, 12 h; (ii) APTES, CH_3CN, RT, 17 h; (iii) CH_3CN, $CHCl_3$, pyridine, RT, 12 h; and (iv) CH_3CN, H_2O, RT, 12 h.

5.8 Polymer Brush-Modified Colloidal Nanopores

5.8.1 PAAM-Modified Colloidal Nanopores

We modified the surface of nanopores in colloidal films assembled from 205 nm silica spheres, with poly(acrylamide) brushes using surface-initiated ATRP. First, silica surface was modified with initiator moieties (Scheme 5.5), followed by the polymerization of acrylamide in the presence of copper(I) chloride and bipyridine (Scheme 5.5). We found that the colloidal crystal lattice remained unperturbed by the polymerization. The polymer brush thickness could be controlled by polymerization time and was monitored by measuring the flux of redox species across the nanopores using cyclic voltammetry. The nanopore size and polymer brush thickness were calculated based on the limiting current change. Polymer brush thickness increased in the course of 26 h of polymerization in a logarithmic manner from 1.3 to 8.5 nm leading to nanopores as small as 7.5 nm.[55]

5.8.2 PNIPAM-Modified Colloidal Nanopores

We prepared polymer brushes of poly(*N*-isopropylacrylamide) (PNIPAM), a well-known temperature responsive polymer,[56] inside the colloidal nanopores using surface-initiated ATRP, as described above and measured the temperature response for the membranes modified at different polymerization time. As can be seen in Fig. 5.12, we observed two types of response.[57] For nanoporous films modified with a thin polymer brush, the limiting current increased with temperature (Fig. 5.12A), while for colloidal films modified with a thick polymer brush a reverse change was observed, where the limiting current decreases with increasing

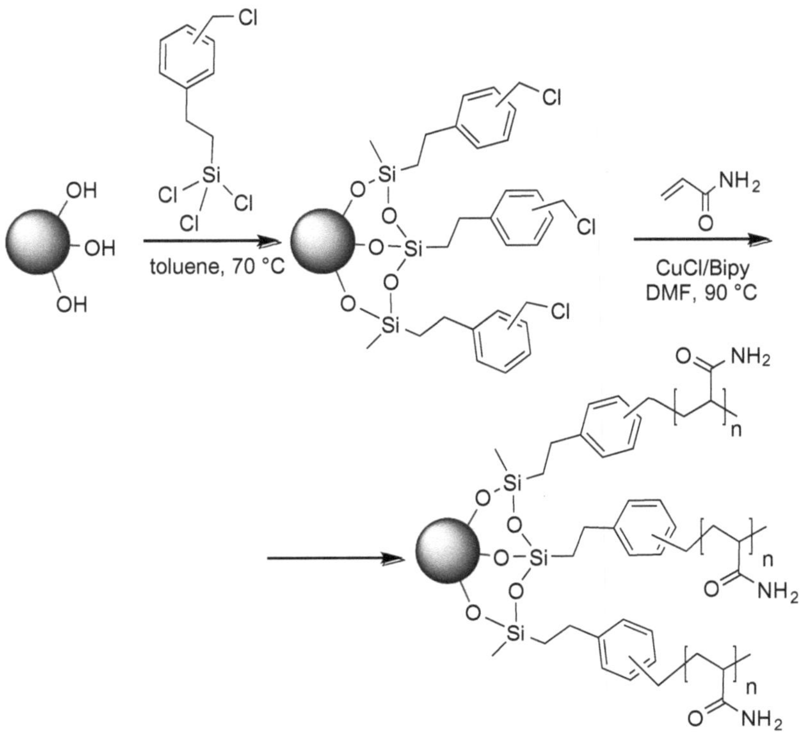

SCHEME 5.5 Preparation of PAAM-modified colloidal nanopores.

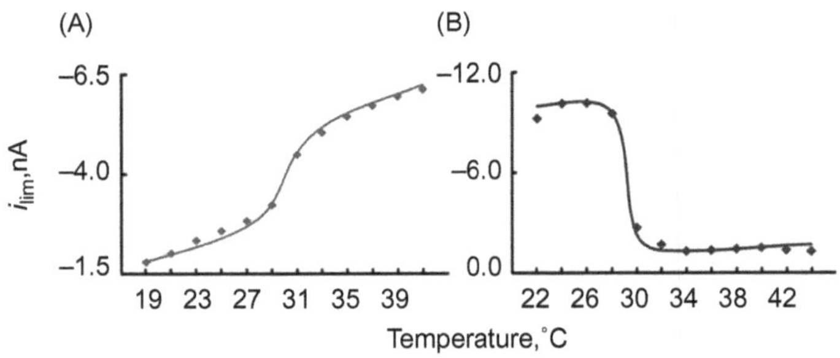

FIGURE 5.12 Limiting current ($Ru(NH_3)_6^{3+}$) as a function of increasing temperature for PNIPAM-opal film Pt electrodes after polymerization for (A) 15 min and (B) 90 min. Source: *Reproduced with permission from reference Schepelina, O.; Zharov, I. PNIPAAM-Modified Nanoporous Colloidal Films with Positive and Negative Temperature Gating. Langmuir 2007, 23, 12704–12709. Copyright 2007 American Chemical Society.*

temperature (Fig. 5.12B). These results are consistent with two types of PNIPAM morphologies inside the nanopores, which lead to two types of molecular transport mechanisms. We believe that when *N*-isopropylacrylamide is polymerized for a short period of time, it forms a dense brush (Fig. 5.13A) and the transport happens in the polymer-free volume of the nanopores. With increasing temperature, the conformation of the polymer chains changes and the brush shrinks, providing a larger volume for diffusion (Fig. 5.13A), resulting in the observed increase

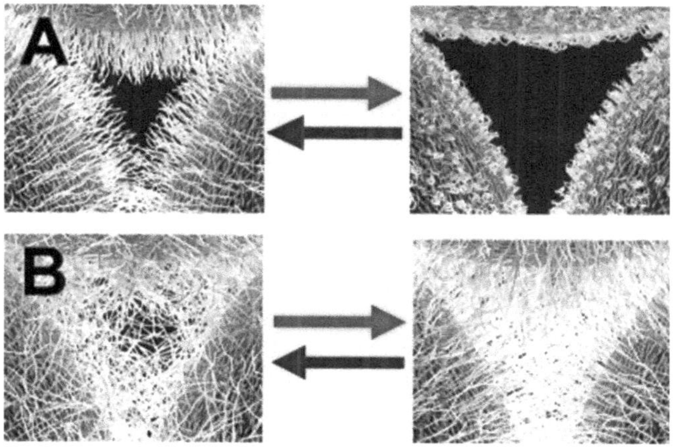

FIGURE 5.13 Schematic representation of the processes that occur upon heating and cooling of a (A) PNIPAM brush (15 min polymerization) and (B) PNIPAM gel (90 min polymerization) inside a colloidal nanopore. Source: *Reproduced with permission from reference Schepelina, O.; Zharov, I. PNIPAAM-Modified Nanoporous Colloidal Films with Positive and Negative Temperature Gating.* Langmuir ***2007***, 23, *12704–12709. Copyright 2007 American Chemical Society.*

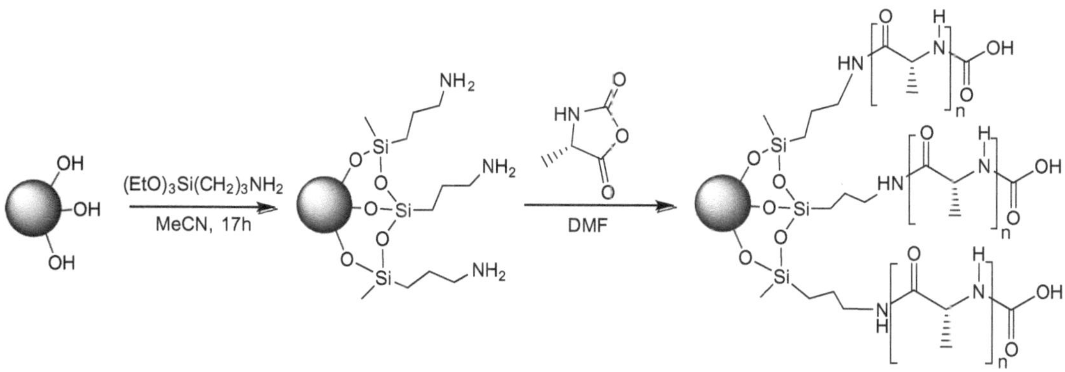

SCHEME 5.6 Preparation of poly(L-alanine)-modified nanopores.

in molecular flux. When polymerization is conducted for a long time, polymer chains become long enough to meet and interpenetrate, leading to a highly porous and permeable hydrogel structure (Fig. 5.13B). When the temperature is increased, the hydrogel does not shrink to open the nanopores but become dehydrated and impermeable (Fig. 5.13B).

5.8.3 Polyalanine-Modified Colloidal Nanopores

We modified the nanopore surface with another temperature-responsive polymer, polyalanine.[58] Poly(L-alanine) brushes of different length were grown on the surface of the nanopores in silica colloidal films (Scheme 5.3) assembled on Pt electrodes.[59]

We investigated the temperature response for poly(L-alanine)-modified colloidal films using cyclic voltammetry and a neutral probe molecule, $Fc(CH_2OH)_2$ (Scheme 5.6). As can

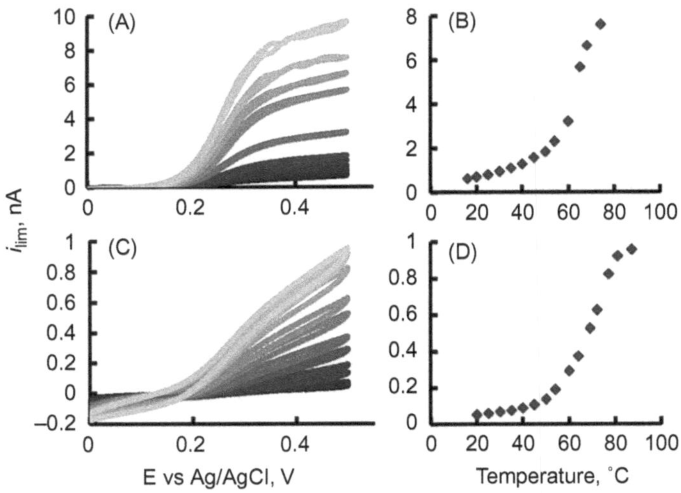

FIGURE 5.14 (A–C) Representative voltammetric responses, and (B–D) plots of $Fc(CH_2OH)_2$ limiting current as a function of temperature for poly(L-alanine)-colloidal membrane Pt electrodes after polymerization for 1 and 3 h, respectively. Source: *Reproduced with permission from reference Abelow, A.E.; Zharov, I. Poly(L-alanine)-Modified Nanoporous Colloidal Films. Soft Matter 2009, 5, 457–462. Copyright 2008 Royal Society of Chemistry.*

be seen in Fig. 5.14, for nanoporous films modified with a thinner polymer brush the limiting current increased with increasing temperature, with a transition temperature of ca. 65°C (Fig. 5.14A and B). For colloidal films modified with thicker polymer brushes, a temperature response was observed at a higher temperature of ca. 75°C (Fig. 5.14C and D). The temperature-dependent change in the limiting current was reversible for both types of polymer brushes.

5.8.4 PDMAEMA-Modified Colloidal Nanopores

We prepared colloidal nanopores modified with polymer brushes of 2-(dimethylamino)ethyl methacrylate (DMAEMA),[60] an environmentally responsive polymer whose behavior is governed by both electrostatic and hydrophobic interactions. The silica surface was aminated, followed by the reaction with 2-bromoisobutyryl bromide, followed by the ATRP of DMAEMA. First, we studied the pH response for PDMAEMA-modified colloidal films using positively charged $Ru(NH_3)_6^{3+}$. As can be seen in Figs. 5.15 and 5.16, the limiting current for PDMAEMA-modified colloidal film electrodes is highly dependent on pH. The limiting current increases with increasing pH by ~80% with an abrupt change at pH ~ 4–5. At high pH, the amine groups of the polymer chains are deprotonated, and the polymer is considered to be neutral. As a result, the polymer chains attain a collapsed conformation due to the hydrophobic interactions, and the diffusion of the positively charged $Ru(NH_3)_6^{3+}$ is not sterically hindered by the polymer brush, nor it is repelled electrostatically. In contrast, at low pH, the polymer brush becomes protonated and stretches away from the surface because of electrostatic repulsions between the charged monomer units and polymer chains.

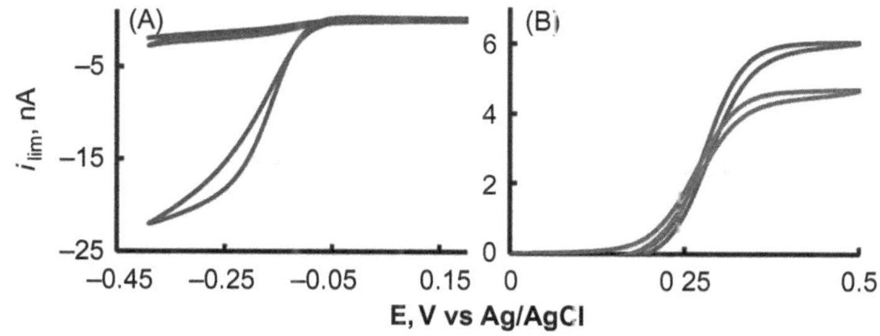

FIGURE 5.15 Representative voltammetric responses for PDMAEMA-colloidal film Pt electrodes (20 h polymerization) at different pH for $Ru(NH_3)_6^{3+}$ (A) and for $Fc(CH_2OH)_2$ (B). Voltammograms recorded above pH 5 are at the bottom in (A) and at the top in (B). Voltammograms recorded below pH 4 are at the top in (A) and at the bottom in (B). Source: *Reproduced with permission from reference Schepelina, O.; Zharov, I. Poly(2-(dimethylamino)ethyl methacrylate)-Modified Nanoporous Colloidal Films with pH and Ion Response. Langmuir 2008, 24, 14188–14194. Copyright 2008 American Chemical Society.*

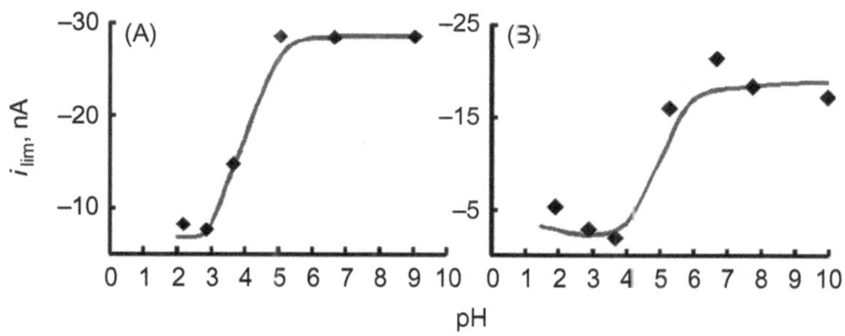

FIGURE 5.16 Limiting current ($Ru(NH_3)_6^{3+}$) as a function of increasing pH for PDMAEMA-colloidal membrane Pt electrodes for 5 h (A) and for 20 h (B) polymerizations. Source: *Reproduced with permission from reference Schepelina, O.; Zharov, I. Poly(2-(dimethylamino)ethyl methacrylate)-Modified Nanoporous Colloidal Films with pH and Ion Response.* Langmuir **2008**, 24, *14188–14194. Copyright 2008 American Chemical Society.*

The diffusion of $Ru(NH_3)_6^{3+}$ in this case is blocked as a result of both electrostatic repulsion from the positively charged polymer chains and the steric hindrance.

To isolate the pH effect on the polymer conformation, we examined the diffusion of a neutral redox-active molecule, $Fc(CH_2OH)_2$. The limiting current of $Fc(CH_2OH)_2$ decreased only by ~30% (Fig. 5.15B) with no abrupt change when pH was lowed from neutral to acidic, which should result exclusively from the conformational changes in the PDMAEMA chains.

The influence of the solution ionic strength on the diffusion across the protonated PDMAEMA-modified colloidal membranes was studied by measuring the limiting current of $Ru(NH_3)_6^{3+}$ and $Fc(CH_2OH)_2$ as a function of KCl concentration. The limiting current increased with increasing KCl concentration. The addition of KCl progressively screens the charge within the polymer brush. As a result, the diffusion of $Ru(NH_3)_6^{3+}$ becomes easier. No significant effect of the salt concentration on the diffusion of $Fc(CH_2OH)_2$ at low pH was observed, which suggests that, under these conditions, the conformation of the PDMAEMA

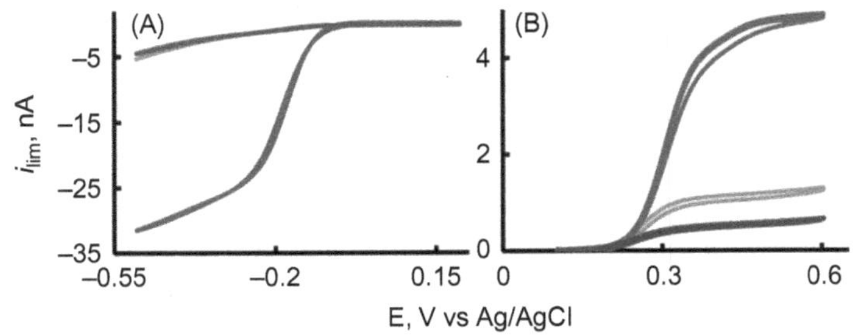

FIGURE 5.17 Representative voltammetric responses of PDMAEMA-colloidal film Pt electrodes for (A) $Ru(NH_3)_6^{3+}$ before quaternization (bottom), after quaternization (top), and after quaternization in the presence of 0.5 M KCl (top) and (B) $Fc(CH_2OH)_2$ before quaternization (top), after quaternization (bottom), and after quaternization in the presence of 0.5 M KCl (middle). Source: *Reproduced with permission from reference Schepelina, O.; Zharov, I. Poly(2-(dimethylamino)ethyl methacrylate)-Modified Nanoporous Colloidal Films with pH and Ion Response.* Langmuir ***2008***, 24, *14188–14194. Copyright 2008 American Chemical Society.*

chains is not significantly affected by the salt concentration. Both pH- and ion-dependent changes in limiting current were fully reversible.

To demonstrate that the "weak" polyelectrolyte PDMAEMA can be converted into a "strong" polyelectrolyte brush with a fixed, pH-independent number of charges, we treated the polymer with ethyl bromide to quaternize the amine groups. The limiting current of both neutral and positively charged species decreased significantly after the quaternization (Fig. 5.17). This suggests that the polymer brush carrying a large number of quaternary ammonium groups blocks the diffusion of ions and molecules almost completely regardless of pH because of both strong electrostatic repulsion and steric hindrance. At 0.5 M concentration of KCl, no significant change in the limiting current for $Ru(NH_3)_6^{3+}$ (Fig. 5.17A) was observed, indicating that the positive charge of the quaternized polymer brush was not significantly screened under these conditions. At the same time, for $Fc(CH_2OH)_2$ a ~10% increase in the limiting current was observed (Fig. 5.17B), suggesting small conformational changes in the polymer chains.

We also prepared PDMAEMA-modified free-standing colloidal membranes.[61] We found that the M_w of PDMAEMA after 16 h of polymerization was ~20 KDa, after 22 h ~23 KDa, and after 44 h ~31 KDa. We used $Fe(bpy)_3^{2+}$ diffusion to determine the responsive behavior of the PDMAEMA-modified membranes. Fig. 5.18 shows the diffusion rate of $Fe(bpy)_3^{2+}$ with and without trifluoroacetic acid present in solution through the PDMAEMA membranes. For the membrane modified with the polymer for 16 h, the diffusion rate of $Fe(bpy)_3^{2+}$ was 42% (2.4 times) lower in the presence of trifluoroacetic acid than in the absence of the acid (Fig. 5.18A). After a longer polymerization time (22 h), the acid-controlled response of the membrane increased greatly (Fig. 5.18B) and the diffusion blockage reached 95%. The switching behavior of the PDMAEMA-modified colloidal membranes was reversible. In addition to the polymer length, increasing the membrane thickness as well as decreasing the nanopore size enhanced the gating behavior and allowed to achieve a complete acid-controlled gating (Fig. 5.19).

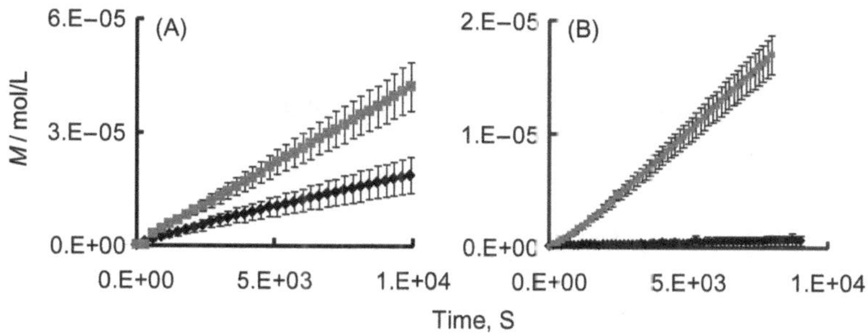

FIGURE 5.18 Diffusion rates of $Fe(bpy)_3^{2+}$ through PDMAEMA-modified colloidal membranes with (black) and without (gray) 50 mM trifluoroacetic acid after (A) 16 h and (B) 22 h of polymerization. Source: *Reproduced with permission from reference Schepelina, O.; Poth, N.; Zharov, I. pH-Responsive Nanoporous Silica Colloidal Membranes. Adv. Funct. Mater. 2010, 20, 1962–1969. Copyright 2010 Wiley, Inc.*

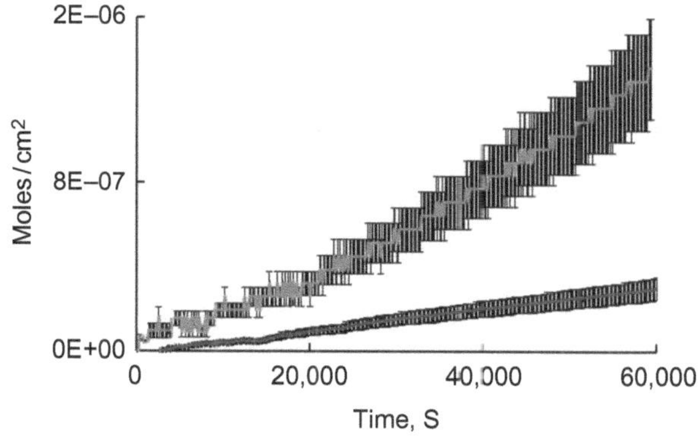

FIGURE 5.19 Representative plots of the molecular flux of ferrocene carboxaldehyde through the SiO_2@Au membranes modified with PMAA via ATRP for 45 min (with TFA, top; without TFA, bottom). Source: *Reproduced with permission from reference Ignacio-de Leon, P.A.; Zharov, I. SiO2@Au Core-Shell Nanospheres Self-Assemble to Form Colloidal Crystals That Can Be Sintered and Surface Modified to Produce pH-Controlled Membranes. Langmuir 2013, 29, 3749–3756. Copyright 2013 American Chemical Society.*

5.8.5 PMAA-Modified Colloidal Nanopores

We modified the surface of gold-coated nanopores with poly(methacrylic acid) (MAA), prepared by surface-initiated ATRP (Scheme 5.7).[39] We studied the flux of a neutral dye through the PMAA-modified nanopores and found that it increased 13-fold upon addition of an acid because the protonated polymer collapsed onto the nanopore surface leading to larger pore size.

5.8.6 PSPM- and PSSA-Modified Colloidal Nanopores

We prepared sintered silica colloidal membranes modified with poly(3-sulfopropyl-methacrylate) (PSPM) and poly(styrenesulfonic acid) (PSSA) brushes covalently attached

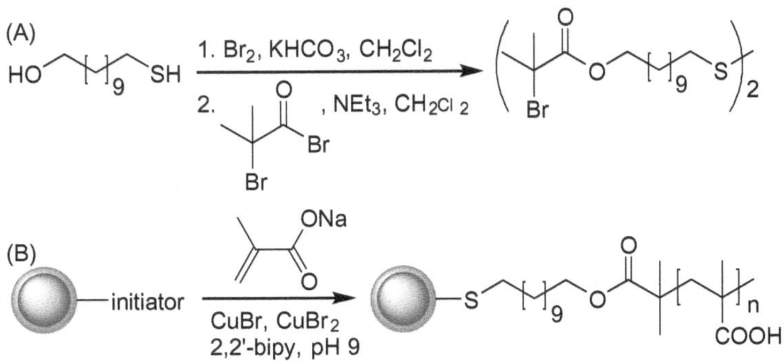

SCHEME 5.7 (A) Synthesis of the initiator molecule for (B) the ATRP of MAA on the surface of the Au-coated SiO_2 nanopores.

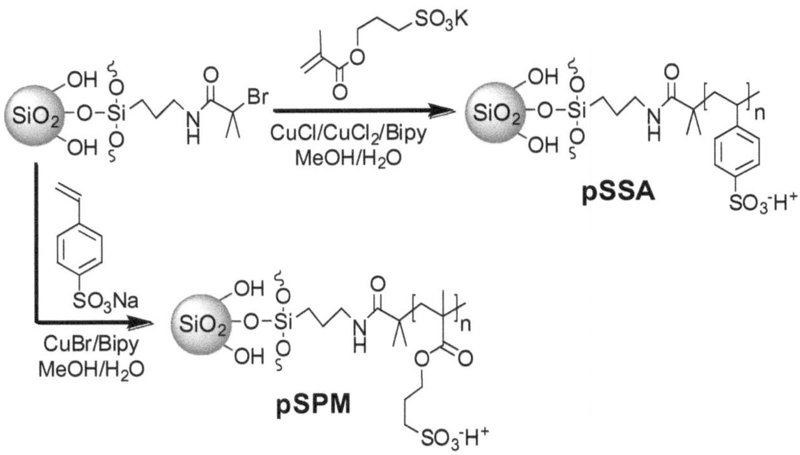

SCHEME 5.8 Preparation of PSPM and PSSA brushes on silica surface.

to the nanopore surface.[62] The sintered membranes ~300 μm thick and ~1 cm across were modified with PSPM- and PSSA-sulfonated polymer brushes by surface-initiated ATRP (Scheme 5.8). Fig. 5.20 shows an SEM image of a pore-filled sintered membrane, comprised of ~300 nm diameter silica spheres. Thermogravimetric analysis (TGA) was used to characterize the polymer-modified sintered membranes. The polymer weight percentage for PSPM membranes is ~8 wt% and the for PSSA membranes is ~6 wt% after 10 h of polymerization. Water uptake of the sintered membranes (Fig. 5.20) was measured by soaking the membranes in water at room temperature for 24 h. Given that the void fraction of the colloidal crystal is 26%, it appears that both polymers are significantly hydrated, with PSPM being hydrated almost completely, and PSSA falling slightly below. The water uptake of PSPM and PSSA sintered silica membranes is lower than that of Nafion (~38% water uptake measured under the same conditions[63]). The sintered polymer-modified membranes did not swell in the course of these experiments.

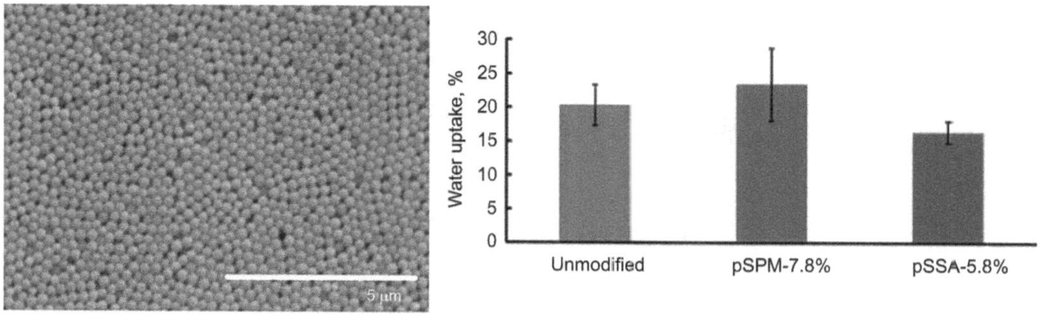

FIGURE 5.20 Left: SEM image of PSPM-filled sintered silica colloidal crystal comprised of ~300 nm diameter silica nanoparticles. Size bar = 5 μm. Right: Water uptake for pore-filled sintered silica membranes after soaking in water at room temperature for 24 h. Source: *Reproduced with permission from reference Smith, J.J.; Zharov, I. Preparation and Proton Conductivity of Self-Assembled Sulfonated Polymer-Modified Silica Colloidal Crystals. Chem. Mater. 2009, 21, 2013–2019. Copyright 2008 American Chemical Society.*

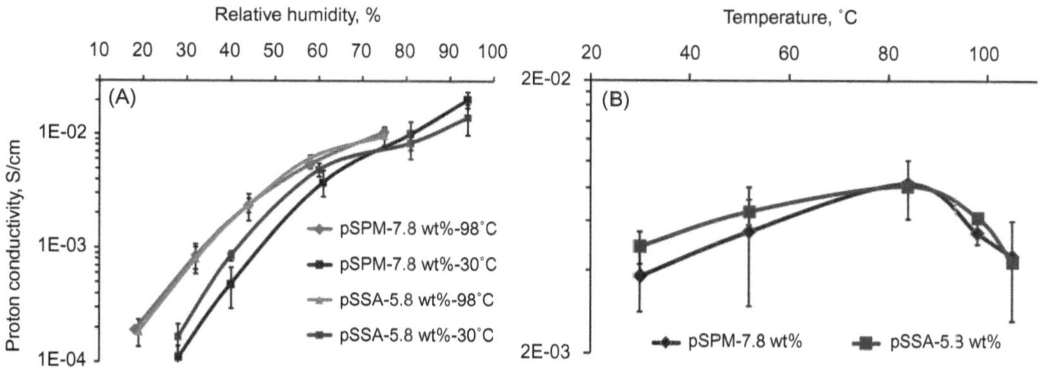

FIGURE 5.21 Proton conductivity of SPM- and SSA-sintered colloidal membranes as a function of (A) RH at 30°C and 98°C and (B) temperature at 60% RH. Source: *Reproduced with permission from reference Smith, J.J.; Zharov, I. Preparation and Proton Conductivity of Self-Assembled Sulfonated Polymer-Modified Silica Colloidal Crystals.* Chem. Mater. ***2009***, 21, *2013–2019. Copyright 2008 American Chemical Society.*

We measured the proton conductivity of the polymer-modified colloidal membranes using electrochemical impedance spectroscopy. Fig. 5.21A shows the proton conductivity of sintered PSPM and PSSA membranes at 30°C and 98°C as a function of RH. It is clear that both membranes have similar proton conductivities which increase with RH, with a maximum value of $\sim 2 \times 10^{-2}$ S/cm achieved at 30°C and 94% RH. The proton conductivity of the polymer-modified sintered colloidal membranes is affected by temperature, as shown in Fig. 5.21B. The proton conductivity for both membranes at 60% RH increases gradually with increasing temperature up to 85°C, and then decreases.

Overall, the proton conductivity values obtained for the sintered polymer-modified colloidal membranes are comparable to those of Nafion (2×10^{-3}–7×10^{-2} S/cm). The most extensively sulfonated polystyrene membranes reported (20 mol% sulfonation) had the proton conductivity of 5×10^{-2} S/cm at room temperature in their fully hydrated state.[64]

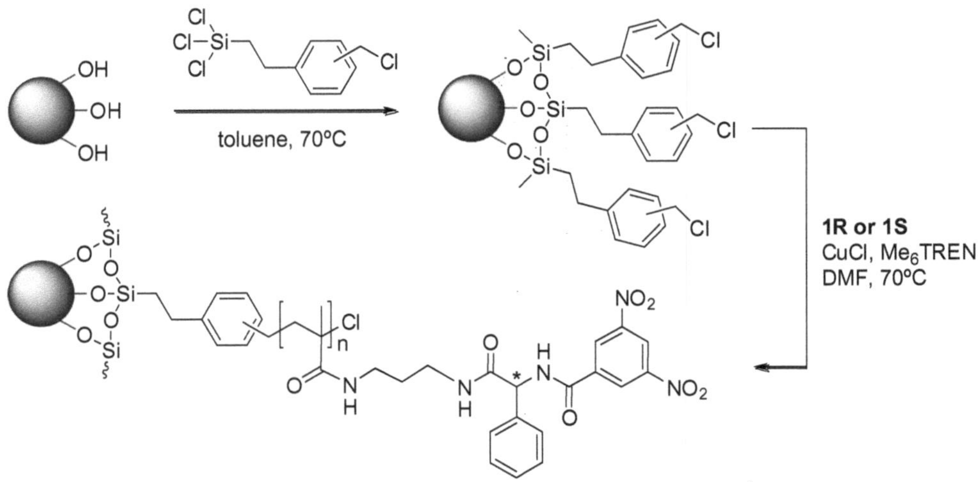

SCHEME 5.9 Preparation of chiral polymer-modified colloidal nanopores by surface-initiated ATRP.

We studied the proton conductivity of these membranes as well as open circuit voltage and linear polarization of the fuel cells prepared using these membranes as a function of sulfonic acid group content in the pore-filling polymer brushes.[65] We found a sigmoidal dependence of the proton conductivity on the amount of sulfonic acid groups. We showed that the proton conductivity of the membrane does not increase significantly after reaching ca. 75% sulfonic acid group content. The fuel cell performance, on the other hand, decreased after reaching ca. 65–70% sulfonic acid group content, which was attributed to the increased methanol permeability of the membranes at higher sulfonic acid group content (Scheme 5.9).

5.8.7 Polymers With Pendant Chiral Groups

We prepared nanoporous silica colloidal films whose nanopores were filled with polymer brushes containing chiral selector moieties in the side chains. These polymers were grown on the surface of the nanopores using surface-initiated polymerization of monomers **8–11** (Chart 5.3). We studied the enantioselective permeation in the resulting pore-filled films and found that they possess a relatively high enantioselectivity. Increasing the number of chiral selectors inside the nanopores and preventing through-solution diffusion by pore-filling with polymers did not lead to increased enantioselectivity compared to that observed for silica colloidal films whose nanopores were surface-modified with a monolayer of chiral selectors. This suggests that the enantioselectivity depends solely on the energy difference between the enantiomer and selector complexes and that through-solution diffusion plays a minor role in the transport of enantiomers, which occurs predominantly through site hopping.

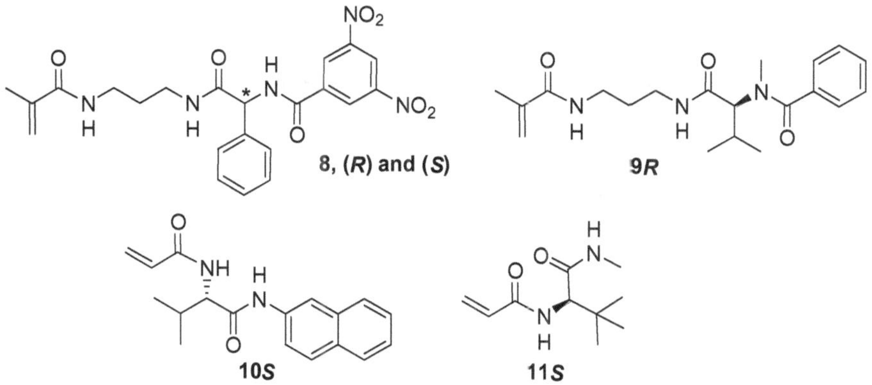

CHART 5.3 Chiral monomers used in ATRP modification of colloidal films and chiral probe used for cyclic voltammetry measurements.

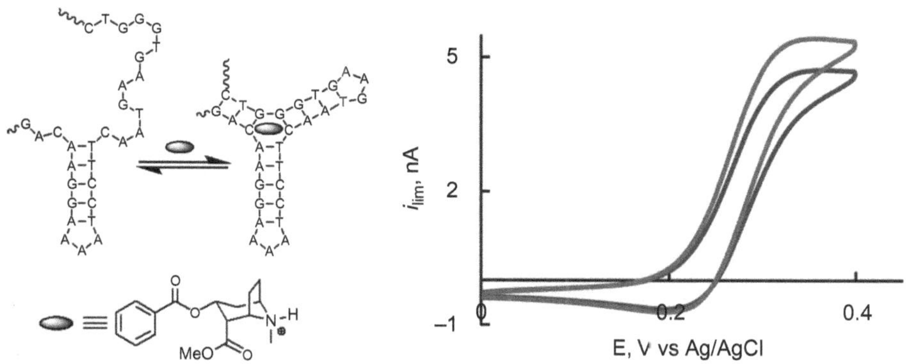

FIGURE 5.22 Left: Cocaine-sensing aptamer binding to cocaine. Right: Representative $Fc(CH_2OH)_2$ voltammetric response for an aptamer-modified opal electrode in the absence (bottom) and in the presence of cocaine (top).

We also prepared free-standing nanoporous silica colloidal membranes surface-modified with chiral polymers, prepared using monomer **8** (Chart 5.3). We measured the diffusion rates of enantiomers through these membranes and calculated the corresponding permselectivity, which was ~1.3.

5.9 Aptamer-Modified Colloidal Nanopores

We explored nanopores whose surface was modified with a polymer whose conformation responds to a small molecule binding. We used a responsive DNA aptamer, which exhibits selective and specific binding toward cocaine.[66] The secondary structure for this 32-base aptamer possesses a three-way junction, in the middle of which there is a cavity which binds the target molecule (Fig. 5.22). In the absence of a target, the aptamer is thought to remain partially unfolded, with only one of the three junctions folded.

The aptamer described above was attached to the surface of silica colloidal film via maleimide-activated chemistry. The transport rates of a redox-active probe molecule (ferrocene dimethanol) through the resulting nanoporous films was measured as a function of cocaine concentration using cyclic voltammetry. As can be seen in Fig. 5.22, the limiting current for the aptamer-modified colloidal film electrodes increased after the addition of cocaine to the solution.[67] The observed reversible 9.0 $\pm$ 3.5% change corresponds to 0.6 nm increase in effective nanopore radius. This behavior was attributed to the conformational change as described above.

We found that the change in limiting current resulting from cocaine binding was affected by the size of the nanopore. For 7.8 nm nanopores, the limiting current increase was ca. 2.6 times higher compared to 22.5 nm nanopores. This observation was rationalized by assuming that the aptamer size change remains constant regardless of the nanopore size, thus having a greater effect for the smaller nanopores.

5.10 Summary

We described the surface modification of silica colloidal nanopores and the transport properties of the corresponding nanoporous thin films and free-standing membranes. The films were formed by vertical deposition from colloidal solution onto various solid supports. They were also suspended in a glass or silicon support. Free-standing silica colloidal membranes were prepared by sintering the colloidal crystals and colloidal glasses at temperatures above 1000°C. The quality of the fcc packing depended on the thickness of the crystal, with thinner films showing much better ordering. Most importantly, in all cases the films and membranes can be prepared without mechanical defects.

We achieved surface modification of silica colloidal nanopores with small molecules using silanol chemistry and formed polymer brushes on the nanopore surface using ATRP and ring-opening polymerization. The transport in the resulting chemically modified nanopores could be controlled by pH and ionic strength, temperature, light, and small molecule binding. We also showed that silica colloidal membranes surface-modified with sulfonated polymer brushes possess high proton conductivity.

Silica colloidal films and membranes described in this chapter expand the growing family of functional inorganic nanoporous materials.[68] There are various potential applications for surface-modified silica colloidal membranes. Due to their ordered nanoporous structure and mechanical durability, they can be applied in nanofluidics, nanofiltration, fuel cells, and as catalyst support. Reversible control of transport in these membranes via external stimuli may be useful in drug-release devices; in size-, charge-, and structure-selective separations; and in microfluidic and sensing devices.

References

1. Davis, M. E. Ordered Porous Materials for Emerging Applications. *Nature* **2002,** *417,* 813–821.
2. Bayley, H.; Martin, C. R. Resistive-Pulse Sensing-From Microbes to Molecules. *Chem. Rev.* **2000,** *100,* 2575–2594.

3. Nishizawa, M.; Menon, V. P.; Martin, C. R. Metal Nanotubule Membranes with Electrochemically Switchable Ion-Transport Selectivity. *Science* **1995,** *268,* 700–702.
4. van Reis, R.; Zydney, A. Bioprocess Membrane Technology. *J. Membr. Sci.* **2007,** *297,* 16–50.
5. Afonso, C. A. M.; Crespo, J. G. Recent Advances in Chiral Resolution through Membrane-Based Approaches. *Angew. Chem. Int. Ed.* **2004,** *43,* 5293–5295.
6. Piruska, A.; Gong, M.; Sweedler, J. V.; Bohn, P. W. Nanofluidics in Chemical Analysis. *Chem. Soc. Rev.* **2010,** *39,* 1060–1072.
7. Orosz, K. E.; Gupta, S.; Hassink, M.; Abdel-Rahman, M.; Moldovan, L.; Davidorf, F. H., et al. Delivery of Antiangiogenic and Antioxidant Drugs of Ophthalmic Interest through a Nanoporous Inorganic Filter. *Mol. Vision* **2004,** *10,* 555–565.
8. Kipke, S.; Schmid, G. Nanoporous Alumina Membranes as Diffusion Controlling Systems. *Adv. Funct. Mater.* **2004,** *14,* 1184–1188.
9. Santini, J. T., Jr.; Cima, M. J.; Langer, R. A Controlled-Release Microchip. *Nature* **1999,** *397,* 335–338.
10. Ito, T.; Sato, Y.; Yamaguchi, T.; Nakao, S. Response Mechanism of a Molecular Recognition Ion Gating Membrane. *Macromolecules* **2004,** *37,* 3407–3414.
11. Liu, G.; Lu, Z.; Duncan, S. Porous Membranes of Polysulfone-*graft*-poly(tert-butylacrylate) and Polysulfone-*graft*-poly(acrylic acid): Morphology, pH-Gated Water Flow, Size Selectivity, and Ion Selectivity. *Macromolecules* **2004,** *37,* 4218–4226.
12. Ulbricht, M. Advanced Functional Polymer Membranes. *Polymer* **2006,** *47,* 2217–2262.
13. Kallus, S.; Condre, J.-M.; Hahn, A.; Golemme, G.; Algieri, C.; Dieudonne, P., et al. Colloidal Zeolites and Zeolite Membranes. *J. Mater. Chem.* **2002,** *12,* 3343–3350.
14. Tong, H. D.; Jansen, H. V.; Gadgil, V. J.; Bostan, C. G.; Berenschot, C. G. E.; van Rijn, C. J. M., et al. Silicon Nitride Nanosieve Membrane. *Nano Lett.* **2004,** *4,* 283–287.
15. Striemer, C. C.; Gaborski, T. R.; McGrath, J. L.; Fauchet, P. M. Charge- and Size-Based Separation of Macromolecules Using Ultrathin Silicon Membranes. *Nature* **2007,** *445,* 749–753.
16. Liu, N. G.; Dunphy, D. R.; Atanassov, P.; Bunge, S. D.; Chen, Z.; Lopez, G. P., et al. Photoregulation of Mass Transport through a Photoresponsive Azobenzene-Modified Nanoporous Membrane. *Nano Lett.* **2004,** *4,* 551–554.
17. Yamaguchi, A.; Uejo, F.; Yoda, T.; Uchida, T.; Tanamura, Y.; Yamashita, T., et al. Self-Assembly of a Silica-Surfactant Nanocomposite in a Porous Alumina Membrane. *Nat. Mater.* **2004,** *3,* 337–341.
18. Blanco, A.; Chomski, E.; Grabtchak, S.; Ibisate, M.; John, S.; Leonard, S. W., et al. Large-Scale Synthesis of a Silicon Photonic Crystal with a Complete Three-Dimensional Bandgap Near 1.5 micrometres. *Nature* **2000,** *405,* 437–440.
19. Bartlett, P. N.; Ghanem, M. A.; Hallag, E.; De Groot, P.; Zhukov, A. Electrochemical Deposition of Macroporous Magnetic Networks Using Colloidal Templates. *J. Mater. Chem.* **2003,** *13,* 2596–2602.
20. Park, S. H.; Xia, Y. Macroporous Membranes with Highly Ordered and Three-Dimensionally Interconnected Spherical Pores. *Adv. Mater.* **1999,** *10,* 1045–1048.
21. Cassagneau, S.; Caruso, F. Semiconducting Polymer Inverse Opals Prepared by Electropolymerization. *Adv. Mater.* **2002,** *14,* 34–38.
22. Wong, S.; Kitaev, V.; Ozin, G. A. Colloidal Crystal Films: Advances in Universality and Perfection. *J. Am. Chem. Soc.* **2003,** *125,* 15589–15598.
23. Crank, J. *The Mathematics of Diffusion,* 2nd ed.; Oxford University Press: New York, 1975.
24. Brenner, H.; Edwards, D. A. *Macrotransport Processes;* Butterworth-Heinemann: Boston, MA, 1993.
25. Cussler, E. L. *Diffusion. Mass Transfer in Fluid Systems,* 2nd ed.; Cambridge University Press, 1997.

26. Stoeber, W.; Fink, A.; Bohn, E. Controlled Growth of Monodisperse Silica Spheres in the Micron Size Range. *J. Colloid Interface Sci.* **1968,** *26,* 62–69.

27. Jiang, P.; Bertone, J. F.; Hwang, K. S.; Colvin, V. L. Single-Crystal Colloidal Multilayers of Controlled Thickness. *Chem. Mater.* **1999,** *11,* 2132–2140.

28. Onclin, S.; Ravoo, B. J.; Reinhoudt, D. N. Engineering Silicon Oxide Surfaces Using Self-Assembled Monolayers. *Angew. Chem. Int. Ed.* **2005,** *44,* 6282–6304.

29. Flink, S.; van Veggel, F. C. J. M.; Reinhoudt, D. N. *J. Phys. Org. Chem.* **2001,** *14,* 407–415.

30. Kanoh, N.; Kumashiro, S.; Simizu, S.; Kandoh, Y.; Hatakeyama, S.; Tashiro, H., et al. *Angew. Chem. Int. Ed.* **2003,** *42,* 5584–5587.

31. Matyjaszewski, K.; Xia, J. Atom Transfer Radical Polymerization. *Chem. Rev.* **2001,** *101,* 2921–2990.

32. Newton, M. R.; Bohaty, A. K.; White, H. S.; Zharov, I. Chemically Modified Opals as Thin Permselective Nanoporous Membranes. *J. Am. Chem. Soc.* **2005,** *127,* 7268–7269.

33. Bohaty, A. K.; Zharov, I. Suspended Self-Assembled Opal Membranes. *Langmuir* **2006,** *22,* 5533–5536.

34. Bohaty, A.; Abelow, A. E.; Zharov, I. Nanoporous Silica Colloidal Membranes Suspended in Glass. *J. Porous Mater.* **2011,** *18,* 297–304.

35. Bohaty, A. K.; Smith, J. J.; Zharov, I. Free-Standing Silica Colloidal Nanoporous Membranes. *Langmuir* **2009,** *25,* 3096–3101.

36. Ignacio-de Leon, P. A.; Zharov, I. Size-Selective Transport in Colloidal Nano-Frits. *Chem. Commun.* **2011,** *47,* 553–555.

37. Jal, P. K.; Patel, S.; Mishra, B. K. *Talanta* **2004,** *62,* 1005–1028.

38. Khabibullin, A.; Zharov, I. Nanoporous Membranes with Tunable Pore Size by Pressing/Sintering Silica Colloidal Spheres. *ACS Appl. Mater. Interfaces* **2014,** *6,* 7712–7718.

39. Ignacio-de Leon, P. A.; Zharov, I. SiO2@Au Core-Shell Nanospheres Self-Assemble to Form Colloidal Crystals That Can Be Sintered and Surface Modified to Produce pH-Controlled Membranes. *Langmuir* **2013,** *29,* 3749–3756.

40. Newton, M. R.; Bohaty, A. K.; Zhang, Y.; White, H. S.; Zharov, I. pH and Ionic Strength Controlled Cation Permselectivity in Amine-Modified Nanoporous Opal Films. *Langmuir* **2006,** *22,* 4429–4432.

41. Smith, J. J.; Zharov, I. Ion Transport in Sulfonated Nanoporous Colloidal Films. *Langmuir* **2008,** *24,* 2650–2654.

42. Smith, J. J.; Abbaraju, R. R.; Zharov, I. Proton Transport in Assemblies of Silica Colloidal Spheres. *J. Mater. Chem.* **2008,** *18,* 5335–5338.

43. Teuka, T.; Tadanaga, K.; Hayashi, A.; Tatsumisago, M. *J. Am. Chem. Soc.* **2006,** *128,* 16470–16471.

44 Gao, G.; Yang, Y.; Lee, D.; Holdcroft, S.; Frisken, B. *Macromolecules* **2006,** *39,* 8060–8066.

45. Subbaraman, R.; Ghassemi, H.; Zawodzinski, T. A., Jr. *J. Am. Chem. Soc.* **2007,** *129,* 2238–2239.

46. Bohaty, A. K.; Newton, M. R.; Zharov, I. Light-Controlled Ion Transport Through Spiropyran-Modified Nanoporous Silica Colloidal Films. *J. Porous Mater.* **2010,** *17,* 465–473.

47. Bertelson, R. C. Spiropyrans. In *Organic Photochromic and Thermochromic Compounds;* Crano, J. C., Guglielmetti, R. J., Eds.; Plenum Publishing Corp.: New York, 1999.

48. Fissi, A.; Pieroni, O.; Ruggeri, G.; Ciardelli, F. *Macromolecules* **1995,** *28,* 302–309.

49. Cichelli, J.; Zharov, I. Chiral Selectivity in Surface-Modified Porous Colloidal Films. *J. Am. Chem. Soc.* **2006,** *128,* 8130–8131.

50. Lee, S. B.; Mitchell, D. T.; Trofin, L.; Nevanen, T. K.; Söderlund, H.; Martin, C. R. Antibody-Based Bio-Nanotube Membranes for Enantiomeric Drug Separations. *Science* **2002,** *296,* 2198–2201.

51. Rmaile, H. H.; Schlenoff, J. B. Optically Active Polyelectrolyte Multilayers as Membranes for Chiral Separations. *J. Am. Chem. Soc.* **2003,** *125,* 6602–6603.
52. Mollard, A.; Ibragimova, D.; Antipin, I. S.; Stoikov, I. I.; Zharov, I. Molecular Transport in Thiacalixarene-Modified Nanoporous Colloidal Films. *Micropor. Nanopor. Mater.* **2010,** *131,* 378–384.
53. Cichelli, J.; Zharov, I. Chiral Permselectivity in Nanoporous Opal Films Surface-Modified with Chiral Selector Moieties. *J. Mater. Chem.* **2007,** *17,* 1870–1875.
54. Mollard, A.; Ibragimova, D.; Antipin, I. S.; Stoikov, I. I.; Zharov, I. Molecular Transport in Thiacalixarene-Modified Nanoporous Colloidal Films. *Micropor. Mesopor. Mater.* **2010,** *131,* 378–384.
55. Schepelina, O.; Zharov, I. Polymer-Modified Opal Nanopores. *Langmuir* **2006,** *22,* 10523–10527.
56. Heskins, M.; Guillet, J. E. Solution Properties of Poly(*N*-isopropylacrylamide). *J. Macromol. Sci. Chem. A.* **1968,** *2,* 1441–1455.
57. Schepelina, O.; Zharov, I. PNIPAAM-Modified Nanoporous Colloidal Films with Positive and Negative Temperature Gating. *Langmuir* **2007,** *23,* 12704–12709.
58. Mart, R. J.; Osborne, R. D.; Stevens, M. M.; Ulijn, R. V. Peptide-Based Stimuli-Responsive Biomaterials. *Soft Matter* **2006,** *2,* 822–835.
59. Abelow, A. E.; Zharov, I. Poly(L-alanine)-Modified Nanoporous Colloidal Films. *Soft Matter* **2009,** *5,* 457–462.
60. Schepelina, O.; Zharov, I. Poly(2-(dimethylamino)ethyl methacrylate)-Modified Nanoporous Colloidal Films with pH and Ion Response. *Langmuir* **2008,** *24,* 14188–14194.
61. Schepelina, O.; Poth, N.; Zharov, I. pH-Responsive Nanoporous Silica Colloidal Membranes. *Adv. Funct. Mater.* **2010,** *20,* 1962–1969.
62. Smith, J. J.; Zharov, I. Preparation and Proton Conductivity of Self-Assembled Sulfonated Polymer-Modified Silica Colloidal Crystals. *Chem. Mater.* **2009,** *21,* 2013–2019.
63. Pereira, F.; Vallé, K.; Belleville, P.; Morin, A.; Lambert, S.; Sanchez, C. *Chem. Mater.* **2008,** *20,* 1710–1718.
64. Carreta, N.; Tricoli, V.; Picchioni, P. *J. Membr. Sci.* **2000,** *166,* 189–197
65. Khabibullin, A.; Minteer, S. D.; Zharov, I. The Effect of Sulfonic Acid Group Content in Pore-Filled Silica Colloidal Membranes on Their Proton Conductivity and Direct Methanol Fuel Cell Performance. *J. Mater. Chem. A.* **2014,** *2,* 12761–12769.
66. Stojanovic, M. N.; Prada, P.; Landry, D. W. Aptamer-Based Folding Fluorescent Sensor for Cocaine. *J. Am. Chem. Soc.* **2001,** *123,* 4928–4931.
67. Abelow, A. E.; White, R. J.; Plaxco, K. W.; Zharov, I. Nanoporous Silica Colloidal Films with Molecular Transport Gated by Aptamers Responsive to Small Molecules. *Coll. Czech Chem. Commun.* **2011,** *76,* 683–694.
68. Abelow, A. E.; Zharov, I. Reversible Nanovalves in Inorganic Materials. *J. Mater. Chem.* **2012,** *22,* 21810–21818.

6

Advanced Modeling of Ion Transport in Polymer and Polyelectrolyte-Modified Nanochannels and Nanopores

M. Tagliazucchi[1], I. Szleifer[2]

[1]UNIVERSIDAD DE BUENOS AIRES–CONICET, BUENOS AIRES, ARGENTINA
[2]NORTHWESTERN UNIVERSITY, EVANSTON, IL, UNITED STATES

CHAPTER OUTLINE

Chemically Modified Nanopores and Nanochannels.

6.1 Introduction

In this chapter, we introduce the reader to a molecular theory that we developed to model chemically modified nanopores and nanochannels. There are a variety of different theoretical approaches that can be used, and have been presented in the literature, to study transport in nanochannels. Thus, the first question that we need to address is why there is a need for a new theory. In order to answer this question, let us consider the different modeling approaches that can be used to describe transport in nanopores and nanochannels (Scheme 6.1). On the one hand, we have particle-based approaches such as molecular dynamics (MD) and Brownian dynamics simulations, which can be either atomistic (all atoms in the system are explicitly excluded) or coarse grained (atoms are grouped together into coarse-grained particles). Atomistic MD is the most detailed approach available to study ion transport in nanopores. This approach has been applied to biological nanopores and nanochannels. However, the length and time scales of interest in solid-state chemically modified nanopores and nanochannels are well beyond those of biological channels and those that can be studied with atomistic MD simulations. For example, the prototypical biological pore, α-hemolysin, has a radius of ~1.3 nm and a length of ~10 nm. If filled with a 1.0 M electrolyte solution, α-hemolysin contains ~60 salt ions and ~1700 water molecules. On the other hand, a nanopore in a Si_3N_4 membrane with a radius of 6 nm and a length of 50 nm[1] filled with 1.0 M KCl contains $\sim 2 \times 10^5$ water molecules and 7×10^4 salt ions (and the situation only gets worse if we consider long nanochannels). For this number of molecules, systematic studies are extremely time consuming (if not impossible) to simulate with atomistic detail even with the availability of supercomputers. On the other hand, solid-state

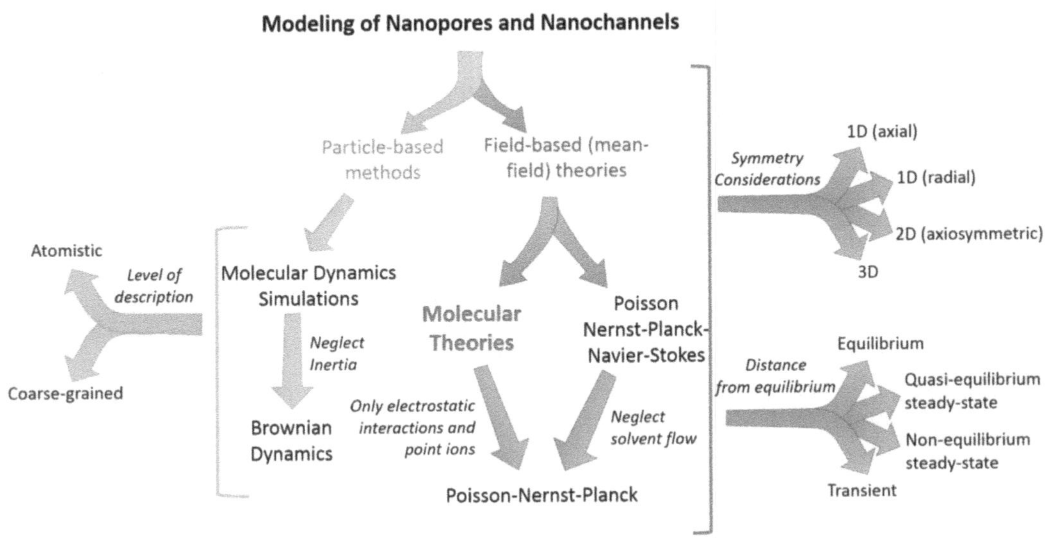

SCHEME 6.1 Roadmap of modeling approaches for nanochannels and nanopores. The approach described in this chapter belongs to the Molecular Theories group.

nanopores and channels can be described in the continuum limit by using field-based (mean-field) methods. For example, the Poisson–Boltzmann (PB) equation for equilibrium systems and the Poisson–Nernst–Planck–Navier–Stokes (PNP-NS) equations in nonequilibrium conditions (see chapter: Theoretical Basis for Structure and Transport in Nanopores and Nanochannels). These continuum approaches have been shown to be valid for channel radii larger than 2 Debye lengths[2,3] and are relatively easy to solve for a variety of conditions. However, they treat the ions as point charges and do not incorporate any molecular details and, therefore, they cannot describe the behavior of polyelectrolytes or bound proteins within chemically modified pores and channels. In order to treat specific molecular systems and to capture the coupling that exits between molecular organization, physical interactions, and chemical state, we have developed a theory that provides a continuum description in terms of the density distributions of the different molecular species as well as the interaction fields and explicitly includes the size, shape, charge distribution, chemical state, and conformations of all molecular species in the system. This molecular theory is an enabling tool to study chemically modified nanochannels and nanopores because it includes chemical details of the molecules in the system, can handle the appropriate length and temporal scales, and can be derived within a formal framework, which allows it to be generalized to many different cases of interest. As we will show in this chapter, the molecular theory reduces to the Poisson–Nernst–Planck equations when only electrostatic interactions are considered and there are no macromolecular species in the system (i.e., there are no molecules with conformational degrees of freedom).

For a given field-based modeling approach, one can resort to different approximations depending on the symmetry of the system and its distance to thermodynamic equilibrium (see Scheme 6.1). Commonly, nanochannels and nanopores have cylindrical symmetry, hence transport through them can be described as a two-dimensional (2D) axiosymmetric problem, where the system is assumed to be homogeneous in the angular (θ) coordinate and inhomogeneous in the axial (z) and radial (r) coordinates. For very long channels, one can assume heterogeneity in the z coordinate as well, which leads to a one-dimensional (1D) radial problem that considers inhomogeneities only in r. Assuming that the system is homogeneous in r, but not in z, leads to a 1D axial problem. Finally, in some cases none of these approximations can be used and one should solve the full three-dimensional (3D) problem.

We can also classify systems according to their distance to thermodynamic equilibrium (see Scheme 6.1) into equilibrium system, quasi-equilibrium steady-state, nonequilibrium steady-state, and transient conditions. Each of these regimes requires a specific theoretical treatment (i.e., a different "flavor" of the molecular theory), and their understanding leads to different levels of physical–chemical intuition. Firstly, we have equilibrium systems, where there are no net ionic currents flowing through the channel or pore. Second, there are steady-state quasi-equilibrium systems, where the driving force is very small and there is a vanishingly small ion current through the pore or channel. In this regime the system is in a steady state, so its properties remain constant in time. Moreover, since the fluxes are very small, they do not perturb the molecular organization of the system (quasi-equilibrium condition), thus the properties of the system (i.e., the local density of the ions and conformations of the soft materials inside the pore) are exactly the same as in equilibrium. The third case is the steady-state nonequilibrium regime, where the driving forces and currents flowing through the system are large and the organization of the system can be quite different from the equilibrium one. Some of the most intriguing effects in nanofluidics (such as current rectification and polarization concentration) occur in the nonequilibrium regime, where the flow and the molecular organization are strongly coupled. Finally, we have the transient regime where the properties of the system vary with time.

The molecular theory that we describe in this chapter was originally developed to treat surfactant self-assembly[4] and later generalized to study neutral grafted polymer layers[5] and more recently polyelectrolytes layers.[6] The basic idea of the molecular theory is to write an approximate expression for the free energy of the system of interest in terms of the spatial distributions of all the molecular species. These distributions include the position-dependent densities as well as the probabilities of the different conformations of the molecules, e.g., polymers and proteins, and the chemical state of the different species. The chemical state refers, for example, to whether an acid group is protonated or deprotonated. All the distributions are clearly position dependent and are unknown a priori. Under the conditions of thermodynamic equilibrium, the distributions of molecular species are those that minimize the model free energy. Namely, from all the possible distributions, the theory yields the one that minimizes the free energy and thus provides the equilibrium structure of the system. The molecular theory was first conceived as an equilibrium theory,[4–7] enabling the treatment of a variety of different systems, each of them requiring the proper free-energy formulation. For example, the equilibrium molecular theory can be used—among many other things—to study the binding of proteins to layers with flexible tethers.[8–10] In the linear response regime, when separation of time scales for different degrees of freedom is possible, the molecular theory can be generalized to determine transport properties of the slow degrees of freedom.[11] In order to study systems in far-from-equilibrium conditions, where the molecular organization changes due to the presence of ionic fluxes, we have recently formulated a nonequilibrium steady-state molecular theory as an extension of the equilibrium theory.[12,13] In the nonequilibrium molecular theory, the free energy of the system is minimized with respect to some degrees of freedom, which either equilibrate on a fast timescale (i.e., the electrostatic potential) or are not subjected to external gradients (i.e., the probability distribution for the conformations of grafted molecules). On the other hand, the free energy is not a minimum of the degrees of freedom that are subjected to external gradients, such as the concentration of anions and cations. These degrees of freedom develop fluxes in the system (e.g., ion fluxes), where the driving forces for these fluxes are the appropriate thermodynamic forces (e.g., the gradients of chemical potential of the ions), which are the derivatives of the free energy (e.g., derivative of the free energy with respect to the concentrations of the ions).

In this chapter, we describe the formulation and implementation of both the equilibrium (Section 6.2.1) and the nonequilibrium (Section 6.2.2) molecular theories as they apply to nanopores and nanochannels. We will also describe the application of these theories to a number of chemical modified nanopores of interest. Sections 6.3 and 6.4 report on the equilibrium structure and transport properties of polyelectrolyte-modified nanopores and nanochannels. Section 6.5 explores the binding of proteins to ligands located within long nanochannels and their effect on the ionic conductance of the channel. We conclude this chapter by summarizing our results and discussing the strength and limitations of the molecular theory as well as future challenges in modeling. Appendix I contains a list of the symbols used throughout this chapter and their meaning.

6.2 Formulation of the Molecular Theory

6.2.1 Equilibrium Molecular Theory

6.2.1.1 Formulation of the Free Energy Functional

The first step in the formulation of the molecular theory is to write the free energy functional that describes the system of interest. It is important to emphasize that nanopores impose physical constrained environments to the molecules. Therefore, the distribution (density) of each molecular species will be position dependent. For this reason, the free energy is a functional, namely, it is a function of the distribution functions from which we can write all the energetic and entropic contributions to the free energy. As an example, we consider first a nanochannel coated by an end-grafted polyelectrolyte, which is a weak polyacid, see Fig. 6.1.

The Helmholtz free energy, F, for the system in Fig. 6.1 is given by the functional:

$$
\begin{aligned}
\beta F = & \sum_{\mathrm{i=A,C,w,H^+,OH^-}} \int \rho_{\mathrm{i}}(\mathbf{r})\left[\ln(\rho_{\mathrm{i}}(\mathbf{r})v_{\mathrm{w}}) - 1 + \mu_{\mathrm{i}}^{0}\right]d\mathbf{r} \\
& + \sum_{j}\sum_{\alpha} P(j,\alpha)\ln(P(j,\alpha)) + \iint \frac{\beta\chi g(|\mathbf{r}-\mathbf{r}'|)}{2}\langle n_{\mathrm{P}}(\mathbf{r})\rangle\langle n_{\mathrm{P}}(\mathbf{r}')\rangle d\mathbf{r}d\mathbf{r}' \\
& + \int\left(\langle\rho_{\mathrm{Q}}(\mathbf{r})\rangle\beta\psi(\mathbf{r}) - \frac{1}{2}\varepsilon\beta(\nabla\psi(\mathbf{r}))^2\right)d\mathbf{r} \\
& + \int\langle n_{\mathrm{P}}(\mathbf{r})\rangle\left[f(\mathbf{r})\ln(f(\mathbf{r})) + (1-f(\mathbf{r}))\ln(1-f(\mathbf{r}))\right]d\mathbf{r} \\
& + \int\langle n_{\mathrm{P}}(\mathbf{r})\rangle\left[f(\mathbf{r})\beta\mu_{\mathrm{A^-}}^{0} + (1-f(\mathbf{r}))\beta\mu_{\mathrm{HA}}^{0}\right]d\mathbf{r}
\end{aligned}
\tag{6.1}
$$

where $\mathbf{r}$ is the position vector (i.e., $\mathbf{r} = (x, y, z)$) and $\beta = 1/k_{\mathrm{B}}T$ is the inverse temperature (k_{B} is Boltzmann's constant and T is temperature). The first term in Eq. (6.1) represents the translational or mixing entropies of the mobile species, which include water molecules (w), anions (A), cations (C), protons (H^+), and hydroxyl ions (OH^-). In this term, $\rho_{\mathrm{i}}(\mathbf{r})$ and μ_{i}^{0} are the position-dependent number density and standard chemical potential of species i (i = w, C, A, H^+, and OH^-), respectively. The second term is the conformational entropy of the polyelectrolyte chains, where $P(j, \alpha)$ is the probability of finding the grafted polyelectrolyte chain j in conformation α, the outer sum runs over all the polyelectrolyte chains in the

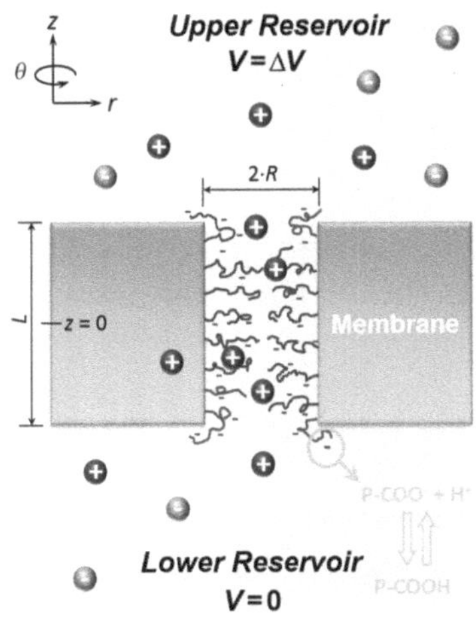

FIGURE 6.1 Schematic representation of polyelectrolyte-modified nanochannel: a nanopore ($R \sim L$) or a nanochannel ($R \ll L$) of radius R and length L is modified by an end-grafted layer of polyelectrolyte chains with a surface density σ. Each chain has N segments. In the case of the figure, each segment is a weak acid that can exist either in its neutral acid form (P—COOH) or as a negatively charged carboxylate species (P—COO$^-$) depending on the local pH. The pore or channel connects two reservoirs containing solutions with the same electrolyte (KCl in concentration C_{salt}) and bulk pH. Each reservoir contains also a reversible electrode. Ion currents result from applying a potential difference ΔV to these electrodes (we use the convention $V=0$ for the lower electrode and $V=\Delta V$ for the upper one).

system and the inner sum runs over all possible polyelectrolyte conformations. The next term in F is the van der Waals (vdW) effective attractive energy between polymer beads which represents the hydrophobicity of the backbone.[14,15] In this term χ is the strength of the effective attractive interactions, i.e., a measure of the hydrophobicity, $g(|\mathbf{r} - \mathbf{r}'|)$ is a distance-dependent van der Waals attractive term and $\langle n_P(\mathbf{r})\rangle$ is the average density of polymer segments at $\mathbf{r}$. Note that χ is an effective segment—segment energy that measures the difference between the segment—segment and segment—solvent attractions; thus the vdW terms for segment—water and water—water interactions do not explicitly appear in F. The average density of segments, $\langle n_P(\mathbf{r})\rangle$ can be calculated as:

$$\langle n_P(\mathbf{r})\rangle = \sum_j \sum_\alpha P(j,\alpha) n(j,\alpha,\mathbf{r}) \tag{6.2}$$

where $n(j, \alpha, \mathbf{r})d\mathbf{r}$ is the number of segments that the chain j has in the volume element between $\mathbf{r}$ and $\mathbf{r} + d\mathbf{r}$ when it is in conformation α. The function $n(j, \alpha, \mathbf{r})$, which completely defines the spatial distribution of segments of the chain j in conformation α, is an input to the theory and it is generated as explained in Section 6.2.3.2.

It is important to mention that the vdW term models only the attractive vdW interactions among polymer segments. We consider intramolecular repulsions in the generation of the function $n(j, \alpha, \mathbf{r})$ by including in this function only self-avoiding conformations. Intermolecular repulsive interactions between all species in the system are modeled as excluded-volume interactions, which are considered through packing constrains, as explicitly described below.

The fourth term in Eq. (6.1) is the electrostatic energy of the system, here $\psi(\mathbf{r})$ is the local electrostatic potential, ε is the dielectric constant, and $\langle\rho_Q(\mathbf{r})\rangle$ is the average charge density at $\mathbf{r}$, given by,

$$\langle\rho_Q(\mathbf{r})\rangle = \sum_{i=A,C,H^+,OH^-} \rho_i(\mathbf{r})q_i + \langle n_P(\mathbf{r})\rangle f(\mathbf{r})q_P \tag{6.3}$$

where the sum runs over all charged mobile species, q_i is the charge of species i in units of the elemental charge (i = A, C, H^+, OH^-, or P for anion, cation, proton, hydroxyl ion, and polymer segment) and $f(\mathbf{r})$ is the fraction of acid groups that are charged (deprotonated) at $\mathbf{r}$. Note that the fraction of charged acid groups depends on the position and therefore segments located in different regions of the system may present different protonation states. The last two contributions to F are the chemical free energy for the acid–base reaction of the acid groups in the polymer:

$$HA + H_2O \underset{}{\overset{K_a}{\rightleftharpoons}} H_3O^- + A^- \tag{6.4}$$

These contributions consider the mixing entropy of the charged and uncharged groups and the difference in standard free energy between A^- and HA through the standard chemical potentials of these species: $\mu^0_{A^-}$ and μ^0_{HA}, respectively.

At this point, we should mention that the theory developed so far for a polyacid can be straightforwardly modified to consider a polybase. The protonation reaction for the polybase is:

$$B + H_2O \overset{K_b}{\rightleftharpoons} OH^- + BH^+ \tag{6.5}$$

The formulation of the theory for the polybase is completely equivalent to that for the polyacid, but replacing A^- by BH^+, HA by B and H^+ by OH^- ($f(\mathbf{r})$ is still the fraction of charged monomers, i.e., A^- for the polyacid and BH^+ for the polybase).

As we mentioned above, the repulsive interactions are separated into intramolecular repulsions, which are exactly considered in the generation of the chain configurations, and the intermolecular repulsions. In both cases they are excluded-volume interactions. The intermolecular excluded-volume interactions are accounted for by packing constraints. Namely, at each position the volume available is fully occupied by molecules, more explicitly in terms of volume fractions:

$$\rho_A(\mathbf{r})v_A + \rho_C(\mathbf{r})v_C + \rho_{OH^-}(\mathbf{r})v_{OH^-} + \langle\phi_P(\mathbf{r})\rangle = 1 \tag{6.6}$$

where v_i (i = A, C, H^+, OH^-) represents the molecular volume of species i and $\langle\phi_P(\mathbf{r})\rangle$ is the volume fraction of the polymer at $\mathbf{r}$, given by:

$$\langle\phi_P(\mathbf{r})\rangle = \sum_j \sum_\alpha P_P(j,\alpha)v_p(j,\alpha,\mathbf{r}) \tag{6.7}$$

where $v_p(j, \alpha, \mathbf{r})d\mathbf{r}$ is the volume that the chain j has in the volume element between $\mathbf{r}$ and $\mathbf{r} + d\mathbf{r}$ when it is in conformation α. The size, shape, and conformations of all the molecular species enter explicitly into the packing constraint, since the way the volume is filled at each position depends on the distribution of molecules and their shape. It is important to note that the integral of the packing constraint over the whole volume shows that the total volume of the system is exactly given by the sum of the number of all the molecular species multiplied by their molecular volumes. This implies that we are reducing the number of thermodynamic independent variables by one. Therefore, from this point forward we define the exchange chemical potentials as the difference between the chemical potential of species i and the chemical potential of the water.[6]

The natural variables of the Helmholtz free energy of the system are the temperature, volume, and number of particles. Namely, at constant temperature, volume, and number of molecules of each species, F is minimal at thermodynamic equilibrium and therefore is a thermodynamic potential. However, the system in Fig. 6.1 is in contact with a bath, thus the mobile species (w, A, C, H^+, and OH^-) can be exchanged between the system and the bath. Therefore their numbers are variable but their chemical potentials are fixed by those given in the bulk solution (reservoirs) and we must consider a semi-grand potential, i.e., a thermodynamic potential with fixed number of polymers but varying number of mobile species. The semi-grand canonical thermodynamic potential is a Legendre transform of F and it is given by:

$$\begin{aligned} \Phi = F - \mu_{\mathrm{A}} \int \rho_{\mathrm{A}}(\mathbf{r})\mathbf{dr} - \mu_{\mathrm{C}} \int \rho_{\mathrm{C}}(\mathbf{r})\mathbf{dr} - \mu_{\mathrm{OH^-}} \int \rho_{\mathrm{OH^-}}(\mathbf{r})\mathbf{dr} \\ - \mu_{\mathrm{H^+}} \int [\rho_{\mathrm{H^+}}(\mathbf{r}) + (1 - f(\mathbf{r}))\langle n_{\mathrm{P}}(\mathbf{r})\rangle]\mathbf{dr} \end{aligned} \tag{6.8}$$

Note that the chemical potential of water is not explicitly included in Eq. (6.8) because the chemical potentials in this equation are exchange chemical potentials as explained above[6] and the exchange chemical potential of the water is $\mu_{\mathrm{w}} \equiv 0$ by definition. It is also important to observe that the last term in Eq. (6.8) contains contributions from the free protons in solution and also from those protons bound to the polymer in order to properly account for all the protons in the system.

In order to find the equilibrium structure of the system, we need to find the minimal free energy. Actually we need to determine the functional extremum of Φ with respect to $\rho_i(\mathbf{r})$, $P(j, \alpha)$, $\psi(\mathbf{r})$, and $f(\mathbf{r})$. The minimization needs to be subjected to a variety of constrains that are required on physical grounds. The first constraint is the packing constraint, which models repulsive interactions among all species in the system, given by Eq. (6.6).

The second constraint to consider is the global electroneutrality, given by:

$$\int \langle \rho_{\mathrm{Q}}(\mathbf{r}) \rangle \mathbf{dr} = 0 \tag{6.9}$$

The third and last constraint is that the sum of the probabilities of all conformations for a given chain must be one (i.e., normalization of $P(\alpha, j)$):

$$\sum_{\alpha} P(\alpha, j) = 1 \tag{6.10}$$

To find the extremum of Φ subjected to the constraints, we introduce Lagrange multipliers. The functional to minimize therefore is:

$$W = \Phi + \int \pi(\mathbf{r}) \left[\sum_{i=A,C,H^+,OH^-} \rho_i(\mathbf{r}) v_i + \langle \phi_P(\mathbf{r}) \rangle v_P - 1 \right] + \lambda \int \langle \rho_Q(\mathbf{r}) \rangle d\mathbf{r} + \xi(j) \left[\sum_{\alpha} P(\alpha, j) - 1 \right] \quad (6.11)$$

where $\pi(\mathbf{r})$, λ, and $\xi(j)$ are the Lagrange multipliers associated to the packing, global electroneutrality, and normalization of conformation probabilities, respectively.

6.2.1.2 *Functional Minimization of the Free Energy*

The equilibrium state of the system is given by the functional extrema of W with respect to the unknown functions that describe the system. Namely,

$$\frac{\delta W[h(\mathbf{r})]}{\delta h(\mathbf{r})} = 0 \quad (6.12)$$

where the functions $h(\mathbf{r})$ are the densities of all mobile species, $\rho_i(\mathbf{r})$ (i = A, C, H^+, OH^-, and w); the fraction of charged monomers, $f(\mathbf{r})$; the electrostatic potential, $\psi(r)$; and the probabilities of each conformation, $P(\alpha, j)$.

The functional minimization with respect to $\rho_i(\mathbf{r})$ yields:

$$\rho_w(\mathbf{r}) v_w = \exp\left[-v_w \beta \pi(\mathbf{r}) - \beta \mu_w^0 \right] \quad (6.13)$$

for the solvent and

$$\rho_i(\mathbf{r}) v_w = \exp\left[-v_i \beta \pi(\mathbf{r}) - q_i \beta (\psi(\mathbf{r}) - \lambda) - \beta \mu_i^0 + \beta \mu_i \right] \quad (6.14)$$

for i = A, C, H^+, and OH^-. Eqs. (6.13) and (6.14) reveal the role of the Lagrange multipliers in the theory. The Lagrange multiplier $\pi(\mathbf{r})$, which enforces the packing constraints, is a position-dependent osmotic pressure. Namely, it is the pressure necessary to keep the chemical potential of the solvent constant under all the different compositions in the inhomogeneous system. The Lagrange multiplier λ, which enforces global electroneutrality, is an additive constant to the electrostatic potential and therefore it has no thermodynamic or physical consequences. Evaluating Eqs. (6.13) and (6.14) in the bulk, allows to relate the position-dependent densities of the mobile species in the system to their bulk values:

$$\rho_w(\mathbf{r}) = \rho_w^{bulk} \exp\left[-v_w \beta (\pi(\mathbf{r}) - \pi^{bulk}) \right] \quad (6.15)$$

for the solvent and

$$\rho_i(\mathbf{r}) = \rho_i^{bulk} \exp\left[-v_i \beta (\pi(\mathbf{r}) - \pi^{bulk}) - q_i \beta (\psi(\mathbf{r}) - \psi^{bulk}) \right] \quad (6.16)$$

for i = A, C, H^+, and OH^-.

The minimization of W with respect to $P(\alpha, j)$, after some algebraic rearrangement, yields:

$$P(\alpha,j) = \frac{1}{q(j)}\exp\left[-\int v_{\mathrm{P}}(\mathbf{r},\alpha,j)\beta\pi(\mathbf{r})d\mathbf{r} - \int n_{\mathrm{P}}(\mathbf{r},\alpha,j)\left(\ln(f(\mathbf{r})) + q_{\mathrm{A}^-}\beta(\psi(\mathbf{r})-\lambda) + \beta\mu^0_{\mathrm{A}^-} + \int \beta\chi g(|\mathbf{r}-\mathbf{r}'|)\langle n_{\mathrm{P}}(\mathbf{r}')\rangle d\mathbf{r}'\right)d\mathbf{r}\right] \tag{6.17}$$

where $q(j)$ is the partition function, which is related to the Lagrange multiplier $\xi(j)$ that enforced the normalization of the probabilities by $q(j) = 1/\exp(1 + \xi(j))$. It is instructing to note that the Boltzmann factor in Eq. (6.17) has contributions from the different interactions considered in the theory: intermolecular repulsions, acid–base equilibrium, electrostatic interactions, and attractive vdW interactions.

The minimum of W with respect to $f(\mathbf{r})$ yields:

$$\ln\left(\frac{f(\mathbf{r})}{1-f(\mathbf{r})}\right) = -\beta\mu^0_{\mathrm{H}^+} - \beta\mu^0_{\mathrm{A}^-} + \beta\mu^0_{\mathrm{HA}} - v_{\mathrm{H}^+}\beta\pi(\mathbf{r}) - \ln(\rho_{\mathrm{H}^+}(\mathbf{r})v_{\mathrm{w}}) \tag{6.18}$$

The last equation can be rewritten to yield the familiar equation for the acid–base equilibrium:

$$K^0_{\mathrm{a}} = \exp\left(-\beta\mu^0_{\mathrm{H}^+} - \beta\mu^0_{\mathrm{A}^-} + \beta\mu^0_{\mathrm{HA}}\right) = \rho_{\mathrm{H}^+}(\mathbf{r})v_{\mathrm{w}}\frac{f(\mathbf{r})}{1-f(\mathbf{r})}\exp(v_{\mathrm{H}^+}\beta\pi(\mathbf{r})) \tag{6.19}$$

where K^0_{a} is the thermodynamic equilibrium constant for the acid–base reaction (Eq. (6.4)), which can be multiplied by the constant factor $\rho_{\mathrm{w}}^{\mathrm{bulk}}/N_{\mathrm{A}}$ (where N_{A} is Avogadro's number) to obtain K_{a}, the equilibrium constant in units of molar concentration. If polybases were present in the system instead of polyacids (i.e., acid–base reaction given by Eq. (6.5)), Eq. (6.19) would be:

$$K^0_{\mathrm{b}} = \exp\left(-\beta\mu^0_{\mathrm{OH}^-} - \beta\mu^0_{\mathrm{BH}^+} + \beta\mu^0_{\mathrm{B}}\right) = \rho_{\mathrm{OH}^-}(\mathbf{r})v_{\mathrm{w}}\frac{f(\mathbf{r})}{1-f(\mathbf{r})}\exp(v_{\mathrm{OH}^-}\beta\pi(\mathbf{r})) \tag{6.20}$$

where K^0_{b} is the thermodynamic equilibrium constant for reaction (6.5) and the subindexes B and BH^+ refer to the deprotonated and protonated base states, respectively.

The extremum of W with respect to $\psi(\mathbf{r})$ yields the Poisson equation:

$$\nabla^2\psi(\mathbf{r}) = -\frac{\langle\rho_{\mathrm{Q}}(\mathbf{r})\rangle}{\varepsilon} \tag{6.21}$$

At this point, it is interesting to note the coupling that exists between the different functions in the molecular theory. For example, the density of the ions (Eq. (6.14)) depends on both the local electrostatic potential and the osmotic pressure. These interaction fields depend, however, on the densities of all species in the system through Eqs. (6.3), (6.6), and (6.21), which effectively creates a highly nonlinear coupling among the distribution of all molecular species, the fraction of protonated acid–base groups and the conformations of the polyelectrolyte. This is the source of the coupling between molecular organization, physical interactions, and chemical state. Note also that because in the molecular theory the Poisson equation is coupled to the molecular organization, the size, shape, and conformation of all molecular species and the chemical state, our theory is more general than the Poisson–Boltzmann equation.[16] Actually, for a system without polymers, the Poisson–Boltzmann equation can be derived from Eq. (6.21) assuming that the ions are simply point charges and the solvent has no volume. In the case of polymer-coated nanopores the application of the Poisson–Boltzmann approximation is not adequate due to its lack of treatment of molecular species.

6.2.1.3 Molecular Theory for Protein Binding

In the previous two sections, we discussed a molecular theory for a system of end-grafted weak polyelectrolytes inside a nanochannel (Fig. 6.1). In that system, the segments in the polyelectrolyte were weak acids or bases and, thus, participated in acid–base chemical equilibria. We consider here a different type of reaction: ligand–receptor binding. Some biological macromolecules, such as proteins, have sites (receptors) with chemical affinity for specific chemical structures (ligands), i.e., the case of streptavidin for biotin.[17] The formation of the ligand–receptor pair can be described in terms of a generic chemical reaction:

$$\mathrm{LR} \rightleftharpoons \mathrm{L} + \mathrm{R} \tag{6.22}$$

Fig. 6.2 shows the system we aim to model: a nanochannel whose inner walls are modified by short polymeric tethers with a functionalized end-ligand that binds to specific receptors located at the surface of spherical proteins. The channel connects two reservoirs containing aqueous solutions with equal concentrations of protein, salt ions, and pH. The proposed Helmholtz free energy functional for this system is:

$$\begin{aligned}
\beta F = & \sum_{i=A,C,w,H^+,OH^-} \int \rho_i(\mathbf{r})\left[\ln\left(\rho_i(\mathbf{r})v_w\right) - 1 + \mu_i^0\right]d\mathbf{r} \\
& + \sum_j \sum_\alpha P(j,\alpha)\ln(P(j,\alpha)) + \iint \frac{\beta\chi g(|\mathbf{r}-\mathbf{r}'|)}{2}\langle n_P(\mathbf{r})\rangle\langle n_P(\mathbf{r}')\rangle d\mathbf{r}d\mathbf{r}' \\
& + \int \left(\langle \rho_Q(\mathbf{r})\rangle \beta\psi(\mathbf{r}) - \frac{1}{2}\varepsilon\beta(\nabla\psi(\mathbf{r}))^2\right)d\mathbf{r} \\
& + \int \int_{S(\mathbf{r})} \rho_{prot}(\mathbf{r},\mathbf{r}')\left[\ln\left(\rho_{prot}(\mathbf{r},\mathbf{r}')v_w\right) - 1 + \beta\mu_{prot}^0\right]d\mathbf{r}'d\mathbf{r} \\
& + \int \langle n_l(\mathbf{r})\rangle \left[f_{unb}(\mathbf{r})\left(\ln(f_{unb}(\mathbf{r})) + \beta\mu^0_{unbound}\right) + \int f_b(\mathbf{r},\mathbf{r}')\left(\ln(f_b(\mathbf{r},\mathbf{r}')) + \beta\mu^0_{bound}\right)d\mathbf{r}'\right]d\mathbf{r}
\end{aligned} \tag{6.23}$$

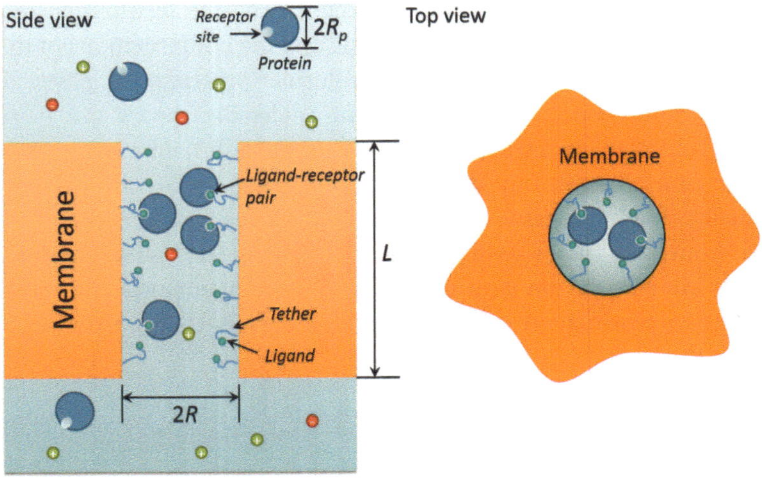

FIGURE 6.2 Scheme of a nanochannel with inner walls modified by tethers terminated in a ligand that binds to receptors on the surface of proteins in solution. The length of the tethers is N and their surface coverage is σ. The proteins, with radius R_{prot} and charge Q_{prot}, can freely rotate and translate in solution. The channel connects two reservoirs that contain solutions of equal concentration of salt ions, pH, and proteins. Ionic currents flow upon applying a potential bias to reversible electrodes located at the reservoirs. Source: *Reproduced from reference Tagliaucchi, M.; Szleifer, I. How Does Confinement Change Ligand-Receptor Binding Equilibrium? Protein Binding in Nanopores and Nanochannels.* J. Am. Chem. Soc. ***2015***, 137, *12539–12551.*

The terms in Eq. (6.23) represent (in order): the translation (mixing) entropies of the ions and solvent, the conformational entropy of the polymer tether, the vdW interactions among tethers, the electrostatic contribution to the free energy, the translational and orientational entropy of the proteins, and the ligand–receptor binding equilibrium. With exception of the last two terms, the other contributions are exactly the same as in Eq. (6.1) and we will not discuss them in detail again. Note that we do not consider the possibility of acid–base equilibrium in the tethers or the proteins, thus the term in Eq. (6.1) accounting for the free energy of acid–base equilibrium was not included in Eq. (6.23).

The next to last term in Eq. (6.23) accounts for the translational and orientational entropy of the proteins, where μ^0_{prot} is the standard chemical potential of proteins and $\rho_{prot}(\mathbf{r}, \mathbf{r}')$ is the density of proteins with center at $\mathbf{r}$ and receptor site at $\mathbf{r}'$. The outer integral runs over the whole space while the inner integral runs only over the domain $S(\mathbf{r})$, which is the set of points on the surface of a protein with center at $\mathbf{r}$ (therefore, the inner integral accounts for the orientational entropy of the protein).

The 10th contribution to the free energy functional is the chemical free energy between the bound and unbound ligands, where $f_{unb}(\mathbf{r})$ is the fraction of unbound ligands at $\mathbf{r}$, $f_b(\mathbf{r}, \mathbf{r}')$ is the fraction of ligands at $\mathbf{r}$ that are bound to a protein with center at $\mathbf{r}'$ and $\mu^0_{unbound}$ and μ^0_{bound} are the standard chemical potentials of the bound and unbound ligands, respectively. Note that $f_{unb}(\mathbf{r})$ and $f_b(\mathbf{r}, \mathbf{r}')$ are related by:

$$f_{unb}(\mathbf{r}) + \int f_b(\mathbf{r}, \mathbf{r}')d\mathbf{r}' = 1 \tag{6.24}$$

In equilibrium, the ions, solvent, and proteins have constant chemical potential, therefore we should consider the semi-grand potential,

$$\Phi = F - \beta\mu_C \int \rho_C(\mathbf{r})d\mathbf{r} - \beta\mu_A \int \rho_A(\mathbf{r})d\mathbf{r} - \beta\mu_{H^+} \int \rho_{H^+}(\mathbf{r})d\mathbf{r} - \beta\mu_{OH^-} \int \rho_{OH^-}(\mathbf{r})d\mathbf{r}$$
$$- \beta\mu_{\text{prot}} \int \left[\int_{S(\mathbf{r})} \rho_{\text{prot}}(\mathbf{r},\mathbf{r}')d\mathbf{r}' + (1 - f_{\text{unb}}(\mathbf{r}))\langle n_l(\mathbf{r})\rangle \right] d\mathbf{r} \quad (6.25)$$

The minimization of Eq. (6.25) requires the same constraints used for the minimization of Eq. (6.8): the packing constraint (which should now include the volume fraction of the proteins), the global electroneutrality constraint, and the normalization constraints. In addition, Eq. (6.24) should be also considered as a constraint and therefore is enforced with a new Lagrange multiplier.

Finding the functional extrema for the density of the ions, water, and electrostatic potential yields the same expressions already discussed in Section 6.2.1.2 for the end-grafted layer of a weak polyacid. The minimization of W with respect to $P(\alpha, j)$ yields,

$$P(\alpha,j) = \frac{1}{q(j)} \exp\left[- \int v_P(\mathbf{r},\alpha,j)\beta\pi(\mathbf{r})d\mathbf{r} \right.$$
$$- \int n(j,\alpha,\mathbf{r})\left(q_{A^-}\beta(\psi(\mathbf{r}) - \lambda) + \int \beta\chi g(|\mathbf{r}-\mathbf{r}'|)\langle n_P(\mathbf{r}')\rangle d\mathbf{r}' \right) d\mathbf{r} \quad (6.26)$$
$$\left. - \int n_l(j,\alpha,\mathbf{r})\left[\ln(f_{\text{unb}}(\mathbf{r})) + \beta\mu^0_{\text{unbound}}\right] \right]$$

which is essentially the same as Eq. (6.17) with the exception that the fraction of charged monomers, $f(\mathbf{r})$, has been replaced by the fraction of ligands that are unbound, $f_{\text{unb}}(\mathbf{r})$. In Eq. (6.26), $n_l(j, \alpha, \mathbf{r})d\mathbf{r}$ is the number of ligands that a chain in conformation α and grafted at j has in the volume element between $\mathbf{r}$ and $\mathbf{r} + d\mathbf{r}$. Note this function stores information about the position of the ligands in the conformations, while $n(j, \alpha, \mathbf{r})$ stores the positions of all segments in the polymeric tethers, not only the ligands.

Minimization of Eq. (6.25) with respect to $\rho_{\text{prot}}(\mathbf{r},\mathbf{r}')$ results in the following expression for the density of the proteins:

$$\rho_{\text{prot}}(\mathbf{r},\mathbf{r}')v_w = \exp\left(- \int \beta\pi(\mathbf{r}'')v_{\text{prot}}(\mathbf{r}'',\mathbf{r})d\mathbf{r}'' - \int \beta\psi(\mathbf{r}'')q_{\text{prot}}(\mathbf{r}'',\mathbf{r})d\mathbf{r}'' + \beta\mu_{\text{prot}} - \beta\mu^0_{\text{prot}} \right) \quad (6.27)$$

where $v_{\text{prot}}(\mathbf{r}, \mathbf{r}')d\mathbf{r}$ and $q_{\text{prot}}(\mathbf{r}, \mathbf{r}')d\mathbf{r}$ are the volume and charge that a protein at $\mathbf{r}'$ has in the volume element between $\mathbf{r}$ and $\mathbf{r} + d\mathbf{r}$.

Finally, the minimization of W with respect to $f_{\text{unb}}(\mathbf{r})$ and $f_b(\mathbf{r}, \mathbf{r}')$ yields,

$$\ln(f_b(\mathbf{r},\mathbf{r}')) + \beta\mu^0_{\text{bound}} - \ln(f_{\text{unb}}(\mathbf{r})) - \beta\mu^0_{\text{unbound}} - \ln(\rho_{\text{prot}}(\mathbf{r}',\mathbf{r})v_w) - \beta\mu^0_{\text{prot}} = 0 \quad (6.28)$$

In this expression, $-\beta\mu^0_{\text{bound}} + \beta\mu^0_{\text{unbound}} + \beta\mu^0_{\text{prot}} = \beta\Delta_d G^0 = \ln(K^0_d)$ is the standard free energy of ligand–receptor binding, Eq. (6.22). Expression (6.28) can be thus rewritten as,

$$\frac{f_{\text{unb}}(\mathbf{r})\rho_{\text{prot}}(\mathbf{r}',\mathbf{r})v_w}{f_b(\mathbf{r},\mathbf{r}')} = K^0_d \tag{6.29}$$

which is nothing else than the equilibrium equation for reaction (6.22) (where K^0_d has units of $(\text{proteins}\cdot v_w)/\text{nm}^3$, so it can be converted to molar units by dividing by $v_w \cdot (10^{-24}\ \text{dm}^3/\text{nm}^3) \cdot 6.02\times 10^{23}$ proteins/mol). In Eq. (6.29), the index $\mathbf{r}'$ is the position of the center of the protein and $\mathbf{r}$ is the position of the ligand–receptor pair. It is instructive to note that the expressions derived for ligand–receptor (Eq. (6.29)) and acid–base equilibrium (Eq. (6.19)) have the same functional form, i.e., despite the very different chemistries involved in acid–base and ligand–receptor equilibria, their thermodynamic treatment by the molecular theory is very similar.

We presented in this chapter a short description of the molecular theory for ligand–receptor binding, stressing the differences from the theory presented in Sections 6.2.1.1 and 6.2.1.2 for a weak polyacid layer. The interested reader can find a more thorough derivation of the theory in the (freely available) Supporting Information of Ref. 10.

6.2.1.4 Conductance in Quasi-Equilibrium Conditions From Equilibrium Calculations

The theories formulated in Sections 6.2.1.1–6.2.1.3 provide equilibrium information (structure and thermodynamics). The ionic conductance of the system can be calculated from this equilibrium organization if the structure under an applied potential and the structure in equilibrium are the same (in other words, if the system does not reorganize in response to the applied potential). This condition is achieved in the quasi-equilibrium limit, i.e., the limit of very small applied potentials.

We will consider the nanochannel in Fig. 6.1 for the condition $L \gg R$. For such long nanochannels, the resistance of the reservoirs is negligible and all the potential drop occurs within the channel. Let us consider the ion currents flowing through a section of the nanochannel with length dL. The mass flux of species, $\mathbf{J}_i(\mathbf{r})$ with i = A, C, H^+, and OH^-, is the product of their drift velocity ($\mathbf{v}_i(\mathbf{r})$) and their concentration, $c_i(\mathbf{r})$,

$$\mathbf{J}_i(\mathbf{r}) = c_i^{\text{eq}}(\mathbf{r})\mathbf{v}_i(\mathbf{r}) \tag{6.30}$$

The total current is the sum of the mass flux of all species integrated across the transversal area of the channel,

$$I = F_C \sum_{i=A,C,H^+,OH^-} q_i \int_A c_i^{\text{eq}}(\mathbf{r})\mathbf{v}_i(\mathbf{r})\hat{\mathbf{n}}\,dA \tag{6.31}$$

where F_C is Faraday's constant and $\hat{\mathbf{n}}$ is a vector normal to the cross section of the channel.

Due to symmetry considerations, the ions move parallel to the channel axis and thus $\mathbf{v}_i$ has only a component in the $\hat{\mathbf{n}}$ direction. Therefore, we can write[18]:

$$\mathbf{v}_i(\mathbf{r}) \cdot \hat{\mathbf{n}} = \frac{D_i q_i F_C}{RT} \frac{dV}{dL} \tag{6.32}$$

where dV/dL is the magnitude of the electric field in the cylindrical section of thickness dL, D_i is the diffusion coefficient of ion i (which we assume to be independent of position) and we have used the Einstein equation for the drift velocity. The conductance of the cylindrical segment of length dL that results from combining Eqs. (6.31) and (6.32) is

$$dG = \frac{I}{dV} = \frac{F_C{}^2}{dLRT} \sum_{i = A,C,H^+,OH^-} q_i{}^2 D_i \int_A c_i^{eq}(\mathbf{r}) dA \tag{6.33}$$

In a cylindrical channel, $c_i^{eq}(\mathbf{r})$ can be assumed to depend only on the radial coordinate r due to the symmetry of the pore.

The total conductance is obtained as the series combination of the conductances of the different cylindrical segments of length dL, namely:

$$G = \left(\int_0^L \left(\frac{1}{dG} \right) \right)^{-1} = \frac{F_C{}^2}{LRT} \sum_{i = A,C,H^+,OH^-} q_i{}^2 D_i \int_A c_i^{eq}(r) dA \tag{6.34}$$

This equation can also be derived by applying Goldman's constant field approximation (CFA),[19,20] which states that the electric field within the channel is constant (i.e., the Goldman CFA implies the quasi-equilibrium condition). Eq. (6.34) allows us to calculate the conductance of a long cylindrical nanochannel in quasi-equilibrium conditions (small ΔV) from the knowledge of the equilibrium concentrations of the ions as a function of the radial coordinate, $c_i^{eq}(r)$. These concentrations can be determined from the molecular theory, e.g., see Eq. (6.14).

What happens if the channel is not cylindrical? In that case, the current lines are no longer parallel to the channel axis and thus $\mathbf{v}_i(\mathbf{r})$ has components in the directions perpendicular to the axis of the channel. Under these conditions Eq. (6.32) is no longer exact, but if the channel radius changes slowly with position, then the components of $\mathbf{v}_i$ perpendicular to the channel axis will be much smaller than the component parallel to the axis. In these cases, Eq. (6.32) will still be approximately valid. Note also that in noncylindrical channels, the cross-sectional area A and the conductance of the cylindrical shells of thickness dL depend on the position within the channel, z:

$$dG(z) = \frac{I}{dV} = \frac{F_C{}^2}{dLRT} \sum_{i = A,C,H^+,OH^-} q_i{}^2 D_i \int_{A(z)} c_i^{eq}(r,z) dA \tag{6.35}$$

so the total conductance now becomes:[21]

$$G = \frac{F_C^2}{RT}\left(\int_0^L \left(\sum_{i=A,C,H^+,OH^-} q_i^2 D_i \int_{A(z)} c_i^{eq}(r,z)\mathrm{d}A\right)^{-1} \mathrm{d}L\right)^{-1} \tag{6.36}$$

In Eq. (6.36) the conductance simply results from the series summation of the resistances of the cylindrical slices of area $A(z)$ and thickness dL in which we decomposed our noncylindrical channel. In order to solve Eq. (6.36), we need the concentration of ions as a function of the radial coordinate for channels of different cross sections A(z). These concentrations can be obtained by solving the molecular theory for infinitely long channels of different cross sections. For example, in order to solve Eq. (6.36) for a conical nanochannel with tip radius R_{tip} and base radius R_{base}, we need to solve the molecular theory for cylindrical nanochannels of different radii between R_{tip} and R_{base} in order to predict the equilibrium concentrations as a function of the radial coordinate for each radius. Then, the resulting concentrations (obtained for infinite cylindrical nanochannels) are used as an input for the approximate Eq. (6.36) to calculate the conductance of a conical channel.

Eq. (6.36) is approximate because we assumed that the current lines in the nanochannel are parallel to the channel axis, which is strictly only valid for an infinitely long cylindrical channel. In practice, this approximation works well for noncylindrical channels if the change in channel radius with position is slow. Namely, as long as the slope of the radius is such that the molecules locally feel as if they were in a cylindrical geometry, the perpendicular component of the velocity will be negligible compared to the parallel one. For example, Fig. 6.3 shows the quasi-equilibrium conductance of a conical nanochannel with $R_{tip} = 5$ nm, $R_{base} = 145$ nm, and $L = 12$ μm (slope of 0.7°) with walls modified by a fixed surface charge with density σ_q. The plot shows that the approximate calculations based on Eq. (6.36) are in very good agreement with the full solutions of the PNP equations using Finite Element calculations (see Appendix I in chapter: Theoretical Basis for Structure and Transport in Nanopores and Nanochannels) for the same system. The advantage of using Eq. (6.36) over the full PNP solution is twofold: (1) the calculation of Eq. (6.36) for very long channels is computational less expensive than the PNP equation, (2) the concentration of ions in Eq. (6.36) is derived from equilibrium molecular theory calculations, which (unlike the PNP equations) can easily incorporate the presence of polymers, coupled acid–base equilibria, non-electrostatic interactions, etc.

6.2.2 Nonequilibrium Molecular Theory

6.2.2.1 Formulation of the Nonequilibrium Molecular Theory

The previous section described methods to calculate the conductance of the channel in the limit of very small biases from the equilibrium structure of the system. For large biases, the system reorganizes in response to the presence of ion fluxes and a full solution of the nonequilibrium

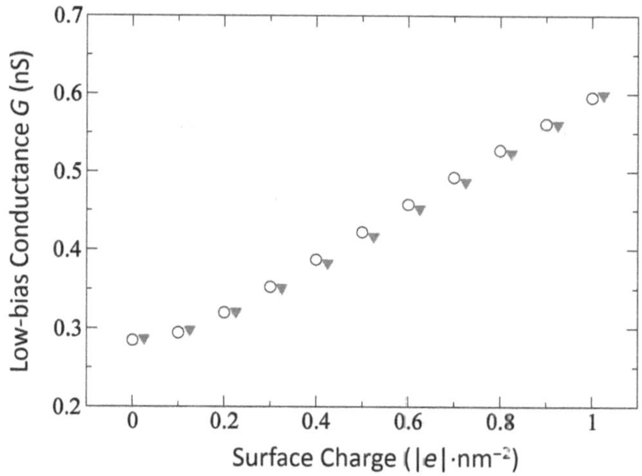

FIGURE 6.3 Low-bias conductance as a function of the surface charge for a conical nanochannel ($R_{tip} = 5$ nm, $R_{base} = 145$ nm, and $L = 12$ μm) modified by surfaces charges only. Solid triangles show the predictions of full PNP calculations and empty circles show the results of Eq. (6.36), where the ion concentrations in equilibrium $c^{eq}(r, z)$ were calculated with the equilibrium molecular theory. Source: *Adapted from reference Gilles, F.M.; Tagliazucchi, M.; Azzaroni, O.; Szleifer, I. Ionic Conductance of Polyelectrolyte-Modified Nanochannels: Nanoconfinement Effects on the Coupled Protonation Equilibria of Polyprotic Brushes. J. Phys. Chem. C.* ***2016****, 120, 4789–4798.*

transport equations is required. The reorganization with the applied bias of the ions is, in fact, responsible of some very interesting effects in ion transport such as current rectification and concentration polarization. In this section, we will derive a nonequilibrium molecular theory for ion transport that allows extending the unique abilities of the theory for modeling polyelectrolyte-modified pores and channels beyond the quasi-equilibrium regime. Our non-equilibrium molecular theory allows to explicitly couple ion transport to the conformations of the polyelectrolyte, electrostatic and non-electrostatic interactions, acid–base equilibria, etc. In fact, we will show that our nonequilibrium molecular theory can be viewed as a generalization of the PNP equations (in analogy, the equilibrium molecular theory in Section 6.2.1 can be thought about as a generalization of the Poisson–Boltzmann equation that includes the size, shape, conformations, and chemical state of the different chemical species).

Our nonequilibrium molecular theory follows the ideas of the local equilibrium approximation,[22,23] which states that all thermodynamic variables in the system can be defined locally and follow the same functional relationships as in equilibrium. The presence of an applied bias generates gradients of chemical potential in the system, which are the driving forces for ion fluxes, according to the equation[23]:

$$\mathbf{J}_i(\mathbf{r}, t) = -D_i \rho_i(\mathbf{r}, t) \nabla \beta \mu_i(\mathbf{r}, t) \tag{6.37}$$

where D_i is the diffusion coefficient of the species i, $\mu_i(\mathbf{r}, t)$ and $\rho_i(\mathbf{r}, t)$ are the chemical potential and the density of the species i at $\mathbf{r}$ and time t, respectively. Eq. (6.37) assumes a linear response between fluxes and driving forces and neglects cross terms between species.[23] This expression can be also derived from the Langevin equations of motion for a system colloidal particles in a bath.[24,25] Note that the chemical potential in Eq. (6.37) depends on the position within the system, unlike those used in Section 6.2.1. In other words, we define a nonequilibrium chemical potential in analogy to the equilibrium one, but, since the system is not in equilibrium, the nonequilibrium chemical potential depends on the position. In the framework of the local equilibrium approximation, we can determine the nonequilibrium chemical potential as the functional derivative of the free energy of the system with respect to the local density of the ions, in analogy to equilibrium thermodynamics.[11] Note that this line of thinking assumes the existence of a length scale where the chemical potential of the ions is well defined. This length scale is of the order of the mean free path and, therefore, in condensed systems under moderated driving forces it is much shorter than any characteristic length scale of the problem (i.e., length of the pore). Hence, the local equilibrium approximation is a rather good approximation for the problem under study.

We will consider here a system simpler than that of Section 6.2.1: a strong polyelectrolyte brush where the presence of protons and hydroxyl ions is neglected (we will discuss the validity of this approximation in the Conclusions section). The free energy for this system is given by:

$$\begin{aligned}\beta F = & \sum_{i=A,C,w} \int \rho_i(\mathbf{r})\left[\ln(\rho_i(\mathbf{r})v_w) - 1 + \mu_i^0\right]d\mathbf{r} \\ & + \sum_j \sum_\alpha P(j,\alpha,t)\ln(P(j,\alpha,t)) + \iint \frac{\beta\chi g(|\mathbf{r}-\mathbf{r}'|)}{2}\langle n_P(\mathbf{r},t)\rangle\langle n_p(\mathbf{r}',t)\rangle d\mathbf{r}d\mathbf{r}' \\ & + \int \left(\langle\rho_Q(\mathbf{r},t)\rangle\beta\psi(\mathbf{r},t) - \frac{1}{2}\varepsilon\beta(\nabla\psi(\mathbf{r},t))^2\right)d\mathbf{r}\end{aligned} \tag{6.38}$$

The terms in this equation have been already described for the equilibrium molecular theory (i.e., see explanation of Eq. (6.1)).

Following the same ideas as for the equilibrium molecular theory, the repulsive interactions among molecular species in the system are modeled as excluded volume repulsions and are taken into account by a packing constraint at each **r**, Eq. (6.39). This packing constraint is enforced at all times and positions.

$$\rho_A(\mathbf{r},t)v_A + \rho_C(\mathbf{r},t)v_C + \rho_w(\mathbf{r},t)v_w + \langle\phi_P(\mathbf{r},t)\rangle = 1 \tag{6.39}$$

Note that due to the presence of the packing constraint, the densities of the ions, solvent, and polyelectrolyte are not thermodynamically independent variables. One can choose the density of the solvent to be a function of the other densities, therefore:

$$\beta\mu_i(\mathbf{r},t) = \frac{\delta\beta F}{\delta\rho_i(\mathbf{r},t)} + \frac{\delta\beta F}{\delta\rho_w(\mathbf{r},t)}\frac{d\rho_w(\mathbf{r},t)}{d\rho_i(\mathbf{r},t)} \tag{6.40}$$

where the derivative

$$\frac{d\rho_w(\mathbf{r},t)}{d\rho_i(\mathbf{r},t)} = -\frac{v_i}{v_w} \tag{6.41}$$

is determined from Eq. (6.39). Replacing F from Eq. (6.38) into Eq. (6.40) yields,

$$\beta\mu_i(\mathbf{r},t) = \mu_i^0 + \ln(\rho_i(\mathbf{r},t)v_w) + v_i\left(\frac{-\ln(\rho_w(\mathbf{r})v_w) - \beta\mu_w^0}{v_w}\right) + q_i\beta\psi(\mathbf{r},t) \tag{6.42}$$

We now can define the position- and time-dependent osmotic pressure in analogy with the relationship found in equilibrium (see Eq. (6.13)), i.e.,

$$\beta\pi(\mathbf{r},t) = \frac{-\ln(\rho_w(\mathbf{r},t)v_w) - \beta\mu_w^0}{v_w} \tag{6.43}$$

which allows us to rewrite Eq. (6.42) as:

$$\beta\mu_i(\mathbf{r},t) = \beta\mu_i^0 + \ln(\rho_i(\mathbf{r},t)v_w) + v_i\beta\pi(\mathbf{r},t) + q_i\beta\psi(\mathbf{r},t) \tag{6.44}$$

We finally insert Eq. (6.44) into Eq. (6.37) to obtain:

$$\mathbf{J}_i(\mathbf{r},t) = -D_i\nabla\rho_i(\mathbf{r},t) - D_i\rho_i(\mathbf{r},t)q_i\nabla\beta\psi(\mathbf{r},t) - D_i\rho_i(\mathbf{r},t)v_i\nabla\beta\pi(\mathbf{r},t) \tag{6.45}$$

In Chapter 2, Theoretical Basis for Structure and Transport in Nanopores and Nanochannels, we showed that the chemical potential of a species i can be split into an ideal (entropic term) and a nonideal contribution, also known as the excess chemical potential or potential of mean force (pmf). The driving force for the flux of the ion i can be therefore expressed as the combination of an entropic part and a mean force (see chapter: Theoretical Basis for Structure and Transport in Nanopores and Nanochannels):

$$\mathbf{J}_i(\mathbf{r},t) = -D_i\nabla\rho_i(\mathbf{r},t) - D_i\rho_i(\mathbf{r},t)\nabla\beta U_{pmf}(\mathbf{r},t) \tag{6.46}$$

where the second term comprises all nonideal contributions to the driving force. Comparison between Eqs. (6.45) and (6.46) shows that the ion fluxes in the molecular theory are driven by two nonideal interactions contained in U_{pmf}: the electrostatic contribution (migration) in the second term of Eq. (6.45) and the nonideal osmotic contribution in the third term of Eq. (6.45). In fact, it is instructive to compare Eq. (6.45) to Fick's first law,

$$\mathbf{J}_i(\mathbf{r},t) = -D_i\nabla\rho_i(\mathbf{r},t) \tag{6.47}$$

which considers only the ideal entropic contribution to the chemical potential and the Nernst–Planck equation,

$$\mathbf{J}_i(\mathbf{r},t) = -D_i\nabla\rho_i(\mathbf{r},t) - D_i\rho_i(\mathbf{r},t)q_i\nabla\beta\psi(\mathbf{r},t) \tag{6.48}$$

which considers both the ideal entropic term and electrostatic interactions, but not the osmotic driving forces. Our theory is a generalization of both the Fick and Nernst–Planck equations because it includes osmotic contributions, but also because it explicitly considers the conformations of grafted macromolecules in the system and their coupling with the fluxes of ions, as we explain below.

We will assume that the field $\psi(\mathbf{r}, t)$ instantaneously responds to any change to the applied bias and thus it corresponds to an extremum of F. The functional extremum of F with respect to $\psi(\mathbf{r}, t)$ yields the Poisson equation:

$$\nabla^2 \psi(\mathbf{r}, t) = -\frac{\langle \rho_Q(\mathbf{r}, t) \rangle}{\varepsilon} \tag{6.49}$$

where $\langle \rho_Q(\mathbf{r}) \rangle$ is the charge density at $\mathbf{r}$ computed from the contribution of all charged species in the system.

At this point we introduce the steady-state condition, which allows some simplifications to the general theory. First, all variables are invariant in time. Second, there is a continuity equation for the ion fluxes:

$$\frac{\partial \rho_i(\mathbf{r})}{\partial t} = -\nabla \cdot \mathbf{J}_i(\mathbf{r}) = 0 \tag{6.50}$$

Third, the steady-state condition will allow us to obtain an expression for the probability of each polyelectrolyte conformation, $P(\alpha, \mathbf{r})$. Note that during a transient response, the conformations of the polyelectrolyte will change with time, but in the steady state their probabilities should remain constant because the polyelectrolyte chains are grafted to the pore walls and thus mass fluxes of polyelectrolytes cannot occur. We thus determine $P(\alpha, \mathbf{r})$ as that which minimizes the functional F, i.e.,

$$\begin{aligned} P_P(\alpha, j) = \frac{1}{q(j)} \exp\Bigg[& -\int v_P(\mathbf{r}, \alpha, j) \beta \pi(\mathbf{r}) d\mathbf{r} \\ & - \int n_P(\mathbf{r}, \alpha, j)(q_{A^-} \beta \psi(\mathbf{r}) + \int \beta \chi g(|\mathbf{r} - \mathbf{r}'|) \langle n_P(\mathbf{r}') \rangle d\mathbf{r}') d\mathbf{r} \end{aligned} \tag{6.51}$$

This equation assumes that the end-grafted polyelectrolyte can relax on the timescale of the experiment, which is always true in steady state.

In summary, the strategy behind the nonequilibrium molecular theory is to propose a free energy functional for the nonequilibrium system. In the steady state, some degrees of freedom ($\psi(\mathbf{r})$ and $P(\alpha, \mathbf{r})$) are assumed to be extrema of the nonequilibrium free energy. Note, however, that this free energy is not the equilibrium one, since there are other degrees of freedom ($\rho_i(\mathbf{r})$) that do not minimize F. The degrees of freedom that do not minimize F (ion concentrations) have associated flows (ion flows) which follow the flux–force relationships, Eq. (6.37). In the flux–force relationships, the driving force is the gradient of the generalized chemical potential, which results from the functional derivative of the free energy with respect to the ion concentrations. As a final comment, we wish to mention that the current formulation of the nonequilibrium approximation does not consider solvent fluxes and the presence of coupled chemical equilibrium, as well as it cannot be used for transient (non-steady-state) systems. We will discuss these approximations and how they can be overcome in the conclusion sections.

6.2.2.2 Boundary Conditions and Current Calculation

In order to use the molecular theory to calculate the conductance of a short nanopore, such as that in Fig. 6.1, it is important to consider the proper boundary conditions. Note that the concentration and/or fixed-potential boundary conditions at the borders of the reservoirs determine the chemical potentials for the ions at these boundaries, which sets the external driving force for ion currents. Fig. 6.4 shows the different boundaries in the system and Table 6.1 shows the equations that describe the typical boundary conditions used at these boundaries.

The boundary condition for the electrostatic potential at the walls assumes that the electric field does not penetrate into the membrane, since its dielectric constant is much smaller than that of the solution. This is actually a very good approximation for typical membrane materials (such as Si_3N_4 and SiO_2, which have dielectric constants of 9.7 and 4.2), see for example SI of Ref. 12.

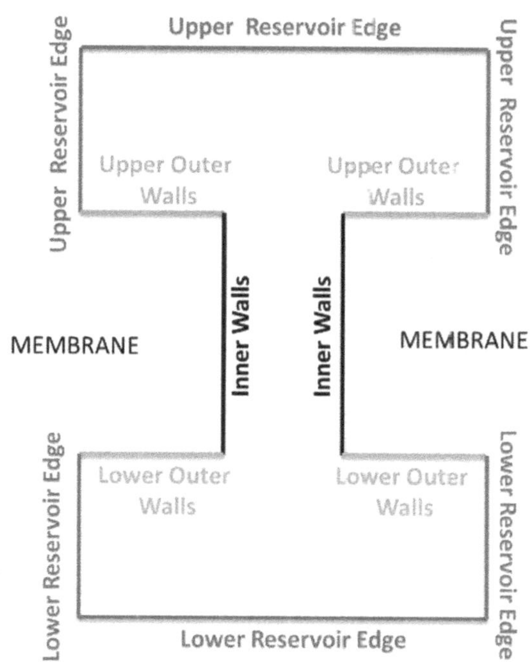

FIGURE 6.4 Scheme showing the appropriate boundary conditions required to use the nonequilibrium molecular theory to calculate the conductance of a short nanopore.

Table 6.1 Boundary Conditions for the Nonequilibrium Molecular Theory

Boundary	Boundary Condition for Electrostatic Potential	Boundary Condition for Ion Concentration
Inner, upper, and lower walls (constant surface charge density)	$\nabla_n \psi = -\frac{\sigma_q}{\varepsilon}$	$\mathbf{J}_i \hat{\mathbf{n}} = 0$
Upper reservoir edge	$\psi = \Delta V$	$c_i = c_i^{\text{upper bulk}}$
Lower reservoir edge	$\psi = 0$	$c_i = c_i^{\text{lower bulk}}$

The boundaries are shown in Fig. 6.4, ∇_n indicates differentiation in the direction normal to the membrane, $\hat{\mathbf{n}}$ indicates a unit vector normal to the surface, σ_q is the charge density of the walls, and ΔV is the potential applied between reservoirs.

The total current through the system must be calculated by integration of the normal components of the ion fluxes on a closed surface. It is possible to choose the boundary between the system and the reservoirs (i.e., the upper or lower reservoir edge in Fig. 6.4) as in our example for Finite Elements in Chapter 2, Theoretical Basis for Structure and Transport in Nanopores and Nanochannels, but in practice this method is not very good for the finite differences scheme used to implement the molecular theory (see Section 6.2.3.2). It is better to consider a plane inside the pore normal to the pore axis (e.g., the plane at $z = 0$). The total current carried by the ion i (i = A or C) across this plane is:

$$i_i = -\int_0^{2\pi}\int_0^{R} |e| q_i \mathbf{J}_i(r,0)\hat{\mathbf{z}} r dr d\theta = -2\pi \int_0^{R} |e| q_i \mathbf{J}_i(r,0)\hat{\mathbf{z}} r dr \qquad (6.52)$$

and the total current flowing through the system is the summation of the currents of the individual ions.

6.2.3 Implementation of the Equilibrium and Nonequilibrium Molecular Theories

6.2.3.1 Symmetry Considerations

Most nanochannels and nanopores have cylindrical (or approximated cylindrical) symmetry, which means that the properties of the system are invariable with respect to rotation around the central axis of the system. In our general formulation of the equilibrium (Section 6.2.1) and nonequilibrium (Section 6.2.2) molecular theories, all functions depend on the coordinate **r** that describes a point in the 3D system. In practice, it is convenient to take advantage of the symmetry of the system and recast the equations of the theory into cylindrical coordinates in order to finally assume homogeneity in the θ (angular) coordinate. The later assumption is exact in some cases (i.e., cylindrical nanochannels in the absence of grafted polyelectrolytes) and a very good approximation in others (i.e., polyelectrolyte-modified cylindrical channels). Moreover, it allows transforming the full 3D physical problem into a 2D numerical problem, which greatly reduces the computational cost of solving it. Finally, we can also approximate very long nanochannels as infinite systems, which are invariant with respect to translations along the channel axis, z. In this case, we can assume that the system is homogenous both in the θ and z coordinates, which reduces the numerical problem to one dimension only (r, the radial coordinate). The reduction of the dimensionality of the problem is one of the biggest advantages of the molecular theory over Monte-Carlo or MD simulations, which cannot straightforwardly exploit symmetry considerations.

Casting the molecular theory into θ-homogeneous cylindrical coordinates is straightforward, but some care is necessary when dealing with the density of polymer segments, Eq. (6.2). If we consider a system that is homogeneous in the θ coordinate, we can no longer assign 3D positions to the grafting points in the system and thus the summation over all chains in Eq. (6.2) cannot be performed. Instead, we should approximate this summation in terms of the probability of finding a chain grafted to an area element on the inner surface of the channel, i.e., the grafting density, σ. Thus Eq. (6.2) becomes:

$$\langle n_{\mathrm{P}}(\mathbf{r})\rangle = \sigma \int \sum_{\alpha} P(\mathbf{r}'(\mathbf{s}'),\alpha)n(\mathbf{r}'(\mathbf{s}'),\alpha,\mathbf{r})\mathbf{ds}' \tag{6.53}$$

where $P(\mathbf{r}'(\mathbf{s}'),\alpha)$ is the probability of finding the chain grafted at $\mathbf{r}'$ in conformation α, $\mathbf{r}'(\mathbf{s}')$ is a parametrization of the channel walls modified by grafted polymers, $n(\mathbf{r}'(\mathbf{s}'),\alpha,\mathbf{r})\mathbf{dr}$ is the number of segments that a chain grafted at $\mathbf{r}'(\mathbf{s}')$ has in the volume element between $\mathbf{r}$ and $\mathbf{r}+\mathbf{dr}$ when it is in conformation α and $\mathbf{ds}'$ the area element. Note that while Eq. (6.2) distinguished between individual chains in the system, Eq. (6.53) is an approximation that assumes that the polymer is homogeneously grafted on the surface of the channel. In order to cast Eq. (6.53) into cylindrical coordinates with a homogeneous θ coordinate, we integrate the number of segments of the chains over all θ:

$$\int_0^{2\pi} n(\mathbf{r}',\alpha,\mathbf{r})r\,\mathrm{d}z\,\mathrm{d}r\,\mathrm{d}\theta = 2\pi r\, n(\mathbf{r}',\alpha,\mathbf{r})\mathrm{d}r\mathrm{d}z = \overline{n}(\mathbf{r}',\alpha,r,z)\mathrm{d}r\mathrm{d}z \tag{6.54}$$

where we define the function $\overline{n}(\mathbf{r}',\alpha,r,z)\mathrm{d}r\mathrm{d}z$ as the number of segments that a chain grafted at $\mathbf{r}'$ and in conformation α has in the cylindrical shell segment between $r+\mathrm{d}r$ and $z+\mathrm{d}z$. Substitution of Eq. (6.54) into Eq. (6.53) gives:

$$\langle n_{\mathrm{P}}(r,z)\rangle = \sigma\frac{1}{2\pi r}\int \sum_{\alpha} P(\mathbf{r}'(\mathbf{s}'),\alpha)\overline{n}(\mathbf{r}'(\mathbf{s}'),\alpha,r,z)\mathrm{d}s' \tag{6.55}$$

In the case where the area modified by polymers are the inner channel walls (e.g., system in Fig. 6.1), can write:

$$\langle n_{\mathrm{P}}(r,z)\rangle = \sigma\frac{1}{2\pi r}\int_0^{2\pi}\int_0^{\infty}\int_{-L/2}^{L/2} \sum_{\alpha} P(r',z',\theta',\alpha)\overline{n}(r',z',\theta',\alpha,r,z)\delta(r'-R)r'\mathrm{d}z'\mathrm{d}r'\mathrm{d}\theta' \tag{6.56}$$

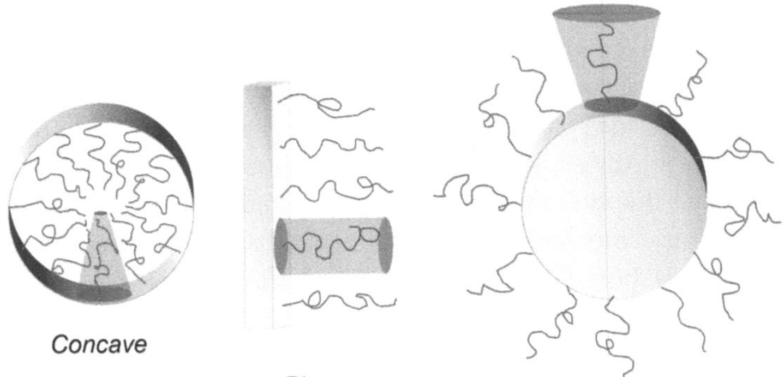

FIGURE 6.5 Scheme showing the effect of substrate curvature over the molecular organization of the end-grafted polymer layer.

where $\delta(r'-R)$ is the Dirac delta function evaluated at $r'-R$. Integration yields:

$$\langle n_P(r,z)\rangle = \sigma\frac{R}{r}\int_{-L/2}^{L/2}\sum_{\alpha}P(z',\alpha)\overline{n}(R,z',\alpha,r,z)\mathrm{d}z' \tag{6.57}$$

where $P(z', \alpha)$ is the probability of finding a chain grafted at z' in conformation α. The factor R/r in this equation is very important, since it relates the ratio of the volumes of the cylindrical shells where the chains are grafted (R) and where the segment density is calculated (r). Because of this factor, grafted layers behave differently on curved and planar surfaces. Let us consider, for example, a grafted layer inside a nanochannel (Fig. 6.5, left), on a planar surface (Fig. 6.5, center), and on the surface of a nanorod (Fig. 6.5, right). In the planar case, the volume element is independent on the distance from the surface. In the case of the nanochannel, R/r increases as one moves away from the surface, and therefore, the available volume decreases as one moves toward the axis of the channel (Eq. (6.57)). Finally, for the rod, the factor R/r decreases as we move away from the surface and thus the available volume increases. For a fixed number of polymer segments (i.e., fixed surface coverage of the polymer), the change in the available volume modulates the change of the density of the polymer with the distance to the surface, which is an important factor that dictates the interactions between polymers and, ultimately, the molecular organization of the whole system. It can be shown that for nanoparticles and spherical holes,[26] the available volume at a distance r from the surface scales as $(r/R)^2$.

6.2.3.2 Numerical Implementation

The sets of integro-differential equations of the equilibrium (Eqs. (6.6), (6.7), (6.13), (6.14), (6.17), (6.18), and (6.21)) and nonequilibrium (Eqs. (6.39), (6.45), (6.49)–(6.51)) molecular theories are highly nonlinear and nonlocal and they have no analytical solution. In order to solve them numerically, we first discretize them using a finite-difference scheme. In other words, the nanopores are divided in cylindrical shells of thickness and length δ and the equations are discretized by replacing integrals by sums and derivatives by finite differences. The final set of nonlinear discretized equations is solved using a Jacobian-Free Newton method with the program Kinsol of the suite Sundials.[27]

In order to solve the theory, we require the function $\overline{n}\,(z', \alpha, r, z)$ (from Eq. (6.57)) as an input. The function $\overline{n}\,(z', \alpha, r, z)\mathrm{d}z\mathrm{d}r$ is the number of segments that a chain in conformation α and grafted at z' has in the cylindrical shell between r and $r - \mathrm{d}r$ and z and $z + \mathrm{d}z$. In other words, the function $\overline{n}\,(z', \alpha, r, z)$ stores all the information of the spatial distribution of the segments of all conformations, i.e., it contains the polymer conformations. While in principle the theory requires the set of all possible conformations, in practice it is enough to include a large sample. Using $\sim 10^6$ conformations usually is enough to achieve convergence of the structural and thermodynamic quantities, although more conformations may be required for very long chains or in the presence of strong fields. In general, we obtain $\overline{n}\,(z', \alpha, r, z)$ by generating random conformations using the rotation isomeric model[28]; although other methods, such as biased sampling techniques[29] or short high-temperature MD simulations,[30] can be used. The only requirements are that the generated conformations are physically allowed (i.e., no overlapping segments) and that they are randomly chosen or that their statistical weight is properly corrected if the generation method is biased.

The other important function that is required as an input for the theory is $g(|\mathbf{r}-\mathbf{r}'|)$, which describes the vdW interactions among segments. Once discretized, this function results in a set of coefficients, which are then determined using numerical integration.[14,31,32]

6.3 Structure of Nanopores and Nanochannels Modified by Polymers and Polyelectrolytes

6.3.1 Structure of Uncharged Polymers in Good Solvent

In the polymer literature, the strength of the effective polymer–polymer attractions is usually described in terms of the quality of the solvent.[28,33] When the polymer–polymer attractions are weaker than the polymer–solvent ones, the solvent is known as a good solvent. On the other hand, the regime of strong effective polymer–polymer attractions is known as the poor-solvent regime.

We will address first the structure of nanopores and nanochannels modified by uncharged polymers in good-solvent conditions, which in our theory corresponds to $\chi=0$. Fig. 6.6 shows the distribution of the polymer in nanopores of different radius and length. The figure shows the volume fraction of the polymer (i.e., the fraction of the volume that is occupied by the polymer) along a plane that includes the axis of the pore. The grafted polymer in this example has a chain length of $N = 75$ and a grafting density of $\sigma = 0.05$ chains $\cdot$ nm^{-2}. The pore dimensions explored in Fig. 6.6 are close to the typical dimensions of pores drilled in SiN membranes ($L \sim 20$–100 nm and $R > 3$ nm).[34]

End-grafted polymer layers inside nanochannels differ from those on planar surfaces due to the combination of two confinement effects: (1) within the nanochannel, the volume

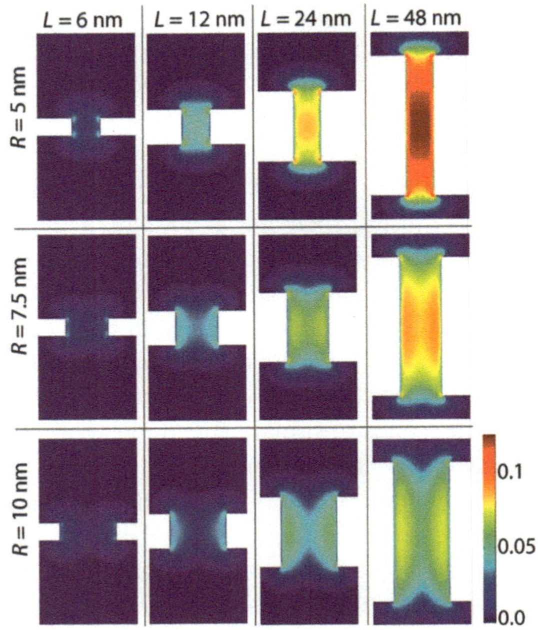

FIGURE 6.6 Volume fraction of the polymer for short nanopores with inner walls modified by a neutral end-grafted polymer in good-solvent conditions. Calculation parameters, $N = 75$, $\sigma = 0.05$ chains·nm^{-2}, and $\chi = 0\ k_BT$. Source: *Reproduced from reference Peleg, O.; Tagliazucchi, M.; Kroeger, M.; Rabin, Y.; Szleifer, I. Morphology Control of Hairy Nanopores.* ACS Nano ***2011***, 5, *4737–4747.*

element decreases as one moves away from the surface, while, for a flat surface, it remains constant (see Section 6.2.3.1 and Fig. 6.5); (2) the number of conformations allowed on a planar surface is greater than that allowed inside a channel, i.e., there are conformations that are allowed on a planar surface but do not fit inside a channel.

Fig. 6.6 shows that the polymers grafted inside the channel stretch in order to place some of their segments outside the pore, leading to an appreciable density of the polymers at the entrances of the pore. This organization is favorable as it minimizes steric repulsions between segments and decreases the osmotic pressure inside the channel. Moreover, this organization can also increase the conformational entropy of the chains, because the segments in the reservoir have more accessible conformations than those within the pore (although there is also an entropic penalty due to the loss of conformational freedom of the stretched part of the chain). The number of polymer segments that are expelled from the pore depends on its radius and length. For example, for $L = 6$ nm and $R = 5$ nm only ~20% of the segments are located inside the channel. As the length and/or radius of the pore increase, the fraction of segments outside the channel decreases.

The results in Fig. 6.6 show the overall density of the polymer layer, but it is also instructive to consider the density of each of the individual chains in the system. Fig. 6.7 shows single chain density profiles for chains grafted at different z positions within a pore of $R = 7.5$ nm and $L = 24$ nm. These profiles show how chains stretch in order to place some

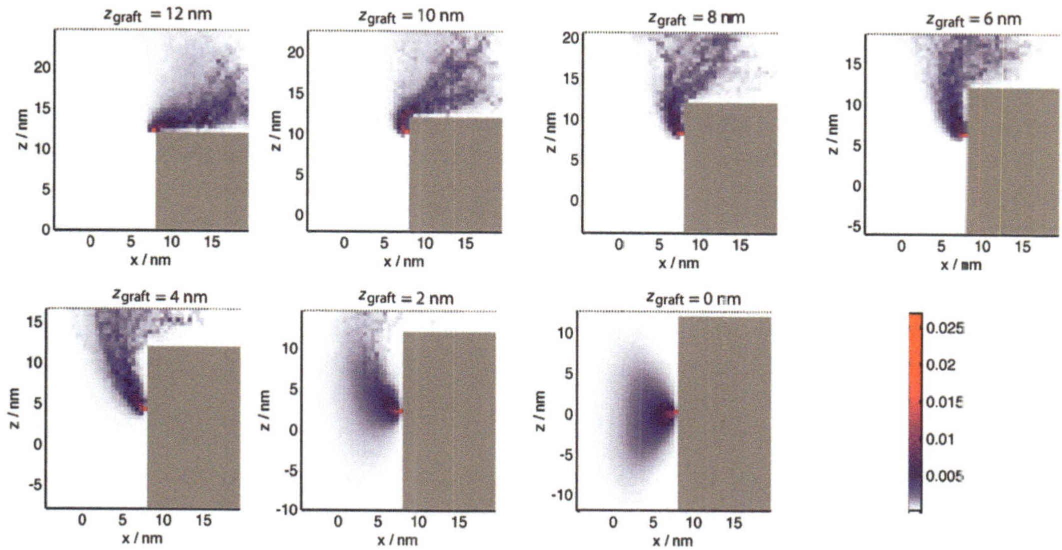

FIGURE 6.7 2D projections of the average density distribution of single chains in the (*x*, *z*) plane for a chain grafted normal to the *y*-axis at different grafting positions (different z_{graft}). The calculation corresponds to a short nanopore with inner walls modified by neutral polymer end-grafted chains in good solvent conditions. Other calculation parameters, $R = 7.5$ nm, $L = 24$ nm, $N = 75$, $\sigma = 0.3$ chains $\cdot$ nm^{-2}, and $\chi = 0\ k_BT$. Source: *Reproduced from reference Peleg, O.; Tagliazucchi, M.; Kroeger, M.; Rabin, Y.; Szleifer, I. Morphology Control of Hairy Nanopores.* ACS Nano ***2011***, 5, *4737–4747.*

of their segments into the reservoirs. For chains that are close to the edge of the pore ($z_{graft} = 8$ and 10 nm), it is entropically favorable to stretch the part of the chain closest to the grafting point in order to grant the rest of the chain access to the reservoirs. On the other hand, the chains that are far from the entrance of the pore ($z_{graft} = 0$ and 2 nm) would need to stretch too much in order to reach the reservoirs and, therefore, they place all their segments within the pore. Fig. 6.7 also shows that in order to reach the reservoirs, the chains must place a large number of segments close to the edge of the pore, which gives rise to the high density regions located at the rim of the pore observed in some of the pores of Fig. 6.6 (e.g., for $L = 24$ nm and $R = 5$ nm). It is also interesting to note that chains with $z_{graft} = 0$ (i.e., chains grafted at the middle of the channel) do not have an ideal mushroom-shape profile, but rather their profile is stretched in the *z*-direction, which results from the confining effect of the pore.[31]

6.3.2 Uncharged Polymers in Poor Solvent: Formation of Self-Organized Aggregates

Polymer solutions in poor-solvent conditions (large χ values in our theory) separate into a solvent-rich phase and a solvent-poor phase, in a process known as phase separation. When the same polymers are end-grafted to a surface, macroscopic phase separation is no longer possible due to the geometric constraint introduced by grafting. However, the chains can still associate and form solvent-poor aggregates, this process is known as "microphase separation".[14,15,35–37] In solution, macroscopic phase separation occurs when the derivative of the chemical potential of the polymer chains (μ_P) with respect to their concentration (ρ_P) becomes equal to zero.[28] This is so, because the chemical potential should always increase with concentration in a thermodynamically stable homogeneous system and thus homogeneous solutions require $\partial\mu_P/\partial\rho_P > 0$. When $\partial\mu_P/\partial\rho_P = 0$, the homogeneous solution becomes unstable and phase separation must occur. One can use a similar criteria, $\partial\mu_P/\partial\sigma = 0$,[15] (where σ is the surface density of the polymers) to estimate the onset of stability of an end-grafted polymer layer with respect to microphase separation.[15]

Fig. 6.8 shows the morphologies of polymer-modified nanopores as a function of χ (polymer–polymer attraction strength) and σ (surface coverage). The solid line corresponds

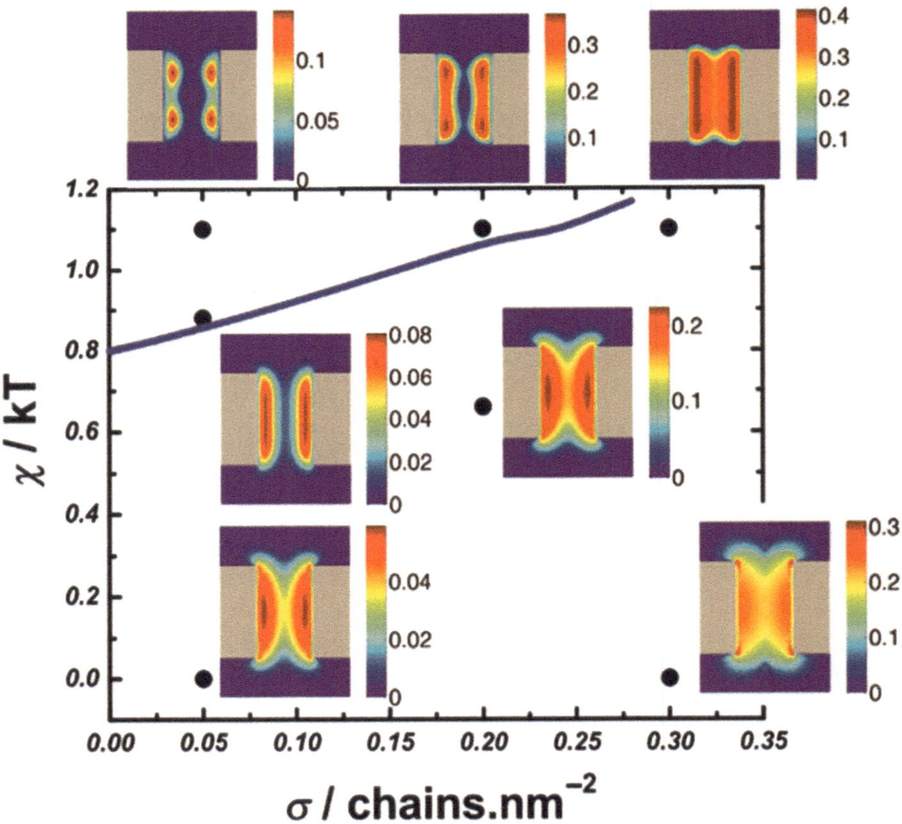

FIGURE 6.8 Volume fraction of the polymer for short nanopores with inner walls modified by a neutral end-grafted polymer for different values of the effective segment–segment attraction strength, χ and grafting density, σ. The solid lines show the onset of thermodynamic stability for the homogeneous layer (calculated for an infinite cylinder), see text. Other calculation parameters, $L = 24$ nm, $R = 7.5$ nm, and $N = 45$. Source: *Reproduced from reference Peleg, O.; Tagliazucchi, M.; Kroeger, M.; Rabin, Y.; Szleifer, I. Morphology Control of Hairy Nanopores.* ACS Nano ***2011***, 5, *4737–4747.*

to $\partial\mu_P/\partial\sigma = 0$ (determined for an infinitely long pore): systems that lie approximately above this line are expected to microphase separate, while the polymer films in systems that lie below the line are expected to be the homogeneous. The molecular theory calculations show, in agreement with the thermodynamic argument, that the systems for $\chi = 1.1\ k_BT$ form aggregates along the z-axis.

Just below the onset of microphase separation (see density maps for $\sigma = 0.3$ chains $\cdot$ nm^{-2}, $\chi = 1.1\ k_BT$ and $\sigma = 0.05$ chains $\cdot$ nm^{-2}, $\chi = 0.88\ k_BT$ in Fig. 6.8), the polymer chains collapse to the walls of the pore and have practically no polymer density in the reservoirs. We refer to the collapse of polymer chains onto the walls of the pore as the *collapse-to-the-walls* mechanism. This mechanism has been recently associated with changes in transport through the pore for stimuli responsive systems.[38] As we will show next, the collapse to the walls is one of the two possible collapse mechanisms in polymer-modified nanopores.

Fig. 6.9 shows the morphology of polymer-modified short pores in poor solvent conditions as a function of the dimensions of the pore. For wide pores ($R < 10$ nm), we observe the formation of aggregates on the walls of the channel, i.e., the *collapse-to-the-walls* mechanism. The number of the aggregates depends on the length of the pore. These aggregates have toroidal shape because we assumed homogeneity in the angular coordinate in this calculations and, therefore, microphase separated structures are symmetric with respect to rotations

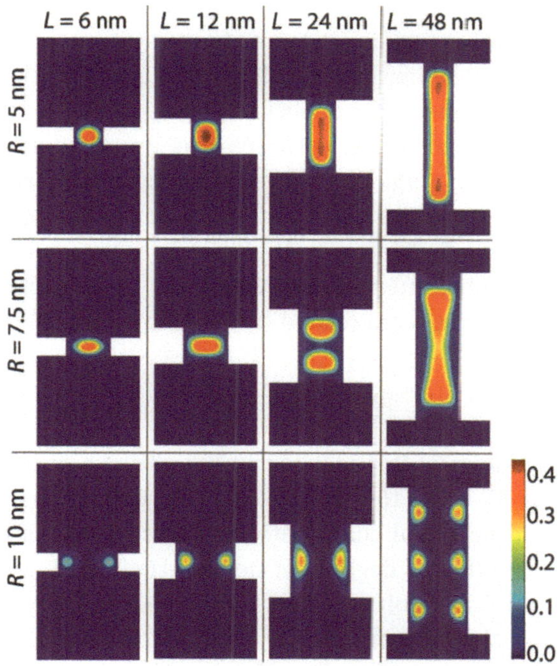

FIGURE 6.9 Color maps showing the volume fraction of the polymer for short nanopores with inner walls modified by a neutral end-grafted polymer in poor-solvent conditions. Calculation parameters, $N = 75$, $\sigma = 0.05$ chains $\cdot$ nm^{-2}, and $\chi = 1.1\ k_BT$. Source: *Reproduced from reference Peleg, O.; Tagliazucchi, M.; Kroeger, M.; Rabin, Y.; Szleifer, I. Morphology Control of Hairy Nanopores.* ACS Nano ***2011***, 5, *4737–4747.*

around the main axis of the pore. In a system where no symmetries are assumed, it would be possible to have inhomogeneities in the angular dimension as well.[39,40] As the radius of the pore decreases, chains can overlap near the pore axis. This process produces the formation of aggregates in the center of the pore (e.g., $R = 5$ nm and $L = 6$ nm) instead of the walls. We refer to the formation of a central plug in poor solvent conditions as the *collapse-to-the-center* mechanism. These two collapse mechanisms have been used in the past to explain the temperature dependence of conductivity in poly(*N*-isopropylacrylamide) modified colloidal films (see chapter: Biomimetic Smart Nanopores and Nanochannels).[38] It is important to note that the concepts of *collapse-to-the-walls* and *collapse-to-the-center* mechanisms are just oversimplified descriptions for the process of the microphase separation that do not capture the details of the rich morphologies observed in Fig. 6.9, such as toroidal aggregates and stacked disk. Experimental realization of these complex morphologies may be useful to gate transport through the pore or for nanofabrication.

6.3.3 Polyelectrolyte-Modified Nanopores and Nanochannels

6.3.3.1 Polymer Organization in Polyelectrolyte-Modified Nanopores and Nanochannels

In this section, we address the structure of nanopores and nanochannels modified by polyelectrolytes (i.e., charged polymers). The polyelectrolyte chains in nanopores may strongly reorganize in the presence of ion currents in the system, thus we will address in this section only the structure in equilibrium (i.e., for a zero applied bias), while the structure out of equilibrium will be discussed later in Sections 6.4.3 and 6.4.4.

Let us consider the system in Fig. 6.10, which was experimentally studied by Yameen et al.[41] This system consists in a single long-aspect-ratio cylindrical nanochannel ($L = 12$ μm and $R = 7.5$ nm) connecting two reservoirs of identical composition. The inner walls of the channel are modified by a grafted layer of 4PVP. Each segment in 4PVP is a weak base, which can exist in a deprotonated neutral state (pyridine) or in a single charged protonated (pyridinium) state. These two states are related by the following acid–base equilibrium:

$$\mathrm{py} + \mathrm{H}^+ \underset{}{\overset{K_a}{\rightleftharpoons}} \mathrm{pyH}^+ \tag{6.58}$$

where the equilibrium constant of this reaction in the absence of interactions (i.e., for free monomers in the bulk) is $K_a = 6.3 \cdot 10^{-6}$ ($\mathrm{p}K_a = 5.2$). Fig. 6.11 shows the volume fraction of the 4PVP as a function of the radial coordinate as predicted by the equilibrium molecular theory (Section 6.2.1), for different degrees of polymerization, N, and pH = 2 and 10. For pH = 2 ($\mathrm{pH} << \mathrm{p}K_a$), the polyelectrolyte is fully protonated and therefore it adopts a highly stretched conformation. In the case of long ($N = 55$) and intermediate ($N = 28$) chains, the length of fully stretched 4PVP chains (11.6 nm for $N = 28$ and 22.8 nm for $N = 55$) is larger than the radius of the pore ($R = 7.5$ nm) and, therefore, the theory predicts that the density of 4PVP is almost homogeneous inside the channel (Fig. 6.11A and C). However, for short chains ($N = 14$), the length of the fully stretched chain (5.8 nm) is shorter than the radius of

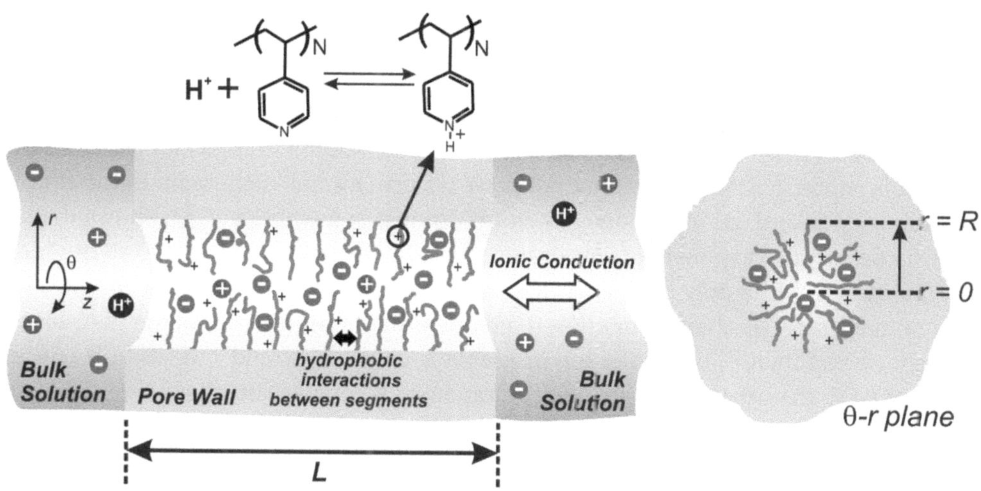

FIGURE 6.10 Scheme of a polyelectrolyte-modified nanochannel. The inner walls of a cylindrical channel of radius *R* and length *L* are modified with end-tethered chains of poly(4-vinyl pyridine) (4PVP). The 4-vinyl pyridine in 4PVP can be either in a positively charged protonated state (pyridinium) or in a neutral deprotonated state (pyridine), which are related by an acid–base equilibrium. The channel connects to two macroscopic reservoirs containing the same concentration of KCl and the same pH. Each reservoir contains a reversible electrode. Ion currents through the system result from applying a potential difference ΔV to these electrodes. Source: *Reproduced from reference Tagliazucchi, M.; Azzaroni, O.; Szleifer, I. Responsive Polymers End-Tethered in Solid-State Nanochannels: When Nanoconfinement Really Matters. J.* Am. Chem. Soc. ***2010***, 132, *12404–12411.*

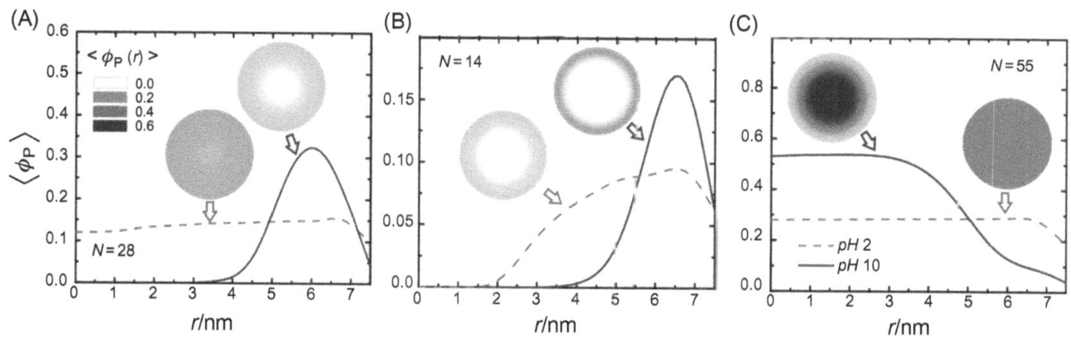

FIGURE 6.11 Polymer volume fraction of 4PVP as a function of the distance from the center of the pore for different chain lengths, *N*, and pH = 2 and 10. Calculation parameters: $\sigma = 0.2$ chains · nm^{-2}, $C_{salt} = 0.1$ M, bulk $pK_a = 5.2$, and $R = 7.5$ nm. Source: *Reproduced from reference Tagliazucchi, M.; Azzaroni, O.; Szleifer, I. Responsive Polymers End-Tethered in Solid-State Nanochannels: When Nanoconfinement Really Matters.* J. Am. Chem. Soc. ***2010***, 132, *12404–12411.*

the channel and, thus, it cannot reach its center (Fig. 6.11B). When the pH is increased to 10 (pH ≫ pK_a), the 4PVP is almost completely uncharged and behaves as a neutral polymer in poor-solvent conditions. In these conditions, the systems with short and intermediate chain lengths ($N = 14$ and 28) collapse to the walls of the channel, while the channel with long chains ($N = 55$) collapses to the center.

6.3.4 Effects of Nanoconfinement on Ion Distribution and Acid-Base Equilibria

In Fig. 6.12 we show the concentration of the ions as a function of the radial coordinate for the case of $N = 28$ and pH = 2 and 10. For pH = 10, the concentration of ions in the system is similar to the bulk concentration of 0.1 M, although we observe a small depression in the concentration of both cations and anions around $r = 6$ nm. Comparison with Fig. 6.11, shows that at this position the density of the polymer is maximum, thus the depletion of ions at $r = 6$ nm is due to steric repulsions with the polymer. When the pH is decreased to 2, 4PVP gets completely protonated. Anions enter the channel in order to compensate for the charges of the polyelectrolyte: Fig. 6.12 shows that the Cl^- concentration inside the channel is around 2 M (its bulk concentration is 0.1 M). On the other hand, the concentration of K^+ inside the channel is decreased to around 0.001 M due to electrostatic repulsions with the polyelectrolyte.

Fig. 6.13A shows the fraction of protonated segments (averaged over the whole channel) as a function of the pH of the solution as predicted by the molecular theory. As a comparison, we show also the results expected for pyridine units in the absence of interactions, calculated using the common formula for acid–base equilibrium in bulk solution:

$$f = \frac{1}{1 + 10^{(\mathrm{pH} - \mathrm{p}K_a)}} \tag{6.59}$$

with the $\mathrm{p}K_a = 5.2$ determined for pyridine in the bulk. We observe that the pyridine units inside the nanochannel require a lower pH than that in the bulk in order to become protonated. We can quantify the effect by defining the apparent $\mathrm{p}K_a$, $\mathrm{p}K_a^{\mathrm{app}}$, as the pH where

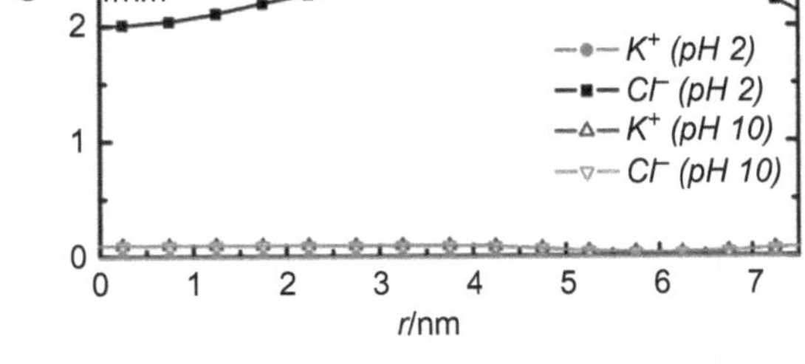

FIGURE 6.12 Molar concentration of K^+ and Cl^- as a function of the distance to the center of the pore for the same system as in Figure 6.11, $N = 28$ and pH = 2 and 10. The inset in the upper right corner shows the concentrations in logarithmic scale. Source: *Reproduced from reference Tagliazucchi, M.; Azzaroni, O.; Szleifer, I. Responsive Polymers End-Tethered in Solid-State Nanochannels: When Nanoconfinement Really Matters.* J. Am. Chem. Soc. ***2010***, 132, *12404–12411.*

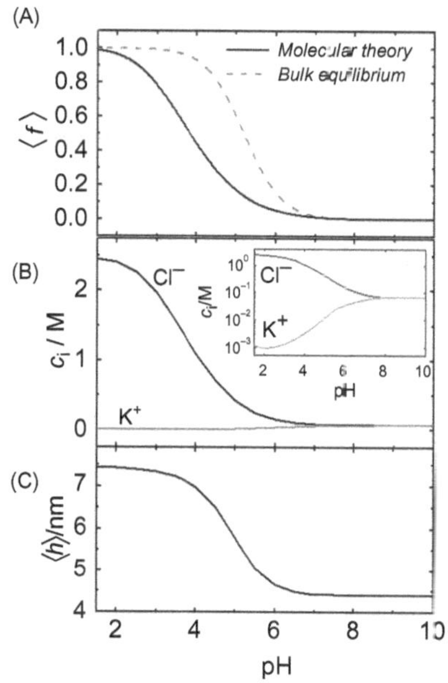

FIGURE 6.13 (A) Fraction of protonated pyridine segments, (B) average concentration of salt ions, and (C) average thickness of the film inside the pore defined according to Eq. (6.60) as a function of the pH of the solution for the system of Fig. 6.11 with $N = 28$. Source: *Adapted from reference Tagliazucchi, M.; Azzaroni, O.; Szleifer, I. Responsive Polymers End-Tethered in Solid-State Nanochannels: When Nanoconfinement Really Matters.* J. Am. Chem. Soc. ***2010***, 132, *12404–12411.*

half of the pyridines inside the nanochannel are protonated. For the example of Fig. 6.13A, we find $\mathrm{p}K_\mathrm{a}^{\mathrm{app}} = 3.83$. The $\mathrm{p}K_\mathrm{a}^{\mathrm{app}}$ is smaller than the bulk $\mathrm{p}K_\mathrm{a}$ because it is energetically more costly to protonate a pyridine unit inside the positively charged 4PVP layer than in the bulk. This process, where the acid–base equilibrium is shifted in order to minimize the electrostatic repulsions, is known as charge regulation (see also chapter: Theoretical Basis for Structure and Transport in Nanopores and Nanochannels) and occurs at the cost of increasing the chemical free energy of the system.

In addition to the charge-regulation mechanism, the 4PVP brush can also reduce electrostatic repulsions by uptake of counterions at the cost of decreasing the translational entropy of the counterions. Fig. 6.13B shows that, in fact, during the protonation transition the amount of counterions in the system increases (and that of co-ions decreases). Finally, electrostatic repulsions can also be reduced by increasing the distance between segments by stretching the brush. We studied the pH dependence of the thickness of the 4PVP, $\langle h \rangle$, which we define from the first moment of the distance from the channel wall $(R - r)$ weighted by the density profile of the polyelectrolyte:

$$\langle h \rangle = \frac{3\int_0^R \langle \phi_\mathrm{P}(r) \rangle (R - r) r \, \mathrm{d}r}{\int_0^R \langle \phi_\mathrm{P}(r) \rangle r \, \mathrm{d}r} \tag{6.60}$$

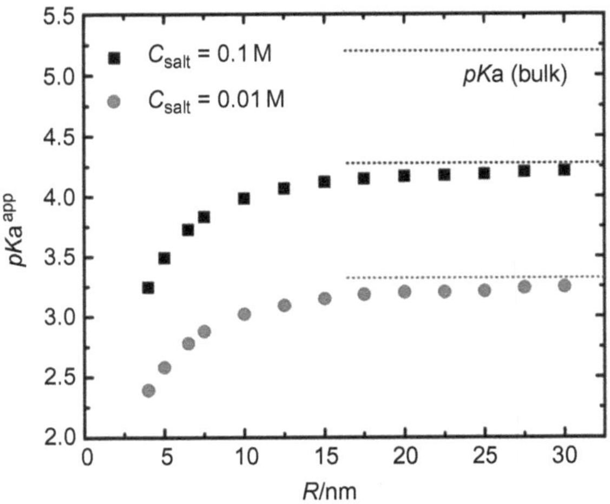

FIGURE 6.14 Apparent pK_a (defined as the pH where half of the pyridine units in the channel are protonated) as a function of channel radius for the same system as in Fig. 6.11 and $N = 28$ for two different salt concentrations. The upper dotted line indicates the pK_a of pyridine in the absence of interactions (p$K_a^{bulk} = 5.2$), the other dot lines show the apparent pK_a as calculated for the planar brushes at $C_{salt} = 0.1$ and 0.01 M, respectively. Source: *Reproduced from reference Tagliazucchi, M.; Azzaroni, O.; Szleifer, I. Responsive Polymers End-Tethered in Solid-State Nanochannels: When Nanoconfinement Really Matters.* J. Am. Chem. Soc. ***2010***, 132, *12404–12411.*

(this definition assures $\langle h \rangle = R$ for a homogeneously filled channel). Fig. 6.13C shows that the thickness of the 4PVP layer increases with decreasing pH. Therefore, the three regulation mechanisms (charge regulation, counterion uptake/release, and swelling/deswelling) are operational at the same time and their relative contribution is the one that minimizes the total free energy of the system.

A very interesting question is how charge regulation operates under nanoconfinement. Our theory predicts that the pK_a^{app} decreases with decreasing channel radius, Fig. 6.14. For wide channels ($R > 30$ nm), the pK_a^{app} of the polyelectrolyte in the pore converges to that on a planar surface (dotted lines). Note that the pK_a^{app} in planar layers is still shifted with respect to the bulk pK_a[15,42,43] due to the charge-regulation effect described above. The effect of the radius of the channel on the pK_a^{app} simply modulates the charge-regulation effect due to the fact that the available volume scales as r/R as one moves away from the surface toward the axis. Therefore, the available volume at a given distance from the surface of the channel decreases as R decreases. This leads to an increase in the density of charged segments within the polyelectrolyte layer, which tends to augment the electrostatic repulsions among segments. These repulsions are partially prevented by the charge-regulation mechanism, hence decreasing the pK_a^{app} as R decreases in order to make it energetically more difficult to protonate a pyridine unit. Decreasing the salt concentration diminishes the importance of the counterion uptake mechanism and augments the electrostatic repulsions between charged segments, thus it boosts charge-regulation effects and decreases the pK_a^{app}.

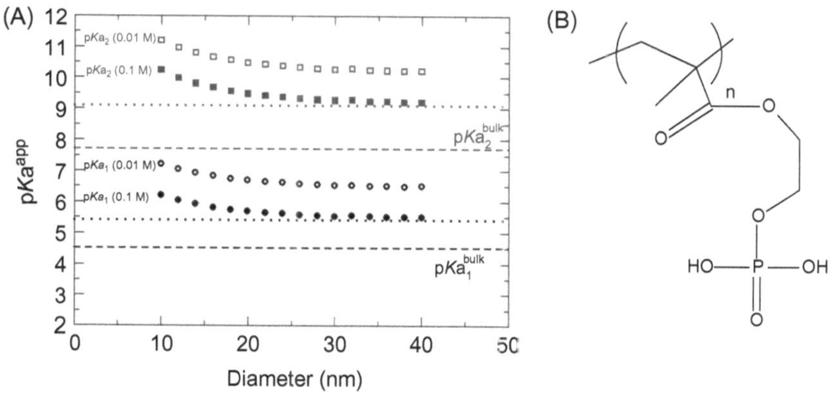

FIGURE 6.15 (A) Apparent pK_as for the phosphate groups in poly(2-(methacryloy oxy)ethyl-phosphate) (PMEP) chains end-grafted to the inner walls of a long cylindrical channel as a function of the diameter of the channel. Each monomer in PMEP (see structure in panel B) has a phosphate group that can exist in three protonation states: diprotonated, monoprotonated, and unprotonated. The pK_{a1}^{app} is defined as the pH where the concentration of diprotonated phosphate groups is equal to that of monoprotonated phosphates. The pK_{a2}^{app} is defined as the pH where the concentration of monoprotonated phosphates is equal to that of fully unprotonated phosphates. The calculations correspond to $C_{salt} = 0.1$ M (solid symbols) or $C_{salt} = 0.01$ M (open symbols). Other calculation parameters: $\sigma = 0.2$ chains/nm^2, degree of polymerization $N = 28$, p$K_{a1}^{bulk} = 4.5$, p$K_{a2}^{bulk} = 7.7$. Source: *Adapted from reference Gilles, F.M.; Tagliazucchi, M.; Azzaroni, O.; Szleifer, I. Ionic Conductance of Polyelectrolyte-Modified Nanochannels: Nanoconfinement Effects on the Coupled Protonation Equilibria of Polyprotic Brushes. J. Phys. Chem. C* ***2016****, 120, 4789–4798.*

In Fig. 6.15A, we present calculation results for a system similar to that in Fig. 6.10 but based on a different end-grafted polyelectrolyte, PMEP,[44] see structure in Fig. 6.15B. PMEP is a diprotic phosphoric acid, whose segments can exist in three different states of charge: neutral (diprotonated phosporic acid), −1 (monoprotonated acid), and −2 (unprotonated phosphate). This system has, therefore, two acid–base transitions:

$$P - PO_3H_2 \overset{K_a}{\rightleftharpoons} P - PO_3H^- + H^+ \tag{6.61}$$

and

$$P - PO_3H^- \overset{K_a}{\rightleftharpoons} P - PO_3^{2-} + H^+ \tag{6.62}$$

The pK_as in bulk for these two transitions are 4.5 and 7.7.[44] Fig. 6.15 shows the same plot as that in Fig. 6.14 (pK_a^{app} vs R) for the PMEP-modified nanochannel. For both transitions, we observe that the pK_a^{app} differs from the bulk pK_a, however while for 4PVP the pK_a^{app} was smaller than that the bulk pK_a, the pK_a^{app} for both protonation reactions in PMEP is actually larger than their bulk counterparts. This difference arises because 4PVP is a polybase and PMEP is a polyacid. The state of protonation in both cases depends on the difference (pH−pK_a^{app}). For 4PVP, decreasing (pH−pK_a^{app}) increases the charge of the polyelectrolyte; therefore, charge regulation works by decreasing pK_a^{app}, so that (pH−pK_a^{app}) increases and the film becomes less charged. On the other hand, the PMEP layer becomes more negatively

Table 6.2 Qualitative Description of the Effect of Surface Curvature and Salt Concentration (C_{salt}) on the Apparent pK_a of Surface-Immobilized Acid–Base Species

	Convex Surface (Sphere or Cylinder)	**Concave Surface (Channel, Pore, or Spherical Hole)**
Polyacid or acid-terminated ligand	$pK_a^{bulk} < pK_a^{app}(R) < pK_a^{app,\ planar}$ $\uparrow R$, $\uparrow pK_a^{app}$ $\uparrow C_{salt}$, $\downarrow pK_a^{app}$	$pK_a^{bulk} < pK_a^{app,\ planar} < pK_a^{app}(R)$ $\uparrow R$, $\downarrow pK_a^{app}$ $\uparrow C_{salt}$, $\downarrow pK_a^{app}$
Polybase or base-terminated ligand	$pK_a^{app,\ planar} < pK_a^{app}(R) < pK_a^{bulk}$ $\uparrow R$, $\downarrow pK_a^{app}$ $\uparrow C_{salt}$, $\uparrow pK_a^{app}$	$pK_a^{app}(R) < pK_a^{app,\ planar} < pK_a^{bulk}$ $\uparrow R$, $\uparrow pK_a^{app}$ $\uparrow C_{salt}$, $\uparrow pK_a^{app}$

charged as (pH−pK_a^{app}) increases; thus the charge regulation mechanism works by displacing pK_a^{app} toward positive values in order to decrease (pH−pK_a^{app}) and shift the acid–base equilibrium toward the neutral layer. As one may expect, the effect of decreasing the radius of PMEP-modified channels is to further increase pK_a^{app}.

In Table 6.2, we summarize the effect of charge regulation on the pK_a^{app} of polyacids or polybases grafted on a concave surface (such as a channel or pore). For completeness, we also include in the table the same information for planar and convex surfaces (such as nanoparticles,[45] nanorods, or nanowires). Regardless of the curvature, the pK_a^{app} is always larger than the bulk pK_a for polyacids and smaller than the bulk pK_a for polybases for the reasons discussed in the previous paragraph. On the other hand, the magnitude of the charge-regulation effect (i.e., the magnitude of the shift of the apparent pK_a from the bulk pK_a) increases with decreasing radius for channels and pores (as shown in Figs. 6.14 and 6.15), but decreases with decreasing radius for nanoparticles and nanowires. The latter behavior is explained by the fact that the available volume increases as r/R as one moves away from a convex surface, so the smaller the radius, the less dense is the polyelectrolyte layer (see Fig. 6.5) and the smaller the electrostatic repulsions among segments. The effect of decreasing salt concentration is in all cases to increase the magnitude of charge regulation and, therefore, increase the difference between the bulk and apparent pK_as.

In addition to the pK_a^{app}, it is also interesting to consider the width of the acid–base transition, as it provides the range of pH required to switch the charge state of the polyelectrolyte within the channel. We define the width of the transition, Δ, as the pH difference required to go from a fraction of protonation of 1/11 to a fraction of protonation of 10/11, namely

$$\Delta = \text{pH}(f = 10/11) - \text{pH}(f = 10/11) \tag{6.63}$$

This definition of Δ guarantees $\Delta = 2$ for the ideal case of an isolated acid–base group in solution (this can be easily proved by combination of Eqs. (6.59) and (6.63)).

The protonation transition for 4PVP (results in Fig. 6.13A) within the pore has $\Delta = 2.97$. This increase in the transition width with respect to the ideal one can be understood by

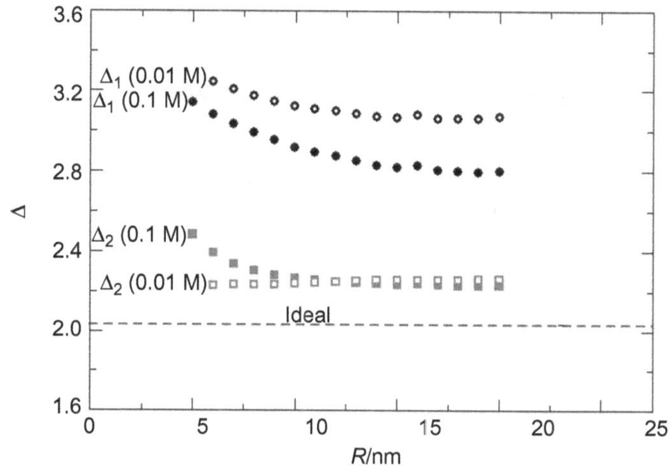

FIGURE 6.16 Width of the transition, Δ, as a function of the radius of the channel, *R* for a nanochannel coated by MEP (same system as Fig. 6.15). The width is defined as the pH range where the degree of dissociation (either for the first, Eq. (6.61), or the second, Eq. (6.62), protonation reaction) changes from 1/11 to 10/11. Solid and open symbols correspond to $C_{salt} = 0.1$ and 0.01 M, respectively. Other calculation conditions were the same as in Fig. 6.15. Source: *Adapted from reference Gilles, F.M.; Tagliazucchi, M.; Azzaroni, O.; Szleifer, I. Ionic Conductance of Polyelectrolyte-Modified Nanochannels: Nanoconfinement Effects on the Coupled Protonation Equilibria of Polyprotic Brushes. J. Phys. Chem. C* ***2016****, 120, 4789–4798.*

recalling the fact that during the protonation transition there is an increase in the charge of the 4PVP layer, which increases the free energy cost of protonating a pyridine unit within the pore. The value of $\Delta = 2$ is the ideal transition width predicted for the bulk, where the free-energy cost of protonation is constant. Inside the pore, the increase in the free-energy cost of protonation with increasing pH stretches the f–pH curve and leads to transition widths larger than the ideal one. Fig. 6.16 shows the widths of the two acid–base transitions of PMEP. In this case, the ideal width is slightly larger than 2 ($\Delta = 2.03$) due to the coupling between equilibria (6.61) and (6.62) even in bulk solution. The values of Δ_1 and Δ_2 for these two dissociation transitions inside the nanochannel are, however, much larger than the ideal value, for the same reasons explained above for 4PVP. Decreasing the salt concentration or the radius of the channel increases the electrostatic interactions in the system and thus widens the acid–base transitions. Another interesting observation is that, for any given radius and salt concentration, Δ_1 is always greater than Δ_2. The reason for this is the following: in the first transition the segments go from a neutral to a -1 state. Thus, the film goes from being completely uncharged to have a large charge density, which triggers very large changes in the thickness and concentration of counter-ions in the film. These changes occur as the pH increases, changing the free energy of protonation and, thus, widening the transition. The same effect occurs during the second transition, but in this case the charge of the segments goes from -1 to -2 and, thus, the reorganization is less drastic than for the first transition, so the width of the transition is smaller.

6.4 Conductance of Polyelectrolyte-Modified Nanopores and Nanochannels

6.4.1 Quasi-Equilibrium (Low-Bias) Conductance of Long Nanochannels

So far, we discussed the equilibrium properties and organization of nanochannels and nanopores. We focus now on study of their conductance by first considering long nanochannels in the quasi-equilibrium (low-bias) regime. In these conditions, the molecular organization of the system is the same as in equilibrium, i.e., that described in the previous section. From the knowledge of the equilibrium structure of a long nanochannel, we can calculate its conductance using Eq. (6.34) for cylindrical channels or Eq. (6.36) for noncylindrical ones (under the additional assumption that the channel cross section changes slowly with z).

Let us first consider a long nanochannel modified by end-grafted PVP chains, i.e., the system in Fig. 6.10. The conductance of this system has been measured experimentally by Yameen et al.[41] Fig. 6.17 shows the experimental pH-dependent conductance for this system (solid symbols), along with the predictions resulting from using Eq. (6.34) with the concentration of ions in equilibrium predicted by the molecular theory. Very good agreement is observed between theory and experiment. It is important to mention that the molecular theory does not require any fitting parameter a priori, although in practice not all parameters of the system are known and one should choose them based on educated guesses and/or fitting. In the case of the experiments of Fig. 6.17, the dimensions of the pore, the pK_a of the pyridine in the bulk and the salt concentration and pH of the measuring solution were well known. The diffusion coefficients of the ions required in Eq. (6.34) were approximated by those in pure solvent[46] because the content of the water of the pore is large (~90%, see Fig. 6.11).

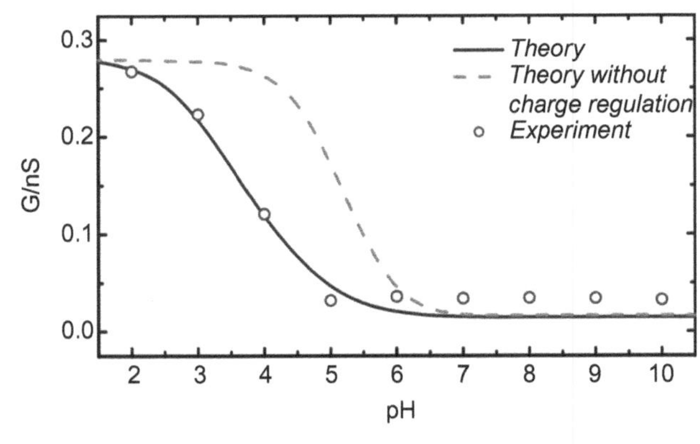

FIGURE 6.17 Comparison between the experimental conductance determined by Yameen et al.[41] and the predictions of the molecular theory and Eq. (6.34) for the system in Fig. 6.10. In the calculations without charge regulation, the fraction of the pyridine segments in 4PVP was fixed to the bulk value, i.e., using Eq. (6.59) (see also bulk equilibrium curve in Fig. 6.13A). Source: *Reproduced from reference Tagliazucchi, M.; Azzaroni, O.; Szleifer, I. Responsive Polymers End-Tethered in Solid-State Nanochannels: When Nanoconfinement Really Matters.* J. Am. Chem. Soc. ***2010***, 132, *12404–12411*.

The surface coverage of 4PVP was chosen to be $\sigma = 0.2$ chains · nm^{-2} from experiments performed on planar surfaces with the same grafting technique.[47] Finally, we chose the polymerization degree of $N = 28$ in order to obtain good agreement with the experiments. This value is very close to the value $N = 32$ measured for polyelectrolyte brushes tethered inside mesoporous metal oxide films.[48] As we will show below, the product $N \cdot \sigma$ controls the current for low pHs; the fact that very good agreement between theory and experiment can be achieved over the whole pH range (in particular the position of the inflection choice) by adjusting only N supports the predictive power of our theoretical approach.

We next analyze the mechanisms that control the change of conductance with the solution pH. The panel for $N = 28$ in Fig. 6.11 showed that 4PVP collapses on the wall of the channel at pH = 10 and homogeneously occupies the pore for pH = 2. Since the channel has low conductance at pH 10 and high at pH 2, we conclude that steric blocking of the channel is not the main mechanism controlling conductance. On the other hand, the G versus pH curve in Fig. 6.17 has a similar shape to that of the fraction of charged pyridines versus pH in Fig. 6.13A, which indicates that the conductance of the channel is governed by an electrostatic mechanism. The conductance of the nanochannel is, in fact, proportional to the concentration of ions inside it, see Eq. (6.34). At high pHs, 4PVP is neutral and the ion concentration inside the channel is similar to that in bulk solution (Fig. 6.13B). As the pH is lowered, the pyridine units protonate and the concentration of anions inside the channel increases in order to compensate for the increasing fixed charge of 4PVP. Note that while the concentration of co-ions decreases in this process, the total ion concentration (anions plus cations) increases, thus, the conductance of the channel also increases.

It is also worthwhile to mention the importance of using a theoretical framework that explicitly includes charge regulation to model the pH-dependent conductance of the system in Fig. 6.17. For instance, neglecting charge regulation and using the bulk value for the fraction of protonated pyridines in the theory (Eq. (6.59)) provide a G versus pH curve that fails to reproduce the experimental results both in the position of the inflexion point and the width of the conductance transition (see dashed line in Fig. 6.17). In other words, the molecular theory provides a way to incorporate the effect of the local environment and the molecular details of the system into the conductance calculations. This tool can be used to propose and validate simplified theories and models. For example, Fig. 6.18 shows the conductance of the system as a function of $\sigma \cdot N \cdot R$ (the product of channel radius, surface coverage, and chain length) for different combinations of N, σ, R, and C_{salt} and pH = 2 (i.e., when the polyelectrolyte is fully protonated). The predictions of the theory show an universal dependence between G and $\sigma \cdot N \cdot R$ that provides a simple design rule. This universal behavior arises because, when all the segments are protonated, the concentration of fixed charges inside the channel scales linearly with the number of segments, which is given by the product of the chains per unit length ($2 \cdot \pi \cdot R \cdot \mathrm{d}L$) and the length of the chains (N). If the most important contribution to conductance comes from the counterions of the polyelectrolyte, then the conductance is proportional to the number of segments in the system. Assuming complete protonation of P4VP moieties, this quantity should be proportional to the number of polymer segments per unit length and thus to $\sigma \cdot N \cdot R$.

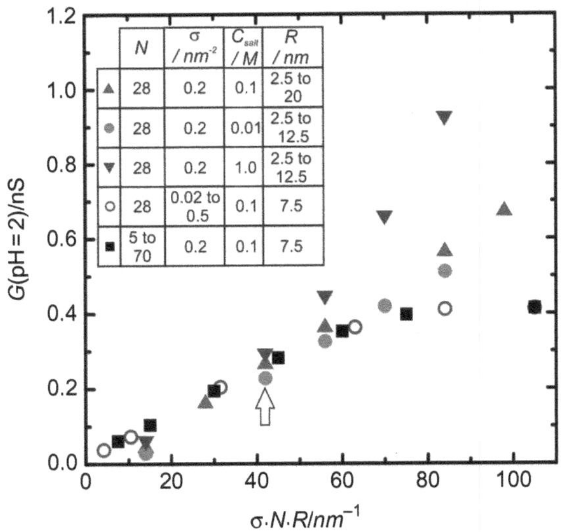

FIGURE 6.18 Conductance of long nanochannels modified by end-grafted 4PVP (system in Fig. 6.10) at pH 2 (completely protonated pyridines) as a function of $\sigma \cdot N \cdot R$. The different combinations of parameters used in the calculations are listed in the inset table. The arrow shows the conductance of the channel of Fig. 6.17.
Source: *Reproduced from reference Tagliazucchi, M.; Azzaroni, O.; Szleifer, I. Responsive Polymers End-Tethered in Solid-State Nanochannels: When Nanoconfinement Really Matters.* J. Am. Chem. Soc. ***2010***, 132, *12404–12411.*

The low-bias conductance of long noncylindrical nanochannels can be calculated with Eq. (6.36). Fig. 6.19 compares the conductance–pH curves predicted for polyelectrolyte-modified nanochannels of different shapes modified by the diprotic polyacid PMEP. The shapes considered in this study are: a cylindrical channel with $R = 5$ nm, a cylindrical channel with $R = 145$ nm, and conical and trumpet-shaped channels with $R_{tip} = 5$ nm and $R_{base} = 145$ nm. The conductance follows the order G (cylindrical, $R = 5$ nm) $< G$ (trumpet-shape) $< G$ (conical) $< G$ (cylindrical, $R = 145$ nm). For the same length, tip, and base radii, the trumpet-shaped channel has a smaller conductance than the conical one because at any position along the channel axis, the cross section of the trumpet-shaped channel is smaller than that of the cylindrical channel. The pK_a^{app} approximately determined from the inflection points in the G versus pH curves also follows the trend: pK_a^{app}(cylindrical $R = 5$ nm) $> pK_a^{app}$ (trumpet-shape) $> pK_a^{app}$ (conical) $> pK_a^{app}$ (cylindrical, $R = 145$ nm) $> pK_a^{bulk}$, because, as explained in Section 6.3.4, decreasing the radius of the channel increases the apparent pK_a of the polyacids.

6.4.2 Quasi-Equilibrium (Low-Bias) Conductance of Short Nanopores

In the previous section we studied quasi-equilibrium through very long channels. Let us consider now the low-bias quasi-equilibrium conductance of short nanopores. In this case, the structure of the system is still that of equilibrium, although the approach to calculate the

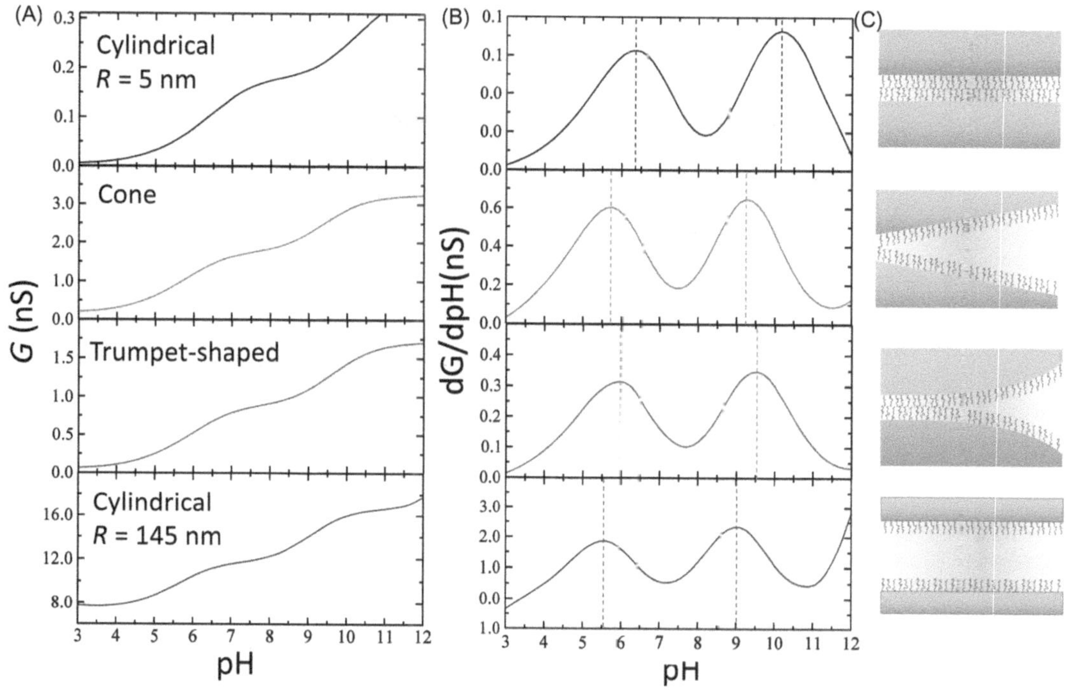

FIGURE 6.19 Conductance of cylindrical (A), conical (B), and trumped-shaped nanochannels (C) modified by end-grafted layers of the diprotic polyacid MEP as a function of solution pH. The dashed lines show the derivatives of the conductance with respect to pH and the vertical dotted lines indicate the position of the maxima of these derivatives (i.e., the inflection points of *G*). Calculations for $R_{tip} = 5$ nm, $R_{base} = 145$ nm, $L = 12$ μm, and $C_{salt} = 0.1$ M. Source: *Adapted from reference Gilles, F.M.; Tagliazucchi, M.; Azzaroni, O.; Szleifer, I. Ionic Conductance of Polyelectrolyte-Modified Nanochannels: Nanoconfinement Effects on the Coupled Protonation Equilibria of Polyprotic Brushes. J. Phys. Chem. C* ***2016****, 120, 4789–4798.*

conductance is complicated by the geometry of the system and the presence of the reservoirs and, thus, simple equations such as (6.34) and (6.36) (valid for long nanochannels) cannot be used anymore. The results presented below were calculated with the nonequilibrium molecular theory (Section 6.2.2) for a small applied bias. Alternatively, it is in principle possible to solve the equilibrium organization of the system performing an equilibrium calculation with an equilibrium molecular theory (Section 6.2.1) and then self-consistently solve for the fluxes (calculated with Eq. (6.37)) that fulfil the continuity condition, Eq. (6.50), while keeping all ionic concentrations fixed to their equilibrium values. In practice, we did not explore this approach as we believed it may be almost as complex as the full solution of the nonequilibrium molecular theory with the disadvantage that it cannot, of course, be extended to the nonlinear regime.

Fig. 6.20A shows the polymer volume fraction and the ionic current lines in short bare nanopores, and polyelectrolyte-modified nanopores where the inner walls (system I) or both

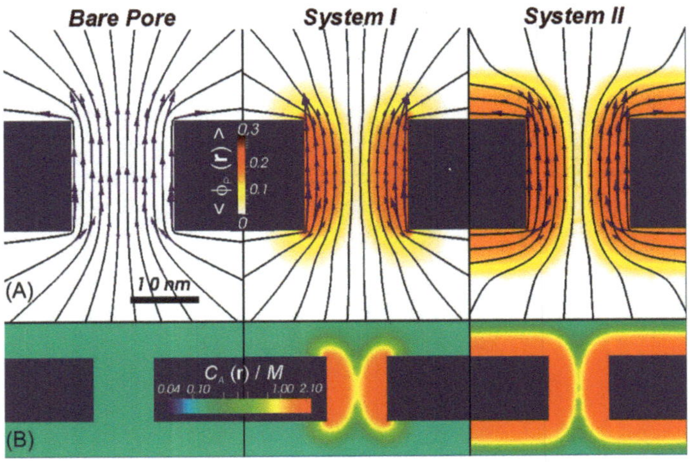

FIGURE 6.20 (A) Volume fraction of the polyelectrolyte inside short nanopores for a cut along a plane containing the axis of the pore. Results are shown for a bare pore, a pore with chains of a strong polycation grafted onto the inner walls only (system I) and a pore with chains of a strong polycation grafted both onto the inner and outer walls (system II). The arrows show the direction and magnitude of ionic currents under quasi-equilibrium conditions (low applied bias) and the solid lines are the flow lines for this vector field. (B) Molar concentration of anions for the systems in (A). Calculation parameters: $N = 20$, $\sigma = 0.5$ chains $\cdot$ nm^{-2}, $R = 7.5$ nm, $L = 16$ nm, $C_{salt} = 0.1$ M, f (charge per segment)= 0.5. Source: *Reproduced from reference Tagliazucchi, M.; Rabin, Y.; Szleifer, I. Ion Transport and Molecular Organization Are Coupled in Polyelectrolyte Modified Nanopores.* J. Am. Chem. Soc. ***2011***, 133, *17753–17763.*

inner and outer walls (system II) are coated by an end-grafted layer of a strong polycation. The distribution of the polyelectrolyte in the system is qualitatively similar to that of neutral polymers in good-solvent conditions, Section 6.3.1, although the chains are more stretched than in the case of a neutral polymer, in order to reduce the electrostatic repulsions among segments.

The arrows in Fig. 6.20A show current vectors that point to the direction of the ion fluxes in the system (the current lines are tangential to these vectors at each point). The size of the arrows is proportional to the magnitude of ion fluxes. The current lines run parallel to the axis of the channel inside the bare nanochannel and with a hemispherical symmetry in the reservoirs. The magnitude of the current vectors is constant inside the channel (all arrows have the same size) but it scales with as r^2 outside the channel, where r is the distance from the mouth of the channel. For large r, the arrows are no longer visible.

In the case of systems I and II, the current fluxes are bent due to the presence of the polyelectrolyte. This process can be understood from the local concentration of charge carriers (ions) in the system. Since the polyelectrolyte is a polycation, the anions are the majority carriers in the system and, thus, we plotted their concentration in Fig. 6.20B. As expected, the distribution of chloride counterions is similar to that of the polycations. The current vectors are larger in the regions containing large concentration of counterions. This organization creates a high-flux region within the polyelectrolyte layer and a low-flux region along the axis of the pore.

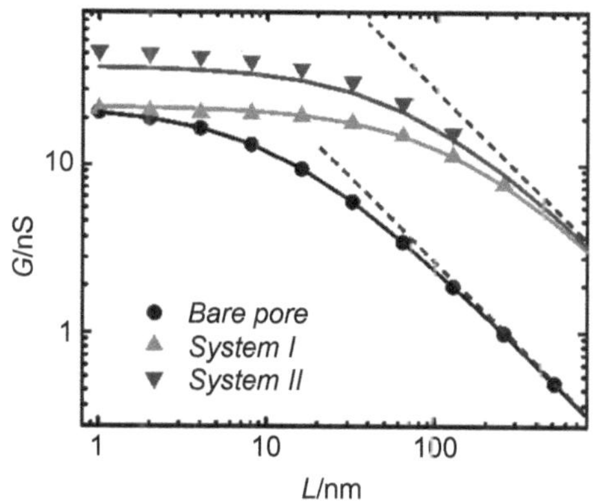

FIGURE 6.21 Quasi-equilibrium (low bias) conductance for short polyelectrolyte-modified nanopores (same system as in Fig. 6.20) as a function of the length of the pore. Symbols show the calculations of the nonequilibrium molecular theory, solid lines are predictions of the simple analytical model Eqs. (6.64)–(6.66), and the dashed line is the conductance expected for a very long nanochannel determined from Eq. (6.34) and an equilibrium molecular theory calculation for an infinitely long channel. Same calculation parameters as for Fig. 6.20. Source: *Adapted from reference Tagliazucchi, M.; Rabin, Y.; Szleifer, I. Ion Transport and Molecular Organization Are Coupled in Polyelectrolyte Modified Nanopores.* J. Am. Chem. Soc. ***2011***, 133, *17753–17763.*

Fig. 6.21 shows the conductance of the systems in Fig. 6.20 as a function of the length of the pore. The symbols are calculations of the nonequilibrium molecular theory, while the solid lines are predictions of a simple model we developed to estimate the low-bias conductance of polyelectrolyte-modified nanochannels. This model is guided by the results of the full molecular theory calculations discussed above and its derivation follows the spirit of that used in Chapter 2, Theoretical Basis for Structure and Transport in Nanopores and Nanochannels, to calculate the conductance of a bare channel, i.e., we approximate the total resistance of the system as a series combination of the resistance of the pore and the reservoirs. If a polyelectrolyte-brush layer is present in a region (either reservoirs or pore), we consider two parallel conductance pathways in that region: through the bulk solution and through the polyelectrolyte layer (i.e., we consider both bulk and surface conductances). The full derivation of the model is rather tedious and we will not reproduce it here as it is freely available online as Supporting Information of Ref. 13. The final equations of the model (used to calculate the solid lines in Fig. 6.21) are as follows:

$$G^{-1} = \frac{1}{\pi\kappa_s}\frac{L}{R^2} + \frac{1}{2R\kappa_s} \tag{6.64}$$

for a bare pore (same equation derived in chapter: Theoretical Basis for Structure and Transport in Nanopores and Nanochannels)

$$G^{-1} = \frac{1}{\pi f N\sigma R\kappa_s}\frac{C_{salt}L}{} + \frac{1}{2R\kappa_s} \tag{6.65}$$

for system I and,

$$G^{-1} = \frac{1}{\pi} \frac{C_{\text{salt}} L}{f N \sigma R \kappa_s} + \frac{1}{\pi \kappa_s} \left[\frac{2C_{\text{salt}}}{f\sigma N} \left(\ln\left(R + \frac{1}{2} \frac{f\sigma N}{C_{\text{salt}}} \right) - \ln(R) \right) \right] + \left(\frac{1}{2} - \frac{1}{\pi} \right) \frac{1}{\kappa_s R} \tag{6.66}$$

for system II, where κ_s is the conductivity of the solution, f is the (fixed) fraction of charge of the segments in the polyelectrolyte, and the rest of the symbols were already defined in this chapter (see also definition of symbols in Appendix I). The derivation of these equations makes the following assumptions (described in further detail in the Supporting Information of Ref. 13): the brush layer inside the pore is completely permselective (i.e., the salt concentration is low) and the thickness of the polyelectrolyte brush is similar to the pore radius. The first term in each of the Eqs. (6.64)–(6.66) represents resistance of the pore, while the other terms arise from the resistance of the reservoirs (the access resistance).

All curves in Fig. 6.21 have two well-defined regions.[49] For $L \ll R$, the total conductance of the system is dominated by the resistance of the reservoirs, which is independent of the length of the pore (see Eqs. (6.64)–(6.66)), and thus G for $L \ll R$ is independent of L. Moreover, the conductance of the bare pore and system I (polyelectrolytes on the inner walls only) converge to the same G as L decreases because in both systems the reservoirs are free of polyelectrolytes (with the exception of the polyelectrolytes that stretch through the mouth of the pore in system I). On the other hand, in the limit $L \gg R$, the conductance of the channel scales as L^{-1}: in this regime the total resistance is dominated by the pore resistance, which scales as L. Systems I and II have the same conductance in this regime because both systems have grafted polyelectrolytes inside the pore. Moreover, in these conditions the channel can be approximated by an infinitely long channel and, thus, its conductance can be determined using Eq. (6.34) in conjunction with a molecular theory calculation considering inhomogeneities only r (i.e., a 1D problem). The G versus L curves calculated with this approximation (dashed lines in Fig. 6.21) show that the approximation is rather good for polyelectrolyte-modified nanochannels with $L > 1$ μm.

In Section 6.4.1, we discussed two mechanisms that change the local ion concentration in the channel: electrostatic interactions and steric repulsions. As we just explained, grafting a polyelectrolyte to a channel increases the conductance (Fig. 6.21) by the electrostatic-interaction mechanism. Fig. 6.22 shows results for a system modified with a neutral polymer. In this case, the polymer interacts with the ions by excluded-volume interactions and, therefore, the concentration of both anions and cations inside the polymer layer is depleted (Fig. 6.22B). Therefore, the currents are smaller near the walls of the pore than near its axis (Fig. 6.22A) and the total conductance of the pore is smaller than that of a bare pore (Fig. 6.22C). This decrease in conductance upon grafting macromolecules into the pore is expected only when their degree of charge is zero (e.g., neutral polymers) or small (e.g., weakly charged proteins[50]) and, therefore, counterions are not incorporated into the channel in large amounts.

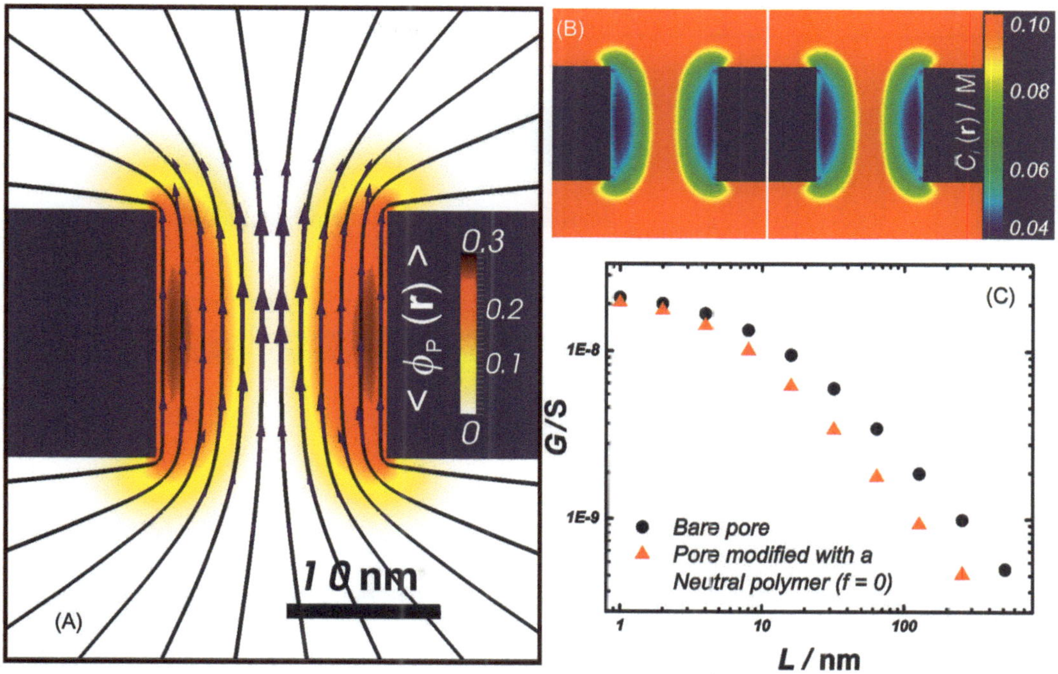

FIGURE 6.22 (A) Volume fraction of the polymer inside short nanopores modified by a neutral polymer for a cut along a plane containing the axis of the pore. The arrows show the direction and magnitude of the ion currents. (B) Color maps for the molar densities of anions (left) and cations (right) for the system shown in (A). (C) Quasi-equilibrium conductances of the pore in panel A and a bare pore as a function of the length of the channel, *L*. Other calculation parameters are the same as in Fig. 6.21. Source: *Reproduced from reference Tagliazucchi, M.; Rabin, Y.; Szleifer, I. Ion Transport and Molecular Organization Are Coupled in Polyelectrolyte Modified Nanopores.* J. Am. Chem. Soc. ***2011***, 133, *17753–17763.*

6.4.3 Nonequilibrium (Large-Bias) Transport in Nanopores: Polyelectrolytes as Electromechanical Gates

So far we have studied systems in the quasi-equilibrium regime, where the potential applied to the channel or pore is small and thus the structure of the system in the steady state is the same as in equilibrium. In this limit, life is linear: the system can be always approximated as a combination of resistances and there is a linear relationship between current and potential (ohmic behavior). If the applied potential is increased beyond the quasi-equilibrium regime, then the structure of the system changes with the applied bias and nonlinear $I-\Delta V$ may arise. What is the threshold of applied potential required to observe this nonlinear behavior? In principle, the quasi-equilibrium limit applies only for $\Delta V \rightarrow 0$, however the system will be approximately linear until the electrostatic energy associated to the applied bias ($|e| \cdot \Delta V$) becomes comparable to the available thermal energy ($k_B T$), in other word, $\Delta V > 25$ mV at room temperature.[12] In Chapter 2, Theoretical Basis for Structure and Transport in Nanopores and Nanochannels, we discussed two effects that arise due to the reorganization

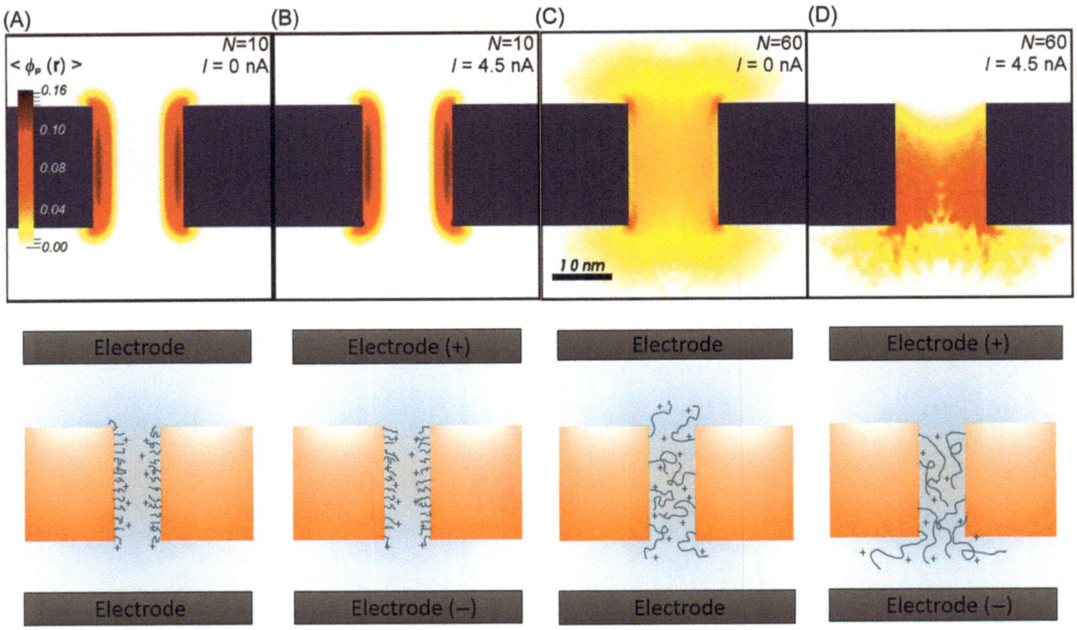

FIGURE 6.23 Volume fraction of the polyelectrolyte for a cut along a plane containing the pore axis for a short nanopore with inner walls modified by an end-grafted layer of a strong polycation for $N = 10$ (panels A and B) and $N = 60$ (panels C and D), both in equilibrium (panels A and C) and under an applied potential bias (panels B and D). Calculation parameters $\sigma = 0.3$ chains · nm^{-2} (for $N = 10$) and 0.05 chains · nm^{-2} (for $N = 60$), $R = 7.5$ nm; $L = 20$ nm; $C_{salt} = 0.1$ M. The lower panel shows schemes of the organization of the polyelectrolyte in each system.
Source: *Adapted from reference Tagliazucchi, M.; Rabin, Y.; Szleifer, I. Ion Transport and Molecular Organization Are Coupled in Polyelectrolyte Modified Nanopores.* J. Am. Chem. Soc. ***2011***, *133, 17753–17763.*

of the ions in the high-bias limit for nanochannels with a fixed surface charge: polarization concentration and current rectification. In this section, we will focus our attention on polyelectrolyte-modified pores in the high-bias limit and show the emergence of new interesting effects that develop due to the reorganization of the polyelectrolyte.

Let us consider again a short pore coated by strong polyelectrolytes in the quasi-equilibrium limit, i.e., System I in Fig. 6.20 of the previous section. Panels A and C of Fig. 6.23 show the volume fraction of the polyelectrolyte in equilibrium conditions ($\Delta V = 0$) for two different lengths of the end-grafted polyelectrolytes, $N = 10$ and $N = 60$. Each of these systems has also different grafting densities ($\sigma = 0.3$ and 0.05 chains · nm^{-2} for $N = 10$ and 60, respectively) in order to keep the product $N \cdot \sigma$ constant, so that they have approximately the same conductance in the quasi-equilibrium limit (see Fig. 6.18 in Section 6.4.1). These equilibrium volume fractions can be qualitatively understood using the same arguments developed for uncharged polymers in good solvent, Section 6.3.1, with the difference that polyelectrolytes stretch more than neutral polymers due to electrostatic repulsions. Panels B and D of Fig. 6.23 show the polyelectrolyte volume fraction under a positive applied potential, such that the current though the pore is 4.5 nA. We observe that the end-grafted layer for $N = 60$ readily reorganizes in the presence of this applied potential and stretches

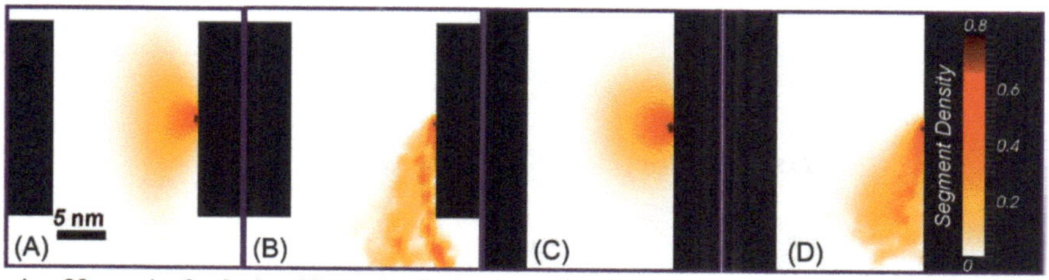

FIGURE 6.24 2D projections of the average density distribution of single chains for polycation chains grafted at the center of either a short nanochannel with $L = 20$ nm (A, B) or a long channel of $L = 128$ nm (C, D) in equilibrium conditions (A, C), or in the presence of ionic currents (B, D). Note that polycation chains are end-grafted on the entire inner surface of the pore (i.e., like in Fig. 6.23), but we show only the maps for a single chain grafted to the center. Other calculations parameters: $N = 60$, $\sigma = 0.05$ chains $\cdot$ nm^{-2}, and $C_{salt} = 0.1$ M. Source: *Reproduced from reference Tagliazucchi, M.; Rabin, Y.; Szleifer, I. Ion Transport and Molecular Organization Are Coupled in Polyelectrolyte Modified Nanopores.* J. Am. Chem. Soc. ***2011***, 133, *17753–17753.*

toward the lower reservoir, where the negatively biased electrode is located. On the contrary, for $N = 10$, the potential-induced deformation is barely observed. In Fig. 6.24, we show the single-chain density profiles for a chain with $N = 60$ grafted to the center of a short pore (panels A and B) and to the center of a very long nanochannel (panels C and D). In the absence of an applied bias, the single chain has a symmetric profile in both cases (panels B and D). This profile is elongated in the z direction for the short nanopore due to the presence of the reservoirs,[31] but not for the long nanochannel. When a potential bias is applied to get a current of $I = 2$ nA, the single chain profiles deform toward the lower reservoir (Fig. 6.24B and D). The effect is larger for the short pore (Fig. 6.24B) than for the long one (Fig. 6.24D) due to the extra volume provided by the reservoirs to place the stretched chain.

It is worthwhile to stress here that the effect of the applied potential on the polyelectrolyte conformation is a truly nonequilibrium effect, which could not occur in the absence of ion currents. Let us consider for a moment replacing the reversible electrodes in the reservoirs by ideally polarizable electrodes. Applying a potential bias to these electrodes will charge their double layers, but, after an initial transient required to charge the electrical double layer (EDL), current will stop flowing since no Faradaic processes are possible. The potential applied to the electrodes will influence only those molecules located within the EDL, i.e., a few Debye lengths from the surface of the electrode. Therefore, the potential cannot influence the polyelectrolyte since the membrane containing the pore is separated from the electrodes by a macroscopic distance. In summary, the reorganization of the polyelectrolyte in the presence of an applied bias requires the presence of ion currents in the system, as these currents sustain the electric field inside the pore that is responsible for stretching the polyelectrolyte chains. Note that the electric field inside the channel is proportional to the ionic current flowing through it and, for this reason, we decided to compare systems with the same ionic current rather than the same applied bias in Fig. 6.23. The electromechanical manipulation of polyelectrolytes through nanopores has been proposed in the past as a mechanism to gate ionic currents through DNA-modified nanochannels.[51]

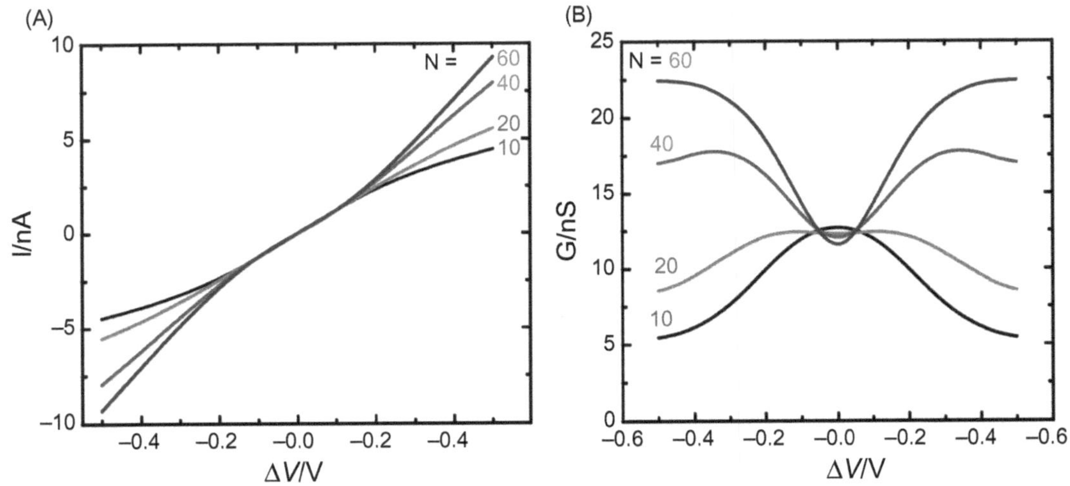

FIGURE 6.25 (A) Current potential curves predicted by the nonequilibrium molecular theory for the system in Fig. 6.23 for different chain lengths (the surface coverages were adjusted to conserve the total number of segments, $N \cdot \sigma$). (B) Differential conductance $G = \partial I / \partial \Delta V$ versus ΔV for the systems in (A). The same calculation parameters as in Fig. 6.23 were used. Source: *Reproduced from reference Tagliazucchi, M.; Rabin, Y.; Szleifer, I. Ion Transport and Molecular Organization Are Coupled in Polyelectrolyte Modified Nanopores.* J. Am. Chem. Soc. ***2011***, 133, *17753–17763.*

Fig. 6.25A shows the I versus ΔV curves for the system of Fig. 6.22 for different chain lengths of the end-grafted polyelectrolyte, N. Once again, we change σ in each case to keep a constant $N \cdot \sigma$. All curves have similar slope around $\Delta V = 0$ due to the fact that they have the same $N \cdot \sigma$ (Section 6.4.1). However, for $|\Delta V| > 0.1$ V, the curves deviate from the ohmic behavior. This behavior is also well illustrated by the differential conductance (defined as the derivative of I with respect to ΔV for each ΔV) in Fig. 6.25B. Interestingly, the conductance for $N = 10$ and $N = 20$ drops with increasing $|\Delta V|$, while it increases for $N = 40$ and $N = 60$. The relationship between this result and the structure of the system discussed in Figs. 6.23 and 6.24 is not simple and requires an in-depth analysis.

Fig. 6.26 analyzes the distribution of the anions in the short nanopores. The anions are the counterions of the end-grafted polycations and thus operate as charge carriers in the system. The steady-state anion concentration shows a depletion region at the lower entrance and an accumulation region at the upper one (panels *iii* and *iv* in Fig. 6.26A), which is characteristic of the concentration polarization effect,[20,52–54] (see discussion in chapter: Theoretical Basis for Structure and Transport in Nanopores and Nanochannels for pores with fixed surface charges). A simple explanation for this effect is based on the fact that the anion conductance of the pore is larger than that of the reservoirs. In the conditions of Fig. 6.26, the anions enter the pore from the lower reservoir and they are transported through the pore (which has a very low resistance for anions) into the positively biased upper reservoir. The conductance of the reservoirs, however, is not enough to supply anions at the lower entrance and to remove them at the upper one, therefore low and high anion-concentration regions evolve at the lower and upper entrances, respectively. Once the system reaches its steady state,

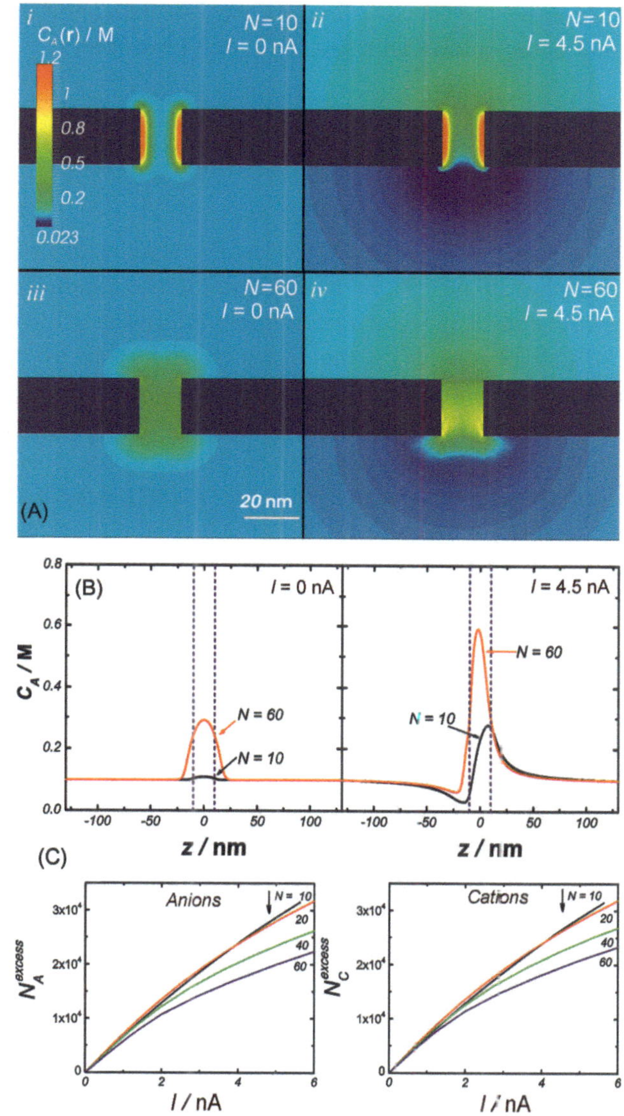

FIGURE 6.26 (A) Map of the molar concentration of anions for the same systems as in Fig. 6.23. (B) Concentration profiles along the axis of the pore ($r = 0$) for the systems shown in (A). The position of the membrane surfaces is shown with dotted lines. (C) Plot of the number excess of anions and cations at $z > 0$ (upper half of the pore) as defined in Eq. (6.67) as a function of the total current through the pore. Source: *Reproduced from reference Tagliazucchi, M.; Rabin, Y.; Szleifer, I. Ion Transport and Molecular Organization Are Coupled in Polyelectrolyte Modified Nanopores.* J. Am. Chem. Soc. ***2011***, 133, *17753–17763.*

the diffusional flux driven by the concentration gradients balance the migration fluxes. The cations, which are the minority carrier, also show depletion and accumulation regions at the lower and upper entrances, respectively, because they need to balance the distribution of anions in order to prevent large deviations from electroneutrality.

In order to better describe the ion distribution, we calculated the excess of anions and cations in the upper half of the system with respect to the lower half as:

$$N_i^{\text{excess}} = 2\pi \left[\int_0^{\infty} \mathrm{d}z \int_0^{\infty} r\,\mathrm{d}r c_i(r,z) - \int_{-\infty}^{0} \mathrm{d}z \int_{-\infty}^{\infty} r\,\mathrm{d}r c_i(r,z) \right] N_A \tag{6.67}$$

where $c_i(r, z)$ is the molar concentration at r and z and N_A is Avogadro's number.

Fig. 6.26C shows a plot of N_i^{excess} as a function of the chain length. Interestingly, the magnitude of the concentration polarization effect decreases with increasing polyelectrolyte length (this is also observed by the fact that the sizes of the anion depletion and enhancement regions in Fig. 6.26A are larger for $N=10$ than for $N=60$). The effect of chain length can be explained by the difference in voltage-driven polyelectrolyte reorganization shown in Fig. 6.23. Long polyelectrolytes reorganize in the presence of the applied bias and stretch toward the lower entrance (see Fig. 6.23D), where the anion depletion region is located. As they move toward the lower entrance, the polyelectrolyte chains carry their counterions with them and, therefore, increase the local conductance of that region. This effect results in a decrease in the magnitude of anion depletion, compared to the system with $N=10$, which, as we shown above (Fig. 6.23C), does not deform in the presence of the applied potential.

In summary, the concentration polarization effect decreases the conductance of the system with respect to the quasi-equilibrium one due to the formation of concentration gradients that oppose to the flow of the majority charge carriers. This effect explains, therefore, the decrease of conductance with increasing $|\Delta V|$ observed in the $I-\Delta V$ curves of Fig. 6.25 for $N=10$ and 20. The increase of conductance with increasing $|\Delta V|$ observed in Fig. 6.25B for $N=40$ and 60 is more challenging to explain. The differences in reorganization of ions and polyelectrolyte observed between $N=10$ and $N=60$ (Figs. 6.23 and 6.26) suggest that polyelectrolyte segments at the lower mouth of the pore play a key role in the effect. Hence, we ascribe the increase of G with ΔV to the combination of three mechanisms. First, the polyelectrolyte segments at the lower entrance decrease the access resistance of the pore, which is the main contribution for short nanopores. Second, the concentration polarization effect is decreased for $N=60$ with respect to $N=10$ due to polyelectrolyte reorganization (Fig. 6.26). Third, stretching the polyelectrolyte out of the pore allows to increase the concentration of the co-ions (cations) and therefore their conductance increases (e.g., for $I=4.5$ nA, we calculate that the cations carry 15% of the total current for $N=10$, but 24% for $N=60$). The complexity of the distribution of ions and polyelectrolyte in the system makes it very difficult to identify the importance of each individual mechanism. The important message is, however, that the reorganization of the polyelectrolyte inside the pore in the presence of the applied bias can feed back into the ion currents: ion transport and polyelectrolyte conformations are two coupled and inseparable processes.

6.4.4 Nonequilibrium Transport in Nanopores With Broken Symmetry and Current Rectification

The $I-\Delta V$ curves in Fig. 6.25A are nonlinear but symmetric, i.e., $I(\Delta V) = -I(-\Delta V)$. This symmetry arises from the fact that the system is symmetric with respect to the plane of the membrane (i.e., the system is symmetric with respect to the plane $z = 0$ in Fig. 6.1). This symmetry can be broken, for example, using asymmetric noncylindrical nanochannels (e.g., conical channels),[5–7,10,12,19–24] by having solutions of different composition in the reservoirs,[55,56] or by having an asymmetric charge distribution in an, otherwise, cylindrical system, e.g., a cylindrical pore where the upper and lower parts of the inner walls have charges of different sign[57–65] (see chapter: Theoretical Basis for Structure and Transport in Nanopores and Nanochannels). We will explore here an example of the latter strategy: a short nanopore that has unmodified inner walls and outer walls modified by polyelectrolyte brushes, Fig. 6.27A. The polyelectrolyte brushes on both sides of the membrane have opposite charge: the upper side is modified by a polycation and the lower one by a polyanion. We choose to study this system because we believe it may be reasonably easy to prepare experimentally: we imagine it could be possible to modify both sides of an ultrathin membrane with the polyelectrolytes first and then drill the pore using an electron beam[34]

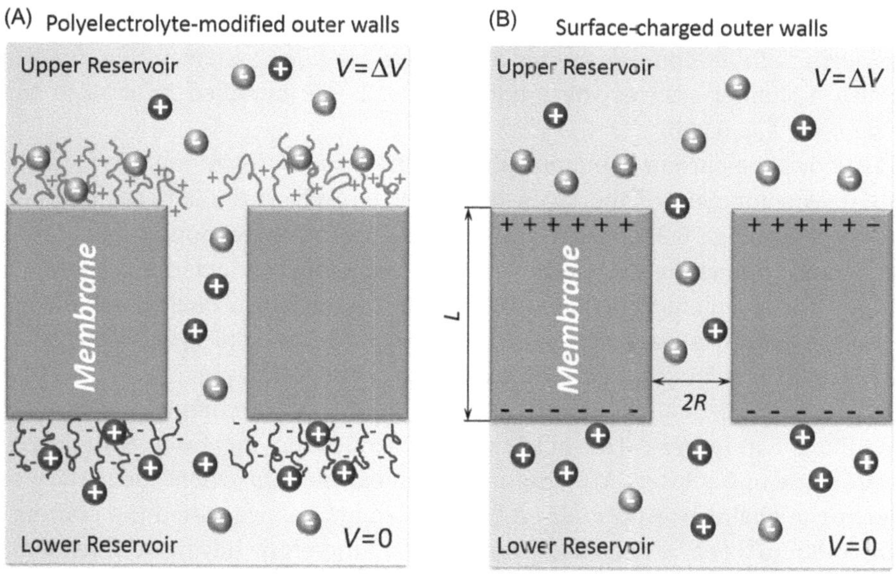

FIGURE 6.27 (A) Scheme of a short nanopore with bare inner walls and outer walls modified by end-grafted layers of strong polyelectrolytes of opposite charge (each segment in the polyelectrolytes has a charge of $f = \pm 1/3$). The pore connects two reservoirs with the same composition and reversible electrodes (we adopt the convention $V = 0$ for the electrode in the lower reservoir). (B) Same as panel A, but the outer walls are modified by surface charges instead of polyelectrolyte layers. Source: *Adapted from reference Tagliazucchi, M.; Rabin, Y.; Szleifer, I. Transport Rectification in Nanopores with Outer Membranes Modified with Surface Charges and Polyelectrolytes.* ACS Nano ***2013****, 7, 9085–9097.*

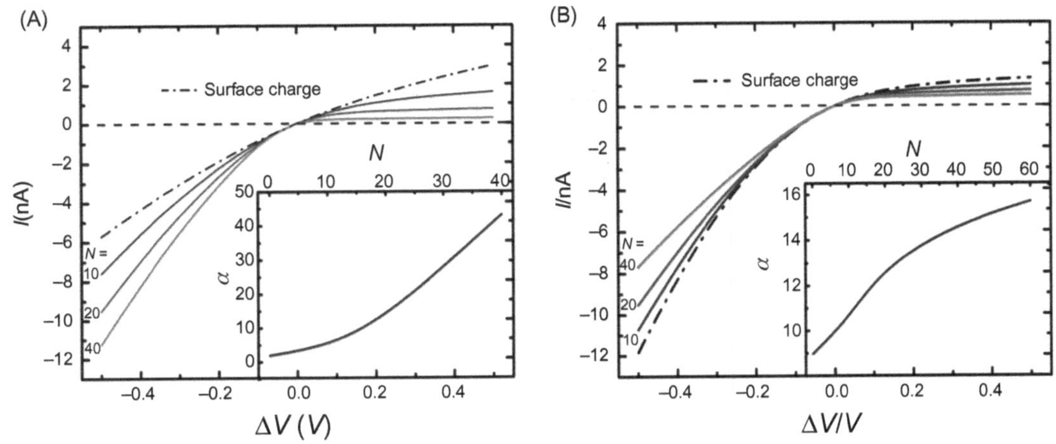

FIGURE 6.28 Current–potential curves predicted by the nonequilibrium molecular theory for the systems schematized in Fig. 6.27 for different lengths of the polyelectrolyte chains (*N*). The insets show the rectification factor α_R as a function of *N*. The grafting density of the polyelectrolytes is fixed in panel A to $\sigma = 0.15$ chains $\cdot$ nm^{-2} ($\sigma_q = 0.15$ |e| $\cdot$ nm^{-2} for the surface-charged outer walls), while in panel B it was varied to keep a constant total surface charge density ($N \cdot \sigma \cdot f$) of 1.0 |e| $\cdot$ nm^{-2} ($\sigma_q = 1.0$ |e| $\cdot$ nm^{-2} for the surface-charged outer walls). Source: *Reproduced from reference Tagliazucchi, M.; Rabin, Y.; Szleifer, I. Transport Rectification in Nanopores with Outer Membranes Modified with Surface Charges and Polyelectrolytes.* ACS Nano ***2013****, 7, 9085–9097.*

or an ion beam.[66] In addition to the polyelectrolyte-modified system in Fig. 6.27A, we will consider also a simpler system where the outer walls are modified by surface charges of opposite sign, see Fig. 6.27B.

Fig. 6.28 shows the current–potential plots for the polyelectrolyte-modified system shown in Fig. 6.27A as a function of the chain length, *N*, as well as the surface-charge-modified channel shown in Fig. 6.27B. All $I-\Delta V$ curves are strongly non-ohmic and show a large conductance branch for $\Delta V<0$ ("on" state) and a low conductance branch for $\Delta V>0$ ("off" state). It is convenient to define a parameter to quantify this rectification effect, we use here the current rectification factor at 0.5 V, $\alpha_R = -I(\Delta V = -0.5\text{ V})/I(\Delta V = 0.5\text{ V})$ ($\alpha_R > 1$ indicates a current-rectifying system, while an ohmic system has $\alpha_R = 1$).

Fig. 6.28A shows results for a constant grafting density (or charge density for surface-charge-modified nanopores). The efficiency of rectification increases with increasing *N* (see α_R vs *N* curve in the inset). We also performed calculations where the surface coverage of polyelectrolyte chains was normalized by their length, i.e., we varied the grafting density with *N* in order to keep a constant product $N \cdot \sigma$; therefore having a constant number of charges per unit area of the external wall. Even after this normalization, we observe that the rectification efficiency increases with *N*. The important questions are, therefore, (1) what the mechanism of current rectification for the systems in Fig. 6.28 is, (2) how the polyelectrolyte chains in the system organize, and (3) how this organization affects current rectification.

Let us consider first the mechanism of current rectification in a surface-charge-modified nanochannel (the same mechanism with small differences apply to the polyelectrolyte-modified

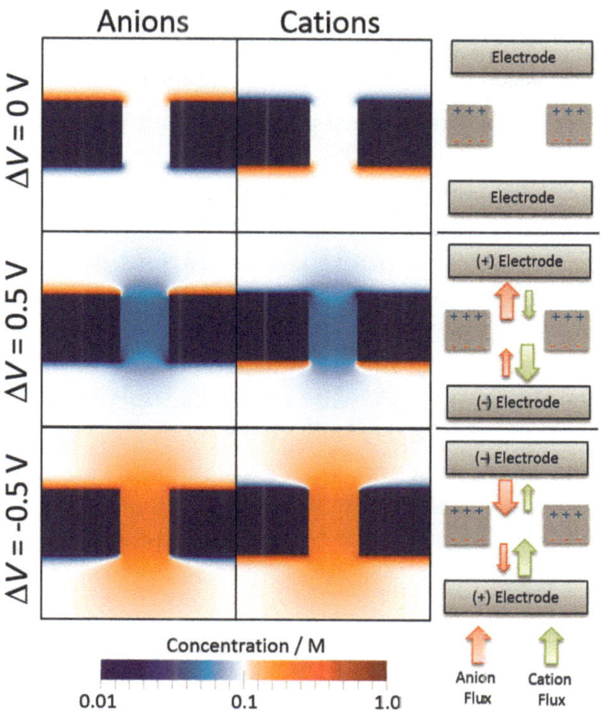

FIGURE 6.29 Left: Concentration of anions and cations in a plane containing the pore axis in equilibrium and for different applied biases. The calculations correspond to a short nanopore ($R = 7.5$ nm, $L = 20$ nm) with outer walls modified by surfaces charges with $\sigma_q = 0.5\ |e| \cdot nm^{-2}$. Other parameters, $C_{salt} = 0.1$ M. Right: Schemes showing the direction and magnitude of the ion fluxes. Source: *Adapted from reference Tagliazucchi, M.; Rabin, Y.; Szleifer, I. Transport Rectification in Nanopores with Outer Membranes Modified with Surface Charges and Polyelectrolytes.* ACS Nano ***2013***, *7, 9085–9097.*

pore, as we explain below). Current rectification is a nonequilibrium process that results from the reorganization of salt ions within the pore due to the applied potential. In Fig. 6.29, we show the concentration of anions and cations for $\Delta V = 0$ V (equilibrium), 0.5 V ("off" state), and -0.5 V ("on" state). In equilibrium ($\Delta V = 0$), the concentration of the ions inside the pore is the same as in the bulk solution as the influence of the charges of the outer wall is limited to a thin region adjacent to the outer walls (the thickness of this region is, of course, of the order of the Debye length, $\lambda_{Debye} \sim 1$ nm for the 0.1 M electrolyte used here). On the other hand, the ion concentration at the center of the pore in nonequilibrium conditions is smaller ($\Delta V = 0.5$ V, "off" state) or larger ($\Delta V = -0.5$ V, "on" state) than that in the bulk. The low and high currents of the "off" and "on" states can be explained by this change in the number of charge carriers in the center of the pore due to the applied potential. However, this observation does not explain why the distribution of the ions in nonequilibrium is different from that in equilibrium.

Let us consider the fluxes of anions and cations in the system as shown in the right panel of Fig. 6.29. For $\Delta V > 0$ V ("off" state) the anions flow toward the positively biased upper electrode. Therefore, the anions enter the pore through the pore entrance with negatively charged outer walls and leave it from the entrance with positively charged outer walls. Thus, the anions experience a large resistance to enter the channel, but a small resistance to leave it, hence their concentration inside the channel decreases. In the case of the cations, the direction of flow is reversed but they also interact with the polyelectrolytes in an opposite way to the anions; thus cations also experience a large resistance to enter the pore and a small one to exit it and their concentration at the center of the pore also drops. For $\Delta V < 0$ V the argument described above is reversed: ions experience a small resistance to enter the channel and a large one to exit it, so that their nonequilibrium concentrations at the center of the pore are larger than in equilibrium.

The mechanism described in the previous paragraph depicts qualitatively the origin of current rectification in the nanopores with chemically modified outer walls shown in Fig. 6.27. It is the basis for a more quantitative analysis that we developed in the form of a simple analytical model for current rectification in these systems. The derivation of the model is lengthy but freely available in the SI of Ref. 12, thus we will not reproduce it here. The model predicts the following nonlinear $I - \Delta V$ relationship:

$$\Delta V = \frac{I}{2}\left(\frac{\Omega^{\text{pore}}}{1 - \frac{I}{2}(\Omega^{=} - \Omega^{\neq})\beta|e|} + \Omega^{=} + \Omega^{\neq}\right) \tag{6.68}$$

where Ω^{i} are resistances for the flux of only one type of ions, i.e., the electrical resistance to the flux of cations or anions (which in the present case are the same due to the symmetry of the system), at zero applied bias. The fact that Ω^{i} are resistances for $\Delta V = 0$ makes it possible to determine them from quasi-equilibrium considerations. In particular, $\Omega^{=}$ is the access resistance for an ion entering the pore though the entrance that has outer walls with the same polarity of charge of the ion (hence the "=" superscript), while $\Omega^{\neq}$ is the access resistance for an ion entering the pore through the side with outer walls with an opposite charge polarity to that of the ion (of course, $\Omega^{\neq} < \Omega^{=}$). Finally, Ω^{pore} is the equilibrium resistance of the pore. Note that Eq. (6.68) is nonlinear and differs from Ohm's law for a system close to the equilibrium by the factor dividing Ω^{pore}. When this factor approaches unity, the system loses its rectification properties and the Ohmic behavior is restored. Eq. (6.68) shows that this can happen if (1) $I \rightarrow 0$, which requires $\Delta V \rightarrow 0$, i.e., the system is in the quasi-equilibrium limit, or (2) $\Omega^{\neq} = \Omega^{=}$, i.e., the ions experience the same resistance at both entrances, which can only occur if the asymmetric charge distribution in the system is removed.

The factor $1 - I/2(\Omega^{=} - \Omega^{\neq})\beta|e|$ in Eq. (6.68) accounts for the change of salt concentration inside the nanopore with ΔV. In other words, the salt concentration inside the pore as a function of the total ion current is given by,

$$c^{\text{pore}} = c^{\text{bulk}}\left(1 - \frac{I}{2}(\Omega^{=} - \Omega^{\neq})\beta|e|\right) \tag{6.69}$$

Note also that Eq. (6.68) is quadratic in the current, thus we can solve for *I*:

$$I = \frac{(\beta|e|\Delta V\Omega^- + \Omega^{\text{pore}} + \Omega^+) - ((\beta|e|\Delta V\Omega^- + \Omega^{\text{pore}} + \Omega^+)^2 - 4\beta|e|\Delta V\Omega^-\Omega^+)^{1/2}}{\beta|e|\Omega^-\Omega^+} \tag{6.70}$$

Finally, Eq. (6.70) can be used to calculate the current rectification factor for a given $|\Delta V|$ as:

$$\alpha_R(|\Delta V|) = \frac{-I(-\Delta V)}{I(\Delta V)} = \frac{\beta|e||\Delta V|\Omega^- - (\Omega^{\text{pore}} + \Omega^+) + ((\beta|e||\Delta V|\Omega^- - (\Omega^{\text{pore}} + \Omega^+))^2 + 4\beta|e||\Delta V|\Omega^-\Omega^+)^{1/2}}{\beta|e||\Delta V|\Omega^- + (\Omega^{\text{pore}} + \Omega^+) - ((\beta|e||\Delta V|\Omega^- + (\Omega^{\text{pore}} + \Omega^+))^2 - 4\beta|e||\Delta V|\Omega^-\Omega^+)^{1/2}} \tag{6.71}$$

where $\Omega^+ = \Omega^= + \Omega^{\neq}$ and $\Omega^- = \Omega^= - \Omega^{\neq}$. In Ref. 12 we have derived expressions for Ω^{pore}, $\Omega^=$, and $\Omega^{\neq}$ for the surface-charge-modified nanopore. These resistances can be calculated using quasi-equilibrium arguments, such as that employed in Chapter 2, Theoretical Basis for Structure and Transport in Nanopores and Nanochannels, to calculate the resistance of a bare short nanopore.

Fig. 6.30 shows systematic comparisons between the predictions of the nonequilibrium molecular theory and that of Eqs. (6.70) and (6.71) for several different sizes of pore (*L* and *R*), surface charge (σ_q), and bulk salt concentration (C_{salt}). These panels show that the simple model captures very well the current rectification effect and in many cases quantitatively agrees with the molecular theory. We have to mention that we did not derive so far expressions for $\Omega^=$ and $\Omega^{\neq}$ for polyelectrolyte-modified pores, although—in principle—such derivation may be done by combining the description of the polyelectrolyte layer used to derive Eqs. (6.64)–(6.66) (see Ref. 13) with the technique used in Ref. 12 to derive single-ion resistances for surface-charged pores.

We now focus again on the polyelectrolyte-modified pore in order to explore how the polyelectrolyte reorganizes in the presence of the applied bias. Fig. 6.31 shows the polyelectrolyte volume fraction in equilibrium ($\Delta V = 0$), in the "off" state ($\Delta V = 0.25$ V) and in the "on" state ($\Delta V = -0.25$ V) for $N = 10$ and $N = 40$. In equilibrium, we observe that long chains ($N = 40$) explore the interior of the pore, which allows them to reduce electrostatic and excluded volume repulsion and gain conformational entropy (see Section 6.3.1).

At this point, we can explain the effect of chain length on the $I - \Delta V$ curves in Fig. 6.28 as the combination of two effects. First, as *N* increases there is an increase in the number of charges per unit area of the outer walls. Increasing this density of surface charges leads to greater electrostatic interactions between the polyelectrolyte and the moving ions and thus, increases the efficiency of current rectification (see also the effect of σ_q on the rectification efficiency in Fig. 6.30C). The second effect of the chain length is through the distribution of polyelectrolyte: an ion entering or leaving the pore will experience much stronger interactions with the polyelectrolyte for N=40 than for N=10 because the long polyelectrolyte occupies a larger volume inside the pore than the short one (see Fig. 6.31 for $\Delta V = 0$ and $N = 40$). Therefore, the difference between the exit and entrance resistances ($\Omega^=$ and $\Omega^{\neq}$) will increase with increasing chain length and the efficiency of current rectification will also increase.

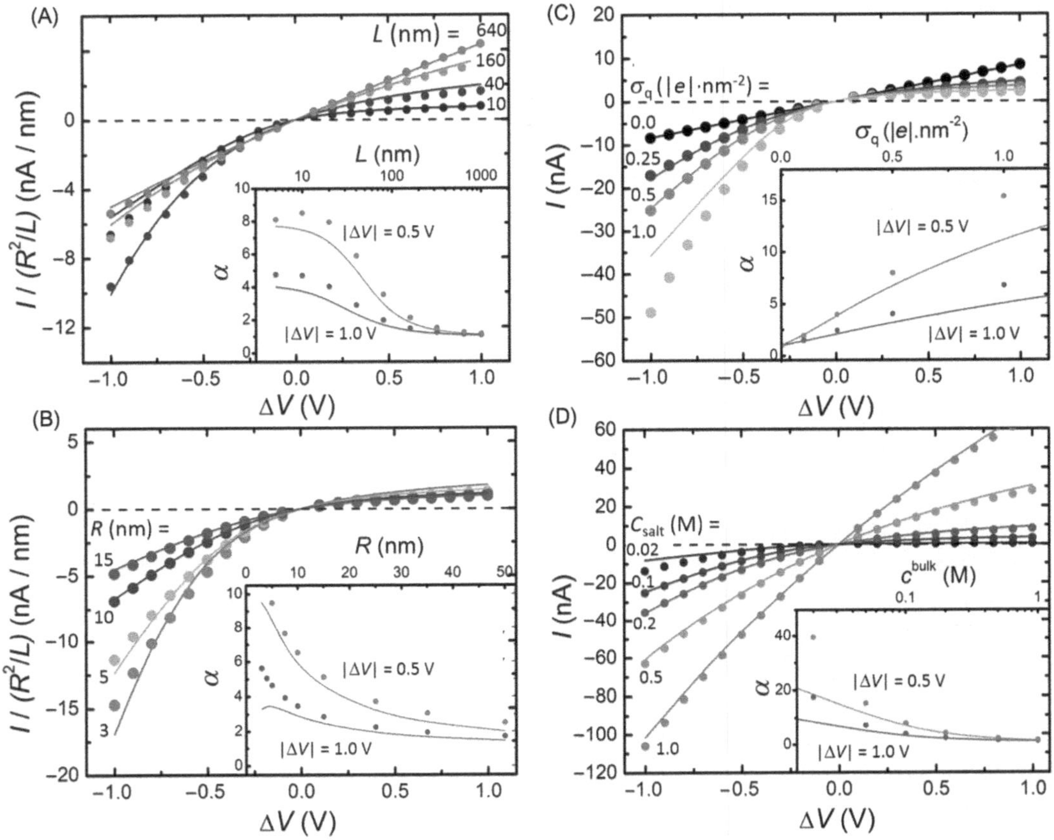

FIGURE 6.30 Current–potential plots for nanopores with surface-charged outer walls for different combinations of L (A), R (B), σ_q (C), and C_{salt} (D). The insets show the rectification factor α_R. Solid symbols show the results of nonequilibrium molecular theory calculations and solid lines show the predictions of the analytical model given by Eq. (6.70). The same parameters as in Fig. 6.29 were used, with the exception of the parameter being scanned in each panel. Source: *Adapted from reference Tagliazucchi, M.; Rabin, Y.; Szleifer, I. Transport Rectification in Nanopores with Outer Membranes Modified with Surface Charges and Polyelectrolytes.* ACS Nano ***2013***, *7, 9085–9097.*

Fig. 6.31 shows that upon applying a potential bias, the polyelectrolyte chains on the outer walls stretch toward the oppositely charged electrode, following the same mechanism that we explained for chains tethered on the inner walls in Fig. 6.23. As expected, long chains ($N=40$) deform much more than short ones ($N=10$). When a positive potential of $\Delta V=0.25$ V is applied (middle panel), the polyelectrolytes chains deform blocking the channel. On the other hand, for $\Delta V=-0.25$ V (bottom panel), the polyelectrolytes deform away from the central axis, leaving a polyelectrolyte-free central channel. Therefore, the pore for $N=40$ works as an electromechanical nanogate with open and close states that coincide with the on and off states for ionic conductance.

Can these electromechanical gates control the transport of something larger than an ion? To answer this question, we determined the potential of mean force (pmf) for the translocation of a

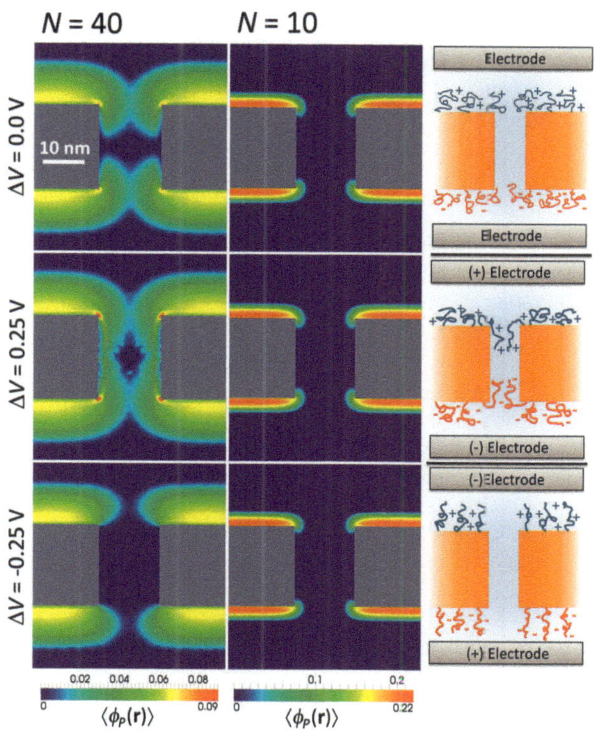

FIGURE 6.31 Volume fraction of the polyelectrolyte for the polyelectrolyte-modified nanopore of Fig. 6 27 for different applied biases. Calculation parameters: $N = 40$, $\sigma = 0.15$ chains $\cdot$ nm^{-2} (left panel), and $N = 10$, $\sigma = 0.6$ chains $\cdot$ nm^{-2} (right panel), other parameters were the same as in Fig. 6.29. The schemes on the right show the organization of the polyelectrolyte chains. Source: *Adapted from reference Tagliazucchi, M.; Rabin, Y.; Szleifer, I. Transport Rectification in Nanopores with Outer Membranes Modified with Surface Charges and Polyelectrolytes.* ACS Nano ***2013***, 7, *9085–9097.*

positively charged sphere along the z axis. The pmf of a particle at $\mathbf{r}$ is the reversible work required to move that particle from a reference position (we choose the bulk solution in the lower reservoir as the reference) to position $\mathbf{r}$ and, therefore, it provides an estimation for the minimum barrier for translocation. In other words, it provides an estimation of whether a process can kinetically occur or not: a large energetic barrier in the pmf indicates that the process will occur too slowly; however, a small barrier (or even the lack of it) does not always mean that the process is fast, because the system may not follow the reversible path considered in the pmf calculation.

We consider the pmf of a test particle (i.e., a particle that does not modify its environment). This is a rough approximation as we know that the particle will strongly interact with the polyelectrolytes as it translocates though the pore[67]; however this calculation will still allow us to obtain a qualitative picture of the translocation process. The pmf is:

$$U^{\mathrm{pmf}}(\mathbf{r}) = \int_V \left(\pi(\mathbf{r}') - \pi^{\mathrm{bulk,lower}}\right) v_{\mathrm{particle}}(\mathbf{r},\mathbf{r}')d\mathbf{r}' + \int_V \left(\psi(\mathbf{r}') - \psi^{\mathrm{bulk,lower}}\right) q_{\mathrm{particle}}(\mathbf{r},\mathbf{r}')d\mathbf{r}' \tag{6.72}$$

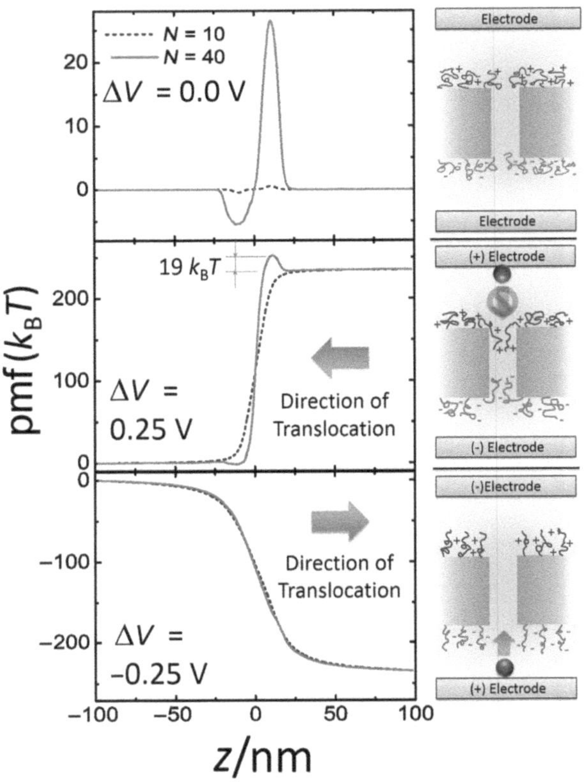

FIGURE 6.32 Potential of mean force, given by Eq. (6.72), for the translocation of a positive spherical particle along the axis of the pore of Fig. 6.31 for different applied voltage and chain length of the polyelectrolyte. By convention, we set the pmf = 0 at the bulk of the lower reservoir, the arrows show the direction of translocation. The negatively and positively charged polyelectrolytes are end-grafted to the outer walls at $z = -10$ nm and $z = 10$ nm, respectively. The scheme at the right shows the organization of the polyelectrolyte. Source: *Adapted from reference Tagliazucchi, M.; Rabin, Y.; Szleifer, I. Transport Rectification in Nanopores with Outer Membranes Modified with Surface Charges and Polyelectrolytes.* ACS Nano ***2013***, *7, 9085–9097.*

where $v_{\text{particle}}(\mathbf{r}, \mathbf{r}')$ and $q_{\text{particle}}(\mathbf{r}, \mathbf{r}')$ are the volume and charge that a particle with its center at $\mathbf{r}$ has in the volume element between $\mathbf{r}'$ and $\mathbf{r}' + d\mathbf{r}'$ and the integrals run over the total volume of the particle.

Fig. 6.32 shows the pmf for the translocation of a positive $(+8|e|$ charges per particle) particle though the pore, both for $N = 40$ and $N = 10$. For $\Delta V = 0$ V and $N = 40$ (solid lines), the particle experiences a large barrier which is located at the position of the positively charged outer wall. As expected, there is also a well in the pmf at the position of the negatively charged outer walls. Note, however, that in the curve for $N = 40$, the barrier is larger than the well, which is explained by the fact that the particle and the polyelectrolyte are repelled by both electrostatic and steric forces in the barrier, while in the well they are attracted by electrostatic forces but still repelled by steric interactions. For the "closed" state of the electrochemical gate ($\Delta V = 0.25$ V), there is a large driving force to drive the particle toward the negatively charged reservoir, but there is also a large barrier for translocation of $\sim 19\ k_BT$. However, this barrier disappears in the "open" state ($\Delta V = 0.25$ V). The pmf of the short chains $N=10$ is always barrierless, confirming the idea that this short chains would make rather poor electromechanical gates.

6.5 Protein Binding into Nanochannels and Applications in Biosensing

Biosensing is arguably one of the most promising applications of chemically modified nanochannels and nanopores. Two main sensing strategies have been developed. On one strategy, the current flowing through a short nanopore under an applied potential bias is continuously monitored. The macromolecules under study (e.g., DNA or a protein) stochastically translocate through the pore driven by the applied potential. Each of these translocation events is recorded as a transient change in the ionic current flowing through the pore, known as a resistive pulse. The shape, duration, and frequency of the resistive pulses are then associated to the chemical structure of the macromolecule and its concentration. The theory and practice of this sensing strategy have been compiled in previous books in literature,[68–70] so we will not comment on it any further. The second binding strategy involves monitoring the steady-state conductance of a nanochannel or nanopore that has specific ligands for a receptor site on a macromolecule in solution. The macromolecule in solution is, therefore, biorecognized by the ligands inside the pore, forming a ligand–receptor (LR) bound pair:

$$\mathrm{L} + \mathrm{R} \rightleftharpoons \mathrm{LR} \tag{6.73}$$

Binding of a macromolecule to the ligands inside the pore causes changes in the steady-state conductance, which are then related to the concentration of the macromolecule in solution as well as its physical–chemical properties (e.g., size and charge). In many cases,[71–74] the ligands inside the nanochannel are connected to the inner walls by flexible oligomeric tethers or molecular spacers (see Fig. 6.2) since long and flexible tethers have been proposed to facilitate the formation of the LR bond.[8,75–77] This binding strategy has been used to measure different analytes, such as proteins,[8,71,74,78–81] DNA,[73,82–85] organic molecules,[86–89] and metal ions.[8,90–93]

In the case where the analyte is a protein, the structure of the bound protein/ligand/tether layer is predicted to be rather complex both on planar surfaces[8] and inside nanochannels.[10] The density of the protein as a function of the distance to the planar substrate[8] or the wall of the channel[10] shows several density peaks due to layering of the bound proteins. The molecular organization of the system is governed by the competition between the binding equilibrium and the steric and electrostatic repulsions between proteins. Due to this competition, the total number of proteins bound to a surface experiences a maximum as a function of the surface density of the ligands on the surface, as predicted by the molecular theory[8] and later observed by experiments.[94,95] Steric repulsions between the oligomeric

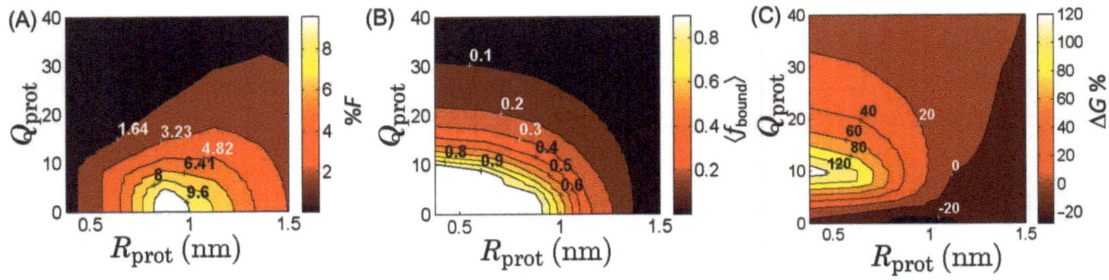

FIGURE 6.33 Contour plot for the percentage of the nanochannel occupied by bound proteins (A), the average fraction of ligands that are bound to the proteins (B) and the predicted change in the conductance upon protein binding (C) as a function of the protein radius, R_{prot}, and charge Q_{prot}. The tethers connecting the ligands to the surface (see Fig. 6.2) are uncharged. Other calculation parameters: $R = 4$ nm, $L \gg R$, N (length of the tethers) = 10, $\sigma = 0.1$ chains · nm^{-2}, $\chi = 0.0$, $K_d = 10^{-15}$ M, $C_{salt} = 0.2$ M and the protein concentration in solution was 500 nM. Source: *Adapted from reference Tagliacchi, M.; Szleifer, I. How Does Confinement Change Ligand-Receptor Binding Equilibrium? Protein Binding in Nanopores and Nanochannels. J. Am. Chem. Soc. 2015, 137, 12539–12551.*

tethers and the proteins inside nanochannels are also predicted to lead to a nonmonotonic dependence of the number of bound proteins with respect to the length of the tethers and the radius of the pore.[10] We will not discuss these effects in detail in this chapter, but rather focus our attention on the predicted changes in conductance as proteins of different size and charge bind to the ligands inside a long nanochannel, see the scheme of the nanochannel in Fig. 6.2.

In Fig. 6.33, we explore the binding of positively charged proteins of different size into a very long nanochannel of $R = 4$ nm. We performed these calculations using the equilibrium molecular theory for protein binding described in Section 6.2.1.3. Panel A shows the percentage of the channel that is filled with bound proteins, $\%F$ (the tethers occupy ~3% of the volume of the channel) as a function of the protein radius (R_{prot}) and charge (Q_{prot}). Panel B shows the average fraction of the ligands that are bound to proteins, $\langle f_{bound} \rangle$. Finally, panel C shows the percent change in channel conductance upon protein binding, which we define as:

$$\Delta G\% = \frac{G - G_0}{G_0} \cdot 100\%. \tag{6.74}$$

where G is the conductance of the channel with bound proteins and G_0 is the conductance of the channel before protein binding. These conductances were estimated in the quasi-equilibrium limit with Eq. (6.34) and the ion concentrations in equilibrium were determined from the molecular theory calculations. We assumed that the diffusion coefficients of the ions were independent on the amount of bound proteins and equal to those in bulk solution.

Note that the three panels of Fig. 6.33 provide complementary information: $\langle f_{bound} \rangle$ (panel B) informs on the number of bound proteins, $\%F$ (panel A) is the percentage of the channel filled by the protein and thus it is proportional to the product of $\langle f_{bound} \rangle$ and the volume of the protein (which depends on R_{prot}), and, finally, ΔG (panel C) depends on the changes in the number of ions inside the channel upon protein binding, which are dictated both by steric and electrostatic interactions. Fig. 6.33B shows that $\langle f_{\mathrm{bound}}\rangle$ decreases both with increasing R_{prot} and Q_{prot}. This result is explained by the decrease in the efficiency of binding due to the increase of the steric and electrostatic repulsions in the system that results from increasing the size and charge of the protein, respectively. Note that even though the concentration of proteins used in the calculation (500 nM) is much larger than the dissociation constant (10^{-15} M), the fraction of bound ligands may be much smaller than one due to the effect of electrostatic and steric repulsions. Fig. 6.33A shows that $\%F$ decreases with Q_{prot}, as expected from increase of electrostatic interactions, but, interestingly, it depends nonmonotonically on R_{prot}. This non-monotonic effect arises because $\%F$ is proportional to both the volume of the protein and the amount of bound protein. Therefore, as R_{prot} increases, the protein volume increases but the number of bound proteins decreases and the maximum of $\%F$ is observed for an intermediate value of R_{prot}.

Fig. 6.33C shows that ΔG is negative (i.e., the conductance decreases upon protein binding) for large and lowly charge proteins, but it is positive (G increases upon binding) for small and highly charged proteins. These results are explained by the competition between the electrostatic and steric interactions between bound proteins and mobile ions. On the one hand, lowly charged proteins reduce the concentration of ions inside the pore due to steric interactions, according the same mechanism shown in Fig. 6.22, where the conductance of a nanopore modified by a neutral polymer was predicted to be smaller than that of a bare pore. On the other hand, highly charged proteins bring their counterions with them as they are incorporated into the channel; thus the ionic concentration inside the pore increases and the conductance of the system is enhanced. The balance between the electrostatic counterion-uptake and the volume-exclusion mechanisms determines if the conductance of the system increases or decreases upon protein binding. This balance is affected by all the parameters in the system (salt concentration, dimensions of the pore, density of ligands, etc.), which may explain the experimental observation that the binding of streptavidin to biotin ligands within nanochannels leads to a decrease of the conductance in some experiments[71] and to an increase in others[78] and even to a decrease or increase in the same experimental system depending on the salt concentration.[74]

Fig. 6.34 explores a variation of the system studied in Fig. 6.33, where the tethers now have a charge of $-|e|$ per segment, i.e., the tethers and the proteins have opposite charge. Panels A and B show that the decrease in protein binding with Q_{prot} is now much weaker than in the case of neutral tethers, because in the scenario of Fig. 6.34, the increase in protein–protein electrostatic repulsions as Q_{prot} increases is counterbalanced by an increase in attractive protein–tether electrostatic attractions. Interestingly, panel C in Fig. 6.34 shows that ΔG % is negative and rather homogeneous for all combinations of Q_{prot} and R_{prot}. This behavior arises because both the steric and electrostatic interactions contribute now to deplete the ion concentration inside the channel upon protein binding. Before binding, the charge of the tethers inside the channel is compensated by mobile counterions that contribute to the conductance. After binding, the charges on the tethers become compensated by those of the oppositely charged protein and, therefore, both the counterions

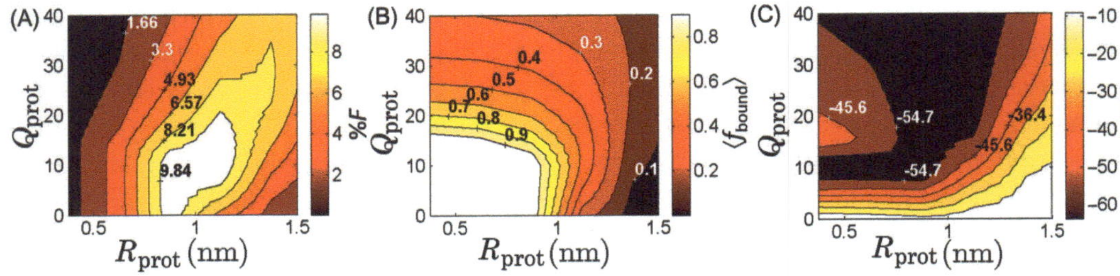

FIGURE 6.34 Same as Fig. 6.33, but the tethers connecting the ligands to the inner surface of the channel have a charge of −1 |e| per segment. *Adapted from reference Tagliaucchi, M.; Szleifer, I. How Does Confinement Change Ligand-Receptor Binding Equilibrium? Protein Binding in Nanopores and Nanochannels.* J. Am. Chem. Soc. ***2015***, *137, 12539–12551.*

of the tethers and the proteins are released, which decreases the conductance of the channel. From the bioanalytical point of view, the changes in conductance predicted in Fig. 6.34 are larger in magnitude and more systematic and homogeneous than those predicted for neutral tethers in Fig. 6.33. Therefore, we believe that using tethers with a charge opposite to that of the macromolecule to be sensed may be an attractive and simple strategy to improve the magnitude and reliability of the sensor response.

6.6 Conclusions and Outlook

We discussed in this chapter our recent efforts to model chemically modified nanopores and nanochannels. Models for these systems are needed in order to understand experimental results, for example, the pH-dependent conductance of the nanochannel of Fig. 6.10, as well as to innovate in the design of nanopore and nanochannel devices. For example, based on our models, we have proposed devices and strategies such as current rectification via modification of the outer walls (Fig. 6.27), the fabrication of polyelectrolyte electromechanical gates (Figs. 6.23 and 6.31) or using ligand–tether molecules with a charge opposite to the proteins in solution to improve biosensor response (Fig. 6.34). Our molecular theory models chemically modified nanopores incorporating chemical and chemical–physical details of the molecules modifying the pore. This methodology stands right in between computer simulations (atomistic or coarse grain) and continuum models (such as the Poisson–Nernst–Planck equations). Computer simulations can easily incorporate molecular details and have been applied to calculate the conductance of biological nanopores,[96] but they cannot cope yet with the large sizes of chemically modified solid-state pores and channels. Moreover, simulations cannot take advantage of the symmetry of the system and they struggle, as well, to deal with chemical equilibria. On the other side, continuum models are very well suited to study nanopores and nanochannels modified by surface charges (even when these charges are pH-dependent[97]); however they cannot incorporate molecular details such as the conformational degree of freedom of polyelectrolytes inside the channel or the steric repulsion effects during protein binding. The molecular theory is especially useful to model chemically modified pores and channels since it incorporates the molecular details of the species in the system into a continuum description of ion fluxes. The fact that the theory is derived from a formal framework is also especially appealing as it allows to easily obtain different "flavors" of the more general theory, i.e., the equilibrium versus nonequilibrium molecular theories (Sections 6.2.1 and 6.2.2, respectively) or different implementations that take advantage of the symmetry of the system (Section 6.2.3.1).

Some of the most interesting predictions discussed in this chapter are related to the structure of the polymer inside the nanopores and nanochannels and its responsiveness to the quality of the solvent and the externally applied potential. We have shown, for example, that polymers in poor solvent can qualitatively display two different collapse mechanisms: *collapse-to-the-walls* and *collapse-to-the-center*. In most previous experimental works dealing with the temperature-driven collapse of poly(isopropylacrylamide) within nanopores,[98,99] an increase in conductance (beyond that simply expected from the increase in ion mobility due to increasing temperature) was observed, which suggests that the *collapse-to-the-walls* mechanism was operative. Interestingly, Zharov and coworkers observed a transition from the *collapse-to-the-walls* to the *collapse-to-the-center* mechanism as the degree of polymerization was increased in experiments with colloidal nanoporous films (see chapter: Chemically Modified Nanopores in Colloidal Crystals).[38] Controlling the two collapse mechanisms may provide an exquisite route for gating ionic conductance using polymers. Moreover, the morphologies that we predict in poor solvent present a level of complexity beyond homogeneous layers collapsed to the walls or to the center of the channel: Fig. 6.9 shows structures such as stacked disks and toroids, whose complexity is expected to increase even further is inhomogeneities in the angular coordinate are allowed.[39,40] If these structures can be experimentally fabricated and stabilized to resist the dissolution of the nanochannel template, one can imagine them opening new avenues in nanofabrication.[40]

Interestingly, the discussion about the *collapse-to-the-center* versus *collapse-to-the-walls* mechanisms has transcended the field of chemically modified solid-state nanopores and it is gaining momentum in the study of the biophysics of the nuclear pore complex. The nuclear pore complex is a large biological nanopore (~30 nm in diameter for yeast, although the size depends on the species) inserted in the double lipid bilayer of the nuclear membrane of every eukaryotic cell.[100–102] This pore is in charge of transporting almost all biological macromolecules (e.g., proteins and m-RNA) between the nucleus and the cytoplasm. Surprisingly, the pore blocks the passage of any macromolecule larger than ~5 nm in size unless they are complexed to specific transport proteins, in which case even macromolecular complexes as large as ~40 nm can translocate through the pore. The selectivity filter is enforced by a family of intrinsically disordered proteins rich in phenylalanine–glycine motifs (the FG-Nups), which form an end-grafted layer on the inner surface of the pore, see Fig. 6.35. The FG domains in the FG-Nups interact attractively, which has led many researchers to believe that the FG-Nups may collapse in vivo. Based on different experiments and/or calculations, some researchers have proposed models where the FG-Nups collapse to the walls,[103,104] collapse to the center,[105] do not collapse,[67] or even display transitions between metastable *collapsed-to-the-center/collapsed-to-the-walls* states.[106] Much work is still

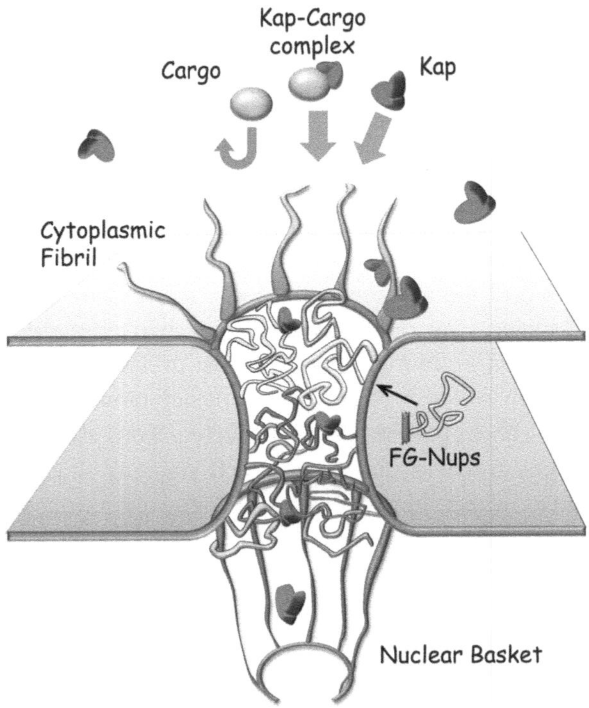

FIGURE 6.35 Schematic representation of the nuclear pore complex.

necessary to unravel the biophysics of the nuclear pore complex; however this biological system stands as an example of what is possible in terms of transport selectivity in nanopores modified by soft gates. In fact, researchers have shown that it is possible to mimic the selectivity of the nuclear pore complex in vitro using solid states channels,[50,107] opening new possibilities for synthetic devices with controlled transport properties. In the future, we believe that theory and simulations should play a very important role not only in helping to understand the biological problem of the NPC, but also in designing synthetic transporters that mimic the function of these biological pores.

Let us now briefly review the advantages of using polyelectrolytes (instead of small charged molecules) for chemical modification of nanopores and nanochannels. First, polyelectrolytes allow to immobilize a much larger amount of fixed charge than surface-confined small molecules. Surface charges are limited to maximum charge densities of $\sigma_q \sim 4.0\ |e| \cdot \mathrm{nm}^{-2}$ (see chapter: Introduction to Chemically Modified Nanochannels and Nanopores), while, for example, the layer of 4PVP chains studied in Sections 6.3.4 and 6.4.1 with $N = 28$ and $\sigma = 0.2$ chains $\cdot\ \mathrm{nm}^{-2}$ has an effective density of surface charges (when fully protonated) of $5.6\ |e| \cdot \mathrm{nm}^{-2}$. The second important advantage of using polyelectrolytes instead of surface-confined small molecules is that polyelectrolytes can stretch toward the center of the channel and, unlike surface charges, influence ion concentration beyond the thickness of the EDL. Due to this effect, polyelectrolyte-modified nanochannels exhibit selectivities greater than those of surface-charge-modified channels, for example, we predict that 99.4% of the ions transported by the PVP-modified long nanochannel in Fig. 6.10 are anions,[32] compared by only 70% for a nanochannel of similar dimensions and a surface charge of 0.5 $|e| \cdot nm^{-2}$.[20] Polyelectrolyte end-grafted layers on the outer walls are also a recurrent theme in our investigations as it is experimentally very challenging to modify only the inner surfaces of pores and channels. We have found that the polyelectrolytes on the outer walls strongly contribute to the conductance of pores in very thin membranes (<100 nm, see Fig. 6.21). However, while the outer walls play an important role in determining the transport properties of thin membranes (i.e., those prepared by electron or ion beam irradiation of thin Si_3N_4 membranes), their effect is negligible in very long channels (e.g., track-etched membranes). In this context, we proposed a strategy for current rectification that involves modifying both sides of a thin membrane with polyelectrolytes of opposite charge and then drilling a hole through it (system in Fig. 6.27). Again, note that this strategy will not work for long nanochannels.

At this point it is important to mention some limitations of our theory as well as to discuss what remains to be done and what we expect to see as the field develops in the near future. A major limitation of our theoretical framework is the lack of hydrodynamic effects, such as electroosmotic flows. In some cases, electroosmotic flows are not expected to be important due to the symmetry of the system, e.g., in the highly symmetric nanopores with chemically modified outer walls of Fig. 6.27 the fluxes of cations and anions are equal in magnitude but have opposite direction, thus completely canceling electroosmotic fluxes inside the channel. In quasi-equilibrium conditions, such as the long nanochannel of Fig. 6.10 and experiments in Fig. 6.17, the ion currents are very small and thus the contribution from electroosmosis will be negligible.[20] However, for short nanopores and high applied biases, electroosmosis is clearly important and we will need to incorporate it into our theory. Another limitation of our theory resides in the choice of the diffusion coefficients, which are inputs for the calculation of conductance both in the quasi-equilibrium (Eq. (6.34)) and the nonequilibrium (Eq. (6.37)) formulations of the molecular theory. In general, we approximate these diffusion coefficients by those in the bulk given the high water content (>80%) of the channels that we study. This approximation seems to work fine considering the good agreement between theory and experiment in Fig. 6.17, however, a consistent method to estimate the diffusion coefficients in polyelectrolyte-modified channels is still highly desirable. An option that we plan to explore in the future involves the use of MD simulations to determine the diffusion coefficient inside the polyelectrolyte layers. Finally, many experimental works probe the conductance of nanopore and nanochannel arrays with electrochemical tools (mainly cyclic voltammetry) in the presence of redox probes instead of the two-reservoir steady-state measurements discussed in this chapter. Recently, we gave some small steps toward the modeling of such systems,[108] but further theoretical work is still needed to provide a sound model for electrochemical measurements.

The field of chemically modified nanochannel and nanopores is still in its initial stages and much systematic work needs to be done to understand how to control transport of ions and larger cargoes by proper modification of solid-state channels with soft and responsive materials. Nature arises as a source of inspiration for the functionality of synthetic channels and pores,[109,110] however the mechanisms used by biological transporters are still far beyond the possibilities of current nanotechnology.[111] Therefore, new solutions—probably based on soft nanotechnology—will be needed to improve transport selectivity and responsiveness for applications in filtration membranes, biosensors, and energy conversion. New approaches will broaden the catalog of chemical and physical stimuli that can modulate transport properties. These new developments will also increase the selectivity of chemical and biochemical transport, which is an area where man-made channels and pores still lag far behind their biological counterparts. In this context, we believe that theory and simulation have the opportunity to propose new designs to accelerate the discovery of this new generation of nanofluidic devices.

Appendix I Symbols Used in This Chapter

Symbol	*Definition*	*Symbol*	*Definition*
$c^{eq}{}_i(\mathbf{r})$	Molar concentration of species i at **r** in equilibrium.	K_b^0	Thermodynamic equilibrium constant for the protonation of a base.
$c_i(\mathbf{r})$	Molar concentration of species i at **r**.	K_b	Equilibrium constant for the protonation of a base (in molar units).
C_{salt}	Salt concentration in the bulk (in molar units).	K_d^0	Thermodynamic equilibrium constant for the dissociation of a Ligand–Receptor pair.
D_i	Diffusion coefficient of species i.	K_d	Equilibrium constant for the dissociation of a Ligand–Receptor pair in molar units.
dL	Length of a segment of a cylindrical nanochannel.	L	Length of the nanochannel or nanopore.
$\|e\|$	Unit of elemental charge.	N_i^{excess}	Excess of anions and cations in the upper half of the system with respect to the lower half.
F	Free-energy functional	N	Number of segments per chain of polymer or polyelectrolyte (chain length).
F_C	Faraday's constant	N_A	Avogadro's number.
$\%F$	Percentage of a channel occupied by bound proteins.	$n_l(j, \alpha, \mathbf{r})d\mathbf{r}$	Number of ligands that a chain in conformation α and grafted at j has in the volume element between **r** and **r** + d**r**
f	Fixed fraction of charge of the segments in a strong polyelectrolyte.	$n(j, \alpha, \mathbf{r})d\mathbf{r}$	Number of segments that the chain j has in the volume element between **r** and **r** + d**r** when it is in conformation α.
$f(\mathbf{r})$	Fraction of segments of a weak polyelectrolyte at **r** that are charged.	$\langle n_p(\mathbf{r})\rangle$	Average density of polymer segments at **r**.
$f_b(\mathbf{r}, \mathbf{r}')$	Fraction of ligands at **r** that are bound to a protein with center at **r**′.	$P(j,\alpha)$	Probability of finding the polyelectrolyte chain j in the system in conformation α.
$f_{unb}(\mathbf{r})$	Fraction of unbound ligands at **r**.	pK_a^{app}	Apparent pK_a (i.e., pK_a that incorporates the effect of local environment).

(Continued)

(Continued)

Symbol	*Definition*	*Symbol*	*Definition*
$\langle f_{\mathrm{bound}} \rangle$	Average fraction of ligands inside a channel that are bound to receptors on proteins.	pK_a^{bulk}	pK_a $(-\log_{10}(K_a))$ of an acid–base reaction for a weak acid–base species in the bulk (i.e., infinite dilution).
G	Conductance of a nanopore or nanochannel.	$q(j)$	Partition function of chain *j*.
$g(\lvert \mathbf{r}-\mathbf{r}' \rvert)$	Distance-dependent van der Waals attractive term.	$q_{\mathrm{particle}}(\mathbf{r}, \mathbf{r}')d\mathbf{r}'$	Charge that a particle with center at **r** has in the volume element between **r**′ and **r**′ + d**r**′.
$\langle h \rangle$	Average thickness of the polymer/polyelectrolyte layer on the inner surface of a nanochannel.	q_i	Charge of species i.
I	Total current through a nanopore or nanochannel.	$q_{\mathrm{prot}}(\mathbf{r}, \mathbf{r}')d\mathbf{r}$	Charge that a protein at **r**′ has in the volume element between **r** and **r** + d**r**.
i_i	Current carried by ion i.	Q_{prot}	Total charge of the proteins in solution.
$\mathbf{J}_i(\mathbf{r})$	Flux of the species i at **r**.	R_{prot}	Radius of the proteins in solution.
K_a^0	Thermodynamic equilibrium constant for the dissociation of an acid.	$\mathbf{r}$	Position in the system.
k_B	Boltzmann's constant.	R	Radius of the nanochannel or nanopore.
		r	Radial coordinate.
K_a	Equilibrium constant for the dissociation of an acid (in molar units).		
R_{tip}, R_{base}	Tip and base radii for a noncylindrical (e.g., conical) nanochannel or nanopore.	$\phi_P(j,\alpha,\mathbf{r})d\mathbf{r}$	Volume that the chain *j* has in the volume element between **r** and **r** + d**r** when it is in conformation α.
T	Temperature.	$\langle \phi_P(\mathbf{r}) \rangle$	Average volume fraction of the polymer at **r**.
U^{pmf}	Potential of mean force for the translocation of a large particle through the pore.	κ_S	Conductivity of the solution.
$v_{\mathrm{particle}}(\mathbf{r}, \mathbf{r}')d\mathbf{r}'$	Volume that a particle with center at **r** has in the volume element between **r**′ and **r**′ + d**r**′.	λ	Lagrange multiplier associated with the global electroneutrality constraint (reference value for the electrostatic potential).
v_i	Molecular volume of species i.	λ_{Debye}	Debye length.
$\mathbf{v}_i(\mathbf{r})$	Drift velocity vector of species i at **r**.	μ_i^0	Standard chemical potential of species i.
$v_{\mathrm{prot}}(\mathbf{r},\mathbf{r}')d\mathbf{r}$	Volume that a protein at **r**′ has in the volume element between **r** and **r** + d**r**.	μ_i	Exchange chemical potential of species i.
$v_p(j,\alpha,\mathbf{r})d\mathbf{r}$	Volume that the chain *j* has in the volume element between **r** and **r** + d**r** when it is in conformation α.	$\pi(\mathbf{r})$	Lagrange multiplier associated to the packing constraint (position-dependent osmotic pressure).
W	Grand canonical potential including minimization constraints.	$\rho_i(\mathbf{r})$	Number density of species i at **r**.
z	Axial coordinate (i.e., parallel to the main axis of the pore or channel).	$\rho_{\mathrm{prot}}(\mathbf{r}, \mathbf{r}')$	Number density of proteins with center at **r** and receptor site at **r**′.
z_{graft}	*z*-position where a chain is grafted.	$\langle \rho_Q(\mathbf{r}) \rangle$	Average charge density at **r**.
α_R	Rectification factor.	σ	Grafting density of polymer/polyelectrolytes (units of chains · nm^{-2})

(*Continued*)

(Continued)

Symbol	Definition	Symbol	Definition
χ	Strength of the effective vdW attractive interactions.	σ_q	Surface charge density (units of $\|e\| \cdot nm^{-2}$).
Δ	Width of an acid–base protonation transition (pH difference required to go from a fraction of protonation of 1/11 to a fraction of protonation of 10/11).	θ	Angular coordinate.
ΔV	Transmembrane potential applied to the pore or channel (i.e., potential applied between the reversible electrodes located at the reservoirs).	$\Omega^{=}$	Single-ion access resistance when the ion and the outer walls of the entrance have the same charge.
ΔG %	Percent change in nanochannel conductance upon binding of proteins to the channel.	$\Omega^{\neq}$	Single-ion access resistance when the ion and the outer walls of the entrance have opposite charge.
ε	Dielectric constant.	Ω^{pore}	Single-ion resistance of the pore.
Φ	Grand canonical potential.	$\xi(j)$	Lagrange multiplier associated with the normalization constraint for the polyelectrolyte conformations of chain j (related to the partition function of chain j).

References

1. Wanunu, M.; Meller, A. Chemically Modified Solid-State Nanopores. *Nano Lett.* **2007,** *7,* 1580–1585.
2. Corry, B.; Kuyucak, S.; Chung, S. H. Tests of Continuum Theories as Models of Ion Channels. II. Poisson-Nernst-Planck Theory Versus Brownian Dynamics. *Biophys. J.* **2000,** *78,* 2364–2381.
3. Corry, B.; Kuyucak, S.; Chung, S. H. Invalidity of Continuum Theories of Electrolytes in Nanopores. *Chem. Phys. Lett.* **2000,** *320,* 35–41.
4. Ben-Shaul, A.; Szleifer, I.; Gelbart, W. M. Chain Organization and Thermodynamics in Micelles and Bilayers. 1. Theory. *J. Chem. Phys.* **1985,** *83,* 3597–3611.
5. Carignano, M. A.; Szleifer, I. Statistical Thermodynamic Theory of Grafted Polymeric Layers. *J. Chem. Phys.* **1993,** *98,* 5006–5018.
6. Nap, R.; Gong, P.; Szleifer, I. Weak Polyelectrolytes Tethered to Surfaces: Effect of Geometry, Acid-Base Equilibrium and Electrical Permittivity. *J. Polym. Sci. Part B: Polym. Phys.* **2006,** *44,* 2638–2662.
7. Szleifer, I.; Carignano, M. A. Tethered Polymer Layers. *Adv. Chem. Phys.* **1996,** *96,* 165–260.
8. Longo, G.; Szleifer, I. Ligand–Receptor Interactions in Tethered Polymer Layers. *Langmuir* **2005,** *21,* 11342–11351.
9. Ren, C.-L.; Carvajal, D.; Shull, K. R.; Szleifer, I. Streptavidin–Biotin Binding in the Presence of a Polymer Spacer. A Theoretical Description. *Langmuir* **2009,** *25,* 12283–12292.
10. Tagliaucchi, M.; Szleifer, I. How Does Confinement Change Ligand-Receptor Binding Equilibrium? Protein Binding in Nanopores and Nanochannels. *J. Am. Chem. Soc.* **2015,** *137,* 12539–12551.
11. Satulovsky, J.; Carignano, M. A.; Szleifer, I. Kinetic and Thermodynamic Control of Protein Adsorption. *Proc. Natl. Acad. Sci. USA* **2000,** *97,* 9037–9041.
12. Tagliazucchi, M.; Rabin, Y.; Szleifer, I. Transport Rectification in Nanopores with Outer Membranes Modified with Surface Charges and Polyelectrolytes. *ACS Nano* **2013,** *7,* 9085–9097.

13. Tagliazucchi, M.; Rabin, Y.; Szleifer, I. Ion Transport and Molecular Organization Are Coupled in Polyelectrolyte Modified Nanopores. *J. Am. Chem. Soc.* **2011,** *133*, 17753–17763.
14. Tagliazucchi, M.; Olvera de la Cruz, M.; Szleifer, I. Self-Organization of Grafted Polyelectrolyte Layers Via the Coupling of Chemical Equilibrium and Physical Interactions. *Proc. Natl. Acad. Sci. USA* **2010,** *107*, 5300–5305.
15. Gong, P.; Genzer, J.; Szleifer, I. Phase Behavior and Charge Regulation of Weak Polyelectrolyte Grafted Layers. *Phys. Rev. Lett.* **2007,** *98*.
16. Israelachvili, J. N. *Intermolecular and Surface Forces*, 3rd ed.; Academic Press: Burlington, MA, 2011.
17. Green, N. M. Avidin. *Adv. Protein Chem.* **1975,** *29*, 85–133.
18. Bard, A. J.; Faulkner, L. R. Electrochemical *Methods*, 2nd ed.; John Wiley & Sons: New York, 2001.
19. Goldman, D. E. Potential, Impedance and Rectification in Membranes. *J. Gen. Physiol.* **1943,** *27*, 37–60.
20. Vlassiouk, I.; Smirnov, S.; Siwy, Z. Ionic Selectivity of Single Nanochannels. *Nano Lett.* **2008,** *8*, 1978–1985.
21. Gilles, F. M.; Tagliazucchi, M.; Azzaroni, O.; Szleifer, I. Ionic Conductance of Polyelectrolyte-Modified Nanochannels: Nanoconfinement Effects on the Coupled Protonation Equilibria of Polyprotic Brushes. *J. Phys. Chem. C* **2016,** 4789–4798.
22. Lebon, G.; Jou, D.; Casas-Vázquez, J. *Understanding Non-equilbirium Thermodynamics;* Springer-Verlag: Berlin, 2008.
23. Prigogine, I. *Introduction to Thermodynamics of Irreversible Processes* Interscience Publishers: New York, 1955.
24. Bettolo Marconi, U. M.; Tarazona, P. Dynamic Density Functional Theory of Fluids. *J. Phys. Condens. Matter* **2000,** *12*, A413–A418.
25. Marconi, U. M. B.; Tarazona, P. Dynamic Density Functional Theory of Fluids. *J. Chem. Phys.* **1999,** *110*, 8032–8044.
26. Tagliazucchi, M.; Szleifer, I. Stimuli-Responsive Polymers Grafted to Nanopores and Other Nano-Curved Surfaces: Structure, Chemical Equilibrium and Transport. *Soft Matter* **2012,** *8*, 7292–7305.
27. Hindmarsh, A. C.; Brown, P. N.; Grant, K. E.; Lee, S. L.; Serban, R.; Shumaker, D. E., et al. Sundials: Suite of Nonlinear and Differential/Algebraic Equation Solvers. *ACM Trans. Math. Software* **2005,** *31*, 363–396.
28. Flory, P. J. *Statistical Mechanics of Chain Molecules* Wiley: New York, 1969.
29. Rosenbluth, M. N.; Rosenbluth, A. W. Monte Carlo Calculation of the Average Extension of Molecular Chains. *J. Chem. Phys.* **1955,** *23*, 356.
30. Longo, G. S.; Olvera De La Cruz, M.; Szleifer, I. pH-Controlled Nanoaggregation in Amphiphilic Polymer Co-Networks. *ACS Nano* **2013,** *7*, 2693–2704.
31. Peleg, O.; Tagliazucchi, M.; Kroeger, M.; Rabin, Y.; Szleifer, I. Morphology Control of Hairy Nanopores. *ACS Nano* **2011,** *5*, 4737–4747.
32. Tagliazucchi, M.; Azzaroni, O.; Szleifer, I. Responsive Polymers End-Tethered in Solid-State Nanochannels: When Nanoconfinement Really Matters. *J. Am. Chem. Soc.* **2010,** *132*, 12404–12411.
33. de Gennes, P.-G. *Scaling Concepts in Polymer Physics* Cornell University, 1979.
34. Krapf, D.; Wu, M. Y.; Smeets, R. M. M.; Zandbergen, H. W.; Dekker, C.; Lemay, S. G. Fabrication and Characterization of Nanopore-Based Electrodes with Radii Down to 2 nm. *Nano Lett.* **2006,** *6*, 105–109.
35. Williams, D. R. M. Grafted Polymers in Bad Solvents—Octopus Surface Micelles. *J. Phys. II* **1993,** *3*, 1313–1318.
36. Pattanayek, S. K.; Pham, T. T.; Pereira, G. G. Morphological Structures Formed by Grafted Polymers in Poor Solvents. *J. Chem. Phys.* **2005,** *122*.

37. Tagliazucchi, M.; Li, X.; Olvera de la Cruz, M.; Szleifer, I. Self-Organized Polyelectrolyte End-Grafted Layers under Nanoconfinement. *ACS Nano* **2014,** *8,* 9998–10008.

38. Schepelina, O.; Zharov, I. PNIPAAM-Modified Nanoporous Colloidal Films with Positive and Negative Temperature Gating. *Langmuir* **2007,** *23,* 12704–12709.

39. Osmanović, D.; Kerr-Winter, M.; Eccleston, R. C.; Hoogenboom, B. W.; Ford, I. J. Effects of Rotational Symmetry Breaking in Polymer-Coated Nanopores. *J. Chem. Phys.* **2015,** *142,* 034901.

40. Wu, Y. Y.; Cheng, G. S.; Katsov, K.; Sides, S. W.; Wang, J. F.; Tang, J., et al. Composite Mesostructures by Nano-Confinement. *Nat. Mater.* **2004,** *3,* 816–822.

41. Yameen, B.; Ali, M.; Neumann, R.; Ensinger, W.; Knoll, W.; Azzaroni, O. Synthetic Proton-Gated Ion Channels via Single Solid-State Nanochannels Modified with Responsive Polymer Brushes. *Nano Lett.* **2009,** *9,* 2788–2793.

42. Dong, R.; Lindau, M.; Ober, C. K. Dissociation Behavior of Weak Polyelectrolyte Brushes on a Planar Surface. *Langmuir* **2009,** *25,* 4774–4779.

43. Israëls, R.; Leermakers, F. A. M.; Fleer, G. J. On the Theory of Grafted Weak Polyacids. *Macromolecules* **1994,** *27,* 3087–3093.

44. Yameen, B.; Ali, M.; Neumann, R.; Ensinger, W.; Knoll, W.; Azzaroni, O. Proton-Regulated Rectified Ionic Transport through Solid-State Conical Nanopores Modified with Phosphate-Bearing Polymer Brushes. *Chem. Commun.* **2010,** *46,* 1908–1910.

45. Wang, D. W.; Nap, R. J.; Lagzi, I.; Kowalczyk, B.; Han, S. B.; Grzybowski, B. A., et al. How and Why Nanoparticle's Curvature Regulates the Apparent p*K*(a) of the Coating Ligands. *J. Am. Chem. Soc.* **2011,** *133,* 2192–2197.

46. CRC Handbook of Chemical and Physics. 72 ed.; CRC Press: Boston, MA, 1991.

47. Jones, D. M.; Brown, A. A.; Huck, W. T. S. Surface-Initiated Polymerizations in Aqueous Media: Effect of Initiator Density. *Langmuir* **2002,** *18,* 1265–1269.

48. Calvo, A.; Fuertes, M. C.; Yameen, B.; Williams, F. J.; Azzaroni, O.; Soler-Illia, G. J. A. A. Nanochemistry in Confined Environments: Polyelectrolyte-Brush Assisted Synthesis of Gold Nanoparticles Inside Ordered Mesoporous Thin Films. *Langmuir* **2010,** *26,* 5559–5567.

49. Grosberg, A. Y.; Rabin, Y. DNA Capture into a Nanopore: Interplay of Diffusion and Electrohydrodynamics. *J. Chem. Phys.* **2010,** *133,* 165102.

50. Kowalczyk, S. W.; Kapinos, L.; Blosser, T. R.; Magalhaes, T.; van Nies, P.; Lim, R. Y., et al. Single-Molecule Transport across an Individual Biomimetic Nuclear Pore Complex. *Nat. Nanotechnol.* **2011,** *6,* 433–438.

51. Harrell, C. C.; Kohli, P.; Siwy, Z.; Martin, C. R. DNA—Nanotube Artificial Ion Channels. *J. Am. Chem. Soc.* **2004,** *126,* 15646–15647.

52. Choi, J. H.; Park, J. S.; Moon, S. H. Direct Measurement of Concentration Distribution within the Boundary Layer of an Ion-Exchange Membrane. *J. Colloid Interface Sci.* **2002,** *251,* 311–317.

53. Holtzel, A.; Tallarek, U. Ionic Conductance of Nanopores in Microscale Analysis Systems: Where Microflu1dics Meets Nanofluidics. *J. Sep. Sci.* **2007,** *30,* 1398–1419.

54. Schoch, R. B.; Han, J. Y.; Renaud, P. Transport Phenomena in Nanofluidics. *Rev. Mod. Phys.* **2008,** *80,* 839–883.

55. Siwy, Z.; Kosińska, I. D.; Fuliński, A.; Martin, C. R. Asymmetric Diffusion through Synthetic Nanopores. *Phys. Rev. Lett.* **2005,** *94.*

56. Ali, M.; Ramirez, P.; Nguyen, H. Q.; Nasir, S.; Cervera, J.; Mafe, S., et al. Single Cigar-Shaped Nanopores Functionalized with Amphoteric Amino Acid Chains: Experimental and Theoretical Characterization. *ACS Nano* **2012,** *6,* 3631–3640.

57. Vlassiouk, I.; Siwy, Z. S. Nanofluidic Diode. *Nano Lett.* **2007,** *7,* 552–556.

58. Cheng, L.-J.; Guo, L. J. Ionic Current Rectification, Breakdown, and Switching in Heterogeneous Oxide Nanofluidic Devices. *ACS Nano* **2009,** *3*, 575–584.
59. Ramírez, P.; Gómez, V.; Cervera, J.; Schiedt, B.; Mafé, S. Ion Transport and Selectivity in Nanopores with Spatially Inhomogeneous Fixed Charge Distributions. *J. Chem. Phys.* **2007,** *126*, 194703.
60. Yan, R.; Liang, W.; Fan, R.; Yang, P. Nanofluidic Diodes Based on Nanotube Heterojunctions. *Nano Lett.* **2009,** *9*, 3820–3825.
61. Kalman, E. B.; Vlassiouk, I.; Siwy, Z. S. Nanofluidic Bipolar Transistors. *Adv. Mater.* **2008,** *20*, 293–297.
62. Daiguji, H. Ion Transport in Nanofluidic Channels. *Chem. Soc. Rev.* **2010,** *39*, 901–911.
63. Daiguji, H.; Oka, Y.; Shirono, K. Nanofluidic Diode and Bipolar Transistor. *Nano Lett.* **2005,** *5*, 2274–2280.
64. Vlassiouk, I.; Smimov, S.; Siwy, Z. Nanofluidic Ionic Diodes. Comparison of Analytical and Numerical Solutions. *ACS Nano* **2008,** *2*, 1589–1602.
65. Zhang, H.; Hou, X.; Zeng, L.; Yang, F.; Li, L.; Yan, D., et al. Bioinspired Artificial Single Ion Pump. *J. Am. Chem. Soc.* **2013,** *135*, 16102–16110.
66. Li, J.; Stein, D.; McMullan, C.; Branton, D.; Aziz, M. J.; Golovchenko, J. A. Ion-Beam Sculpting at Nanometre Length Scales. *Nature* **2001,** *412*, 166–169.
67. Tagliazucchi, M.; Peleg, O.; Kröger, M.; Rabin, Y.; Szleifer, I. Effect of Charge, Hydrophobicity, and Sequence of Nucleoporins on the Translocation of Model Particles through the Nuclear Pore Complex. *Proc. Natl. Acad. Sci. USA* **2013,** *110*, 3363–3368.
68. Muthukumar, M. *Polymer Tranlocation;* Taylor & Francis Inc: Bosa Roca, 2011.
69. Edel, J.; Albretch, T. *Engineered Nanopores for Bioanalytical Applications;* Elsevier, 2013.
70. Gracheva, M. E. *Nanopore-Based Technology;* Humana Press, 2012.
71. Ali, M.; Yameen, B.; Neumann, R.; Ensinger, W.; Knoll, W.; Azzaroni, O. Biosensing and Supramolecular Bioconjugation in Single Conical Polymer Nanochannels. Facile Incorporation of Biorecognition Elements into Nanoconfined Geometries. *J. Am. Chem. Soc.* **2008,** *130*, 16351–16357.
72. Ali, M.; Nasir, S.; Nguyen, Q. H.; Sahoo, J. K.; Tahir, M. N.; Tremel, W., et al. Metal Ion Affinity-Based Biomolecular Recognition and Conjugation Inside Synthetic Polymer Nanopores Modified with Iron-Terpyridine Complexes. *J. Am. Chem. Soc.* **2011,** *133*, 17307–17314.
73. Ali, M.; Neumann, R.; Ensinger, W. Sequence-Specific Recognition of DNA Oligomer Using Peptide Nucleic Acid (PNA)-Modified Synthetic Ion Channels: PNA/DNA Hybridization in Nanoconfined Environment. *ACS Nano* **2010,** *4*, 7267–7274.
74. Karnik, R.; Castelino, K.; Fan, R.; Yang, P.; Majumdar, A. Effects of Biological Reactions and Modifications on Conductance of Nanofluidic Channels. *Nano Lett.* **2005,** *5*, 1638–1642.
75. Balamurugan, S.; Obubuafo, A.; McCarley, R. L.; Soper, S. A.; Spivak, D. A. Effect of Linker Structure on Surface Density of Aptamer Monolayers and Their Corresponding Protein Binding Efficiency. *Anal. Chem.* **2008,** *80*, 9630–9634.
76. Ravan, H.; Kashanian, S.; Sanadgol, N.; Badoei-Dalfard, A.; Karami, Z. Strategies for Optimizing DNA Hybridization on Surfaces. *Anal. Biochem.* **2014,** *444*, 41–46.
77. Del Valle, E. M. M.; Galán, M. A. Effect of the Spacer Arm in Affinity Chromatography: Determination of Adsorption Characteristics and Flow Rate Effect. *Ind. Eng. Chem. Res.* **2002,** *41*, 2296–2304.
78. Umehara, S.; Karhanek, M.; Davis, R. W.; Pourmand, N. Label-Free Biosensing with Functionalized Nanopipette Probes. *Proc. Natl. Acad. Sci. USA* **2009,** *106*, 4611–4616.
79. Siwy, Z.; Trofin, L.; Kohli, P.; Baker, L. A.; Trautmann, C.; Martin, C. R. Protein Biosensors Based on Biofunctionalized Conical Gold Nanotubes. *J. Am. Chem. Soc.* **2005,** *127*, 5000–5001.
80. Ali, M.; Nasir, S.; Ramirez, P.; Cervera, J.; Mafe, S.; Ensinger, W. Carbohydrate-Mediated Biomolecular Recognition and Gating of Synthetic Ion Channels. *J. Phys. Chem. C* **2013,** *117*, 18234–18242.

81. Ali, M.; Ramirez, P.; Tahir, M. N.; Mafe, S.; Siwy, Z.; Neumann, R., et al. Biomolecular Conjugation Inside Synthetic Polymer Nanopores via Glycoprotein-Lectin Interactions. *Nanoscale* **2011,** *3,* 1894–1903.

82. Fu, Y.; Tokuhisa, H.; Baker, L. A. Nanopore DNA Sensors Based on Dendrimer-Modified Nanopipettes. *Chem. Commun.* **2009,** 4877–4879.

83. Jágerszki, G.; Gyurcsányi, R. E.; Höfler, L.; Pretsch, E. Hybridization-Modulated Ion Fluxes through Peptide-Nucleic-Acid-Functionalized Gold Nanotubes. A New Approach to Quantitative Label-Free DNA Analysis. *Nano Lett.* **2007,** *7,* 1609–1612.

84. Wang, X.; Smirnov, S. Label-Free DNA Sensor Based on Surface Charge Modulated Ionic Conductance. *ACS Nano* **2009,** *3,* 1004–1010.

85. Tian, Y.; Hou, X.; Wen, L.; Guo, W.; Song, Y.; Sun, H., et al. A Biomimetic Zinc Activated Ion Channel. *Chem. Commun.* **2010,** *46,* 1682–1684.

86. Sun, Z.; Han, C.; Wen, L.; Tian, D.; Li, H.; Jiang, L. pH Gated Glucose Responsive Biomimetic Single Nanochannels. *Chem. Commun.* **2012,** *48,* 3282–3284.

87. Han, C.; Hou, X.; Zhang, H.; Guo, W.; Li, H.; Jiang, L. Enantioselective Recognition in Biomimetic Single Artificial Nanochannels. *J. Am. Chem. Soc.* **2011,** *133,* 7644–7647.

88. Song, M.; Sun, Z.; Han, C.; Tian, D.; Li, H.; Jiang, L. Design and Fabrication of a Biomimetic Nanochannel for Highly Sensitive Arginine Response in Serum Samples. *Chemistry* **2014,** *20,* 7987–7993.

89. Kumar, B. V.; Rao, K. V.; Sampath, S.; George, S. J.; Eswaramoorthy, M. Supramolecular Gating of Ion Transport in Nanochannels. *Angew. Chem. Int. Ed.* **2014,** *126,* 13289–13293.

90. Zhao, C.; Li, X. S.; Li, L. Y.; Gong, X.; Chang, Y.; Zheng, J. Mimicking the Binding and Unbinding of Fe^{3+} with Transferrin Using a Single Biomimetic Nanochannel. *Chem. Commun.* **2013,** *49,* 9317–9319.

91. Tian, Y.; Zhang, Z.; Wen, L.; Ma, J.; Zhang, Y.; Liu, W., et al. A Biomimetic Mercury(II)-Gated Single Nanochannel. *Chem. Commun.* **2013,** *49,* 10679–10681.

92. Ali, M.; Nasir, S.; Ramirez, P.; Cervera, J.; Mafe, S.; Ensinger, W. Calcium Binding and Ionic Conduction in Single Conical Nanopores with Polyacid Chains: Model and Experiments. *ACS Nano* **2012,** *6,* 9247–9257.

93. Brunsen, A.; Díaz, C.; Pietrasanta, L. I.; Yameen, B.; Ceolín, M.; Soler-Illia, G. J., et al. Proton and Calcium-Gated Ionic Mesochannels: Phosphate-Bearing Polymer Brushes Hosted in Mesoporous Thin Films as Biomimetic Interfacial Architectures. *Langmuir* **2012,** *28,* 3583–3592.

94. Zhao, Z.; Matsui, H. Accurate Immobilization of Antibody-Functionalized Peptide Nanotubes on Protein-Patterned Arrays by Optimizing Their Ligand-Receptor Interactions. *Small* **2007,** *3,* 1390–1393.

95. Fasoli, E.; Reyes, Y. R.; Guzman, O. M.; Rosado, A.; Cruz, V. R.; Borges, A., et al. Para-Aminobenzamidine Linked Regenerated Cellulose Membranes for Plasminogen Activator Purification: Effect of Spacer Arm Length and Ligand Density. *J. Chromatogr. B* **2013,** *930,* 13–21.

96. Aksimentiev, A.; Schulten, K. Imaging α-Hemolysin with Molecular Dynamics: Ionic Conductance, Osmotic Permeability, and the Electrostatic Potential Map. *Biophys. J.* **2005,** *88,* 3745–3761.

97. Yeh, L. H.; Zhang, M.; Qian, S. Ion Transport in a pH-Regulated Nanopore. *Anal. Chem.* **2013,** *85,* 7527–7534.

98. Yameen, B.; Ali, M.; Neumann, R.; Ensinger, W.; Knoll, W.; Azzaroni, O. Ionic Transport through Single Solid-State Nanopores Controlled with Thermally Nanoactuated Macromolecular Gates. *Small* **2009,** *5,* 1287–1291.

99. Guo, W.; Xia, H.; Xia, F.; Hou, X.; Cao, L.; Wang, L., et al. Current Rectification in Temperature-Responsive Single Nanopores. *ChemPhysChem* **2010,** *11,* 859–864.

100. Peleg, O.; Lim, R. Y. H. Converging on the Function of Intrinsically Disordered Nucleoporins in the Nuclear Pore Complex. *Biol. Chem.* **2010,** *391,* 719–730.

101. Rout, M. P.; Aitchison, J. D. The Nuclear Pore Complex as a Transport Machine. *J. Biol. Chem.* **2001,** *276*, 16593–16596.

102. Wente, S. R.; Rout, M. P. The Nuclear Pore Complex and Nuclear Transport. *Cold Spring Harb. Perspect. Biol.* **2010,** *2*.

103. Kapinos, L. E.; Schoch, R. L.; Wagner, R. S.; Schleicher, K. D.; Lim, R. Y. H. Karyopherin-Centric Control of Nuclear Pores Based on Molecular Occupancy and Kinetic Analysis of Multivalent Binding with FG Nucleoporins. *Biophys. J.* **2014,** *106*, 1751–1762.

104. Peters, R. Translocation through the Nuclear Pore: Kaps Pave the Way. *Bioessays* **2009,** *31*, 466–477.

105. Bestembayeva, A.; Kramer, A.; Labokha, A. A.; Osmanović, D.; Liashkovich, I.; Orlova, E. V., et al. Nanoscale Stiffness Topography Reveals Structure and Mechanics of the Transport Barrier in Intact Nuclear Pore Complexes. *Nat. Nanotechnol.* **2014,** *10*, 60–64.

106. Ando, D.; Zandi, R.; Kim.; Yong, W.; Colvin, M.; Rexach, M., et al. Nuclear Pore Complex Protein Sequences Determine Overall Copolymer Brush Structure and Function. *Biophys. J.* **2014,** *106*, 1997–2007.

107. Jovanovic-Talisman, T.; Tetenbaum-Novatt, J.; McKenney, A. S.; Zilman, A.; Peters, R.; Rout, M. P., et al. Artificial Nanopores That Mimic the Transport Selectivity of the Nuclear Pore Complex. *Nature* **2009,** *457*, 1023–1027.

108. Andrieu-Brunsen, A.; Micoureau, S.; Tagliazucchi, M.; Szleifer, I.; Azzaroni, O.; Soler-Illia, G. J. Mesoporous Hybrid Thin Film Membranes with Pmetac@Silica Architectures: Controlling Ionic Gating through the Tuning of Polyelectrolyte Density. *Chem. Mater.* **2015,** *27*, 808–821.

109. Hou, X.; Zhang, H. C.; Jiang, L. Building Bio-Inspired Artificial Functional Nanochannels: From Symmetric to Asymmetric Modification. *Angew. Chem. Int. Ed.* **2012,** *51*, 5296–5307.

110. Guo, W.; Tian, Y.; Jiang, L. Asymmetric Ion Transport through Ion-Channel-Mimetic Solid-State Nanopores. *Acc. Chem. Res.* **2013,** *46*, 2834–2846.

111. Tagliazucchi, M.; Szleifer, I. Transport Mechanisms in Nanopores and Nanochannels: Can We Mimic Nature? *Mater. Today* **2015,** *18*, 131–142.

Index

Note: Page numbers followed by "*f*," "*t*," and "*b*" refer to figures, tables, and boxes, respectively.

Q

FSC
www.fsc.org
MIX
Papier | Fördert
gute Waldnutzung
FSC® C083411

Kolibri 360 GmbH
Ferdinand-Jühlke-Straße 7
99095 Erfurt, Deutschland
produktsicherheit@kolibri360.de